basic
ROBOTICS

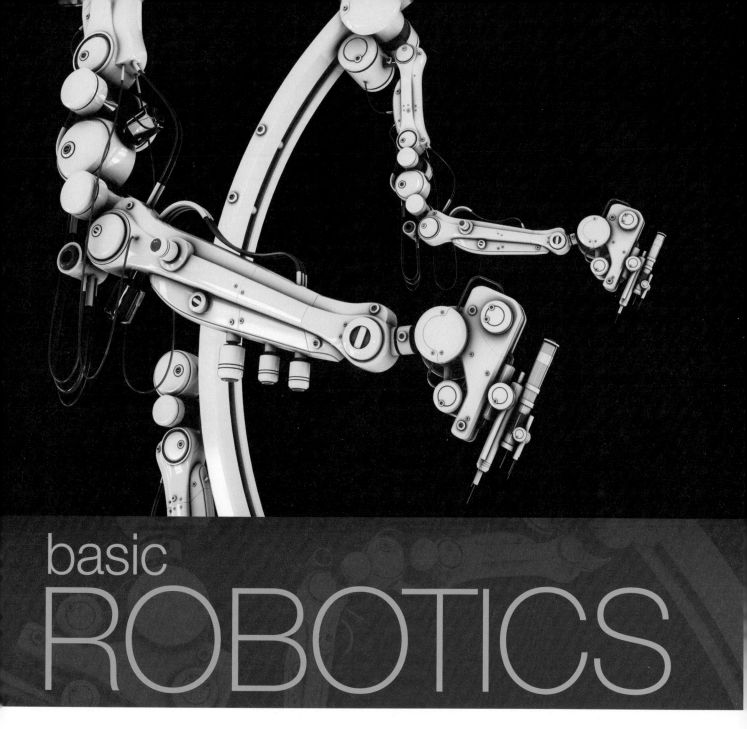

basic ROBOTICS

KEITH DINWIDDIE
Ozarks Technical Community College

CENGAGE
Learning·

Australia • Brazil • Mexico • Singapore • United Kingdom • United States

Basic Robotics
Keith Dinwiddie

SVP, GM Skills & Global Product Management:
Dawn Gerrain

Product Team Manager: Erin Brennan

Associate Product Manager: Nicole Sgueglia

Senior Director, Development:
Marah Bellegarde

Senior Product Development Manager:
Larry Main

Senior Content Developer: Sharon Chambliss

Product Assistant: Jason Koumourdas

Vice President, Marketing Services:
Jennifer Ann Baker

Marketing Manager: Kelsey L. Hagan

Senior Production Director: Wendy Troeger

Production Director: Andrew Crouth

Senior Content Project Manager: James Zayicek

Content Production Management and
Art Direction: Lumina Datamatics, Inc.

Cover image(s): Robotic Arm: ©Alexey Dudoladov/
iStock/Thinkstock; Two robotic hand tools:
©Ociacia/iStock/Thinkstock; Aldebaran
Robotics: © AFP/Getty Images; Dextre, the
Canadian Space Agency's robotic handyman:
© Stocktrek Images/Getty Images

Interior Image(s): Creative brain vector:
©Ideas_supermarket/Shutterstock.com; Tech
vector background with a circuit board texture:
©Kun Kunko/Shutterstock.com

Library of Congress Control Number: 2014943299

ISBN-13: 978-1-133-95019-6

Cengage Learning
20 Channel Center Street
Boston, MA 02210
USA

Cengage Learning is a leading provider of customized learning solutions with office locations around the globe, including Singapore, the United Kingdom, Australia, Mexico, Brazil, and Japan. Locate your local office at:
www.cengage.com/global

Cengage Learning products are represented in Canada by Nelson Education, Ltd.

To learn more about Cengage Learning, visit **www.cengage.com**

Purchase any of our products at your local college store or at our preferred online store **www.cengagebrain.com**

Printed in the United States of America
Print Number: 01 Print Year: 2014

Contents

chapter **5**
End-Of-Arm Tooling   120

chapter **6**
Sensors and Vision   144

chapter **7**
Peripheral Systems   167

chapter **8**
Robot Operation   186

chapter **9**
Programming and File Management 208

chapter **11**
Repairing the Robot 278

chapter **10**
Troubleshooting 238

chapter **12**
Justifying the Use of a Robot 300

preface

About This Book

Robotics is an exciting and growing field in the modern world. Where robots were once the realm of science fiction, they are now a crucial part of the industrial world and spreading into our everyday lives as the field continues to evolve, expand, and mature. Those interested in the field of robotics have many avenues and potential job opportunities, but where does one begin learning about robotics? The purpose of this textbook is to answer that question. There are many great books out there to help one drill down into a specific area of robotics or even specific types of robots, but it is hard to find a good introductory text that is not outdated or tailored to a specific brand of robots. The driving idea behind this book is to fill that gap and give those interested in robotics the tools they need to understand the deep-drilling texts while helping them understand the mechanical marvel that is the modern robot.

This text provides students access to my years of experience working in industrial maintenance as well as working with and teaching various robotic systems to my students. I offer stories of what can go wrong, tips, precautions, best practices, and knowledge to give students a wealth of information without having to spend the years needed to learn this on their own. This knowledge is accompanied by a large number of images to help drive home the points and give students a glimpse into the world of robotics that might otherwise be difficult, if not impossible, to get on their own. Here is a taste of what they will learn.

We begin with an exploration of the fascinating series of events necessary for the birth of the modern robot as well as milestone events in robotics. The timeline in Chapter 1 is one of the most complete timelines you will find relating to robotics and not a reprint of previous timelines, some of which contain inaccurate facts. The discussion of the timeline helps students realize that highly complex machines existed and flourished well before the birth of the digital computer. From there, we discuss what a robot is and the various definitions that are out there, plus some of the reasons why a large range of things fall under the term robot. We look at the two driving theories behind robot creation and finish up with a look at what the future of robotics might hold.

In Chapter 2, we discuss how to work safely with robots as well as the systems that keep people safe when using them. Given the fact that many classes include lab activities, it is important for students to understand the dangers involved. The main point of this chapter is to drive home the need for safety and the dire consequences that can happen if one disregards the rules. No discussion on robotic safety is complete without discussing the danger of electricity. This chapter devotes an entire section to just that. We conclude with a discussion that benefits everyone on how to deal with emergencies.

In Chapter 3, students get to know the robot and its main components. We cover a large number of terms that are confusing to those new to robotics as well as explore the job of the main robot components. Within the power supply section we dig down into electricity a bit so the student can understand the differences between AC and DC, how electricity flows, amp hours, and the difference between single and three-phase power. In the hydraulic and pneumatic portions of this chapter, we review the history of the power sources, how they work, and the differences between the two. This in-depth style of examination continues with a look at the controller, teach pendant, manipulator, and various base types.

Chapter 4 covers the various ways to power robots and how to group them accordingly. We look at motion types for various robots and the type of work envelope this creates. In the section on classifying robots by their drives, we dive down into the direct-drive and indirect-drive systems and take a closer look at belts, chains, and gears. Given the possibility that many of those reading this text may end up creating robots in the field, we look at some of the important math concepts associated with drive systems as well. We close this chapter with a look at how the ISO classifies robots, both industrial and nonindustrial, as they are a standards organization recognized the world over.

Chapter 5 takes students through the diverse world of robot tooling. Tooling is what allows the robot to interact with the world around it and perform the multitude of tasks we desire. During the exploration of tooling, we take look at the most common types, the basics of their operation, the way in which we deal with misalignments, and the need for multiple tooling in the field. We also explore how tooling affects the payload of the robot as well as the different forces of motion that can affect part movement.

Without sensors, the robot has zero information about the world around it, so in Chapter 6, we focus in on these vital information-gathering tools. We start out looking at the basics, such as limit and proximity switches, and work our way up to the complex vision systems in use today. In each section of the chapter, we look at how various sensors operate and what we use them for in industry. Are you curious about how an encoder works? Have you ever thought of using sound to help a robot "see?" How does a vision system find the part? You'll find the answers to these questions and more in Chapter 6.

We often think of the robot as an army of one, but in truth, the robot is often a team player and depends on other machines to get its job done. In Chapter 7, we look at some of the equipment the robot works with and depends on. While we do discuss the work cell and how the robot fits in; the primary focus of the chapter is the equipment to help and work with the robot, including external positioners, safety systems, and equipment to assist with the tooling side of the robot. We wrap up with a look at communications, since all this extra equipment is basically useless if we do not have a way to control it as needed and get the pertinent information to and from these devices.

In Chapter 8, I share my experience from years of working with and teaching the operation of robots and provide the basic information needed when someone plans to run robots. Even though the systems differ, there are fundamental basics that transcend make and model; these are the basis of Chapter 8. From powering the robot up to getting ready to run for the day, this chapter has great tips on what to do and what to watch for. This chapter includes a great guide on what to do when things go wrong and the robot crashes. This happens from time to time; however, few, if any, of the other texts out there talk about how to deal with this stressful situation. By breaking crashes down into simple, easy-to-remember steps, this section gives students the tools they need to weather the storm of a robot crash.

Chapter 9 is about writing programs to control the actions of the robot. We start with a look at the various levels of languages out there and the responsibilities of the programmer when working with each. From there, we go through the process of getting a program ready from the planning stage to normal operation. This chapter does not include the specifics needed to write a program for one particular robot, but it does have all the basic steps and examples of programs from various systems. This is where I have taken my background in programming different systems and distilled a process that works no matter the model. Of course, no programming section is complete without a discussion of the various logic filters used in programming as well as of testing and verifying the program. To round out this chapter, we look at proper file maintenance to ensure that our working program of today survives the complete loss of power tomorrow.

Chapter 10 is all about fixing the robot when something goes wrong. Troubleshooting is a common part of working with the robot, yet so many texts ignore this fact all together. In this chapter, students will learn what troubleshooting is, explore the various ways to gather information about the problem, and then explore ways to filter this information to create a plan of action to fix the problem. Here, students have the chance to tap into my years of knowledge to further them along the path of becoming successful troubleshooters. The chapter finishes with a section on what to do if the troubleshooting process does not work or problems remain, as this will happen from time to time.

A follow-up to Chapter 10, Chapter 11 provides tips and valuable information on repairing and maintaining the robot. We start with a look at preventative maintenance in an effort to avoid breakdowns and then progress to useful tips for fixing the system. We also look at the important difference between part swapping and fixing the robot. Those who master truly fixing equipment are the ones that get the high-paying jobs in the field. We also look at precautions to take before trying a repaired robot as well as what to do when the repair is finished, both crucial parts of the repair process that is at times overlooked, to the detriment of all.

We finish up with a look at how we justify the use of a robot in Chapter 12. There are many reasons to use a robot, and we look at the common ones here, giving the student a better understanding of how we match robot use to cost. We look specifically at return on investment (ROI) and the math involved. For those who end up working in industry, this chapter could mean the difference between getting the right robot for the job at hand and getting the cheapest robot for which the company received a quote. We also address the issue of robots replacing human workers, a topic that can be and has been a heated subject.

This is just a quick snapshot of what you will find in this textbook. I encourage you to take some time and dig deeper to see what other treasures you may find. This work is the culmination of my years of experience coupled with my general dissatisfaction with other introductory robotics textbooks. I have reviewed many and taught from several, but I never found one I really liked. This was the driving force behind the text before you today. It is my hope that you will find this work the answer to your needs as it is for mine.

Accompanying Activities Manual

An activities manual has been specifically designed to accompany this text. Each section in the manual matches up with the corresponding chapter in the book and includes various tasks to strengthen student comprehension. Activities include math problems, matching, multiple-choice, and short answer questions along with short writing assignments, research activities, and suggested labs. Already have your labs planned? There are blank lab forms for students to convey what they did and what they learned. The manual provides options for the classroom in addition to serving as a useful tool to help students retain the key points of each chapter.

Included are several low-cost labs that students can do on their own or as part of a classroom exercise, complete with pictures to help guide the way.

SUPPLEMENTS

The Instructor Companion Website includes the following components to help minimize instructor prep time and engage students:

- PowerPoint—Chapter outlines with images from the book for each textbook chapter.
- Computerized Test Bank in Cognero—Modifiable questions for exams, quizzes, in-class work, or homework assignments, in an online platform.
- Image Library—Images from the textbook that can be used to easily customize the PowerPoint outlines.
- Answer Key—Answers to the end-of-chapter review questions and the quizzes in the Activities Manual.

MINDTAP for Basic Robotics

The MindTap for Basic Robotics includes a fully interactive robotic arm simulation that is controlled by a highly realistic teach pendant. Set in a factory setting, the simulation has a variety of engaging practical applications that provide students with real-world experience in a safe, controlled environment. Animations bring other facets of the robotics industry to life, including a harmonic drive gear, cylindrical robot, and more. Gradable learning activities are integrated throughout the unique learning path to assess student skill mastery and progress.

MindTap is a personalized teaching experience with relevant assignments that guide students to analyze, apply, and improve thinking, allowing instructors to measure skills and outcomes with ease.

- *Personalized Teaching:* Becomes *yours* with a *learning path* that is built with key student objectives. Control what students see and when they see it—match your syllabus exactly by hiding, rearranging, or adding your own content.
- *Guide Students:* Goes beyond a traditional "lift and sift" model by creating a unique learning path of relevant readings, multimedia, and activities that move students up the learning taxonomy from basic knowledge and comprehension to analysis and application.
- *Measure Skills and Outcomes:* Analytics and reports provide a snapshot of class progress, time on task, engagement, and completion rates.

About the Author

I have been a fan of science fiction and robots for as long as I can remember. As a kid, I had two of the jumbo Poppy robots, which have survived more or less intact to this day, and a battery powered Rotate-O-Matic, which did not fare as well (though I do still have some of that robot's pieces). I have been captivated by the potential of the robot since that early age, and I am excited to see the realm of robotics advance to the point where the science fiction of my youth is becoming the science fact of today! Even though I have always loved the robot, it was not until later in life that I truly began my exploration of the field.

Growing up, I enjoyed taking things apart and trying to figure out how they worked. It was not surprising that my military aptitude test pegged me for work in the repair field, and I ended up working on the Huey Helicopter. This was my first official training in the mechanical repair field and it served me well years down the road when I decided to try my hand at industrial maintenance. I tried the engineering route for a while in industry, but found that my passion was for the repair of industrial equipment instead of working on the process of making products.

Once I made this decision, I took the classes needed to get my associate's degree in industrial maintenance. I then worked my way into the maintenance crew at a manufacturer in my town. This is where I first laid hands on the industrial robot, though these robots were older Cartesian-type units with only three axes and error messages written in German. It was not until the company bought newer FANUC robots that I had the chance to work with a modern industrial robot. Unfortunately, my experience at that point was primarily on the repair side, as I had very little chance to play with the programming on the newer robots.

During my time in maintenance (over seven years total at the facility), I had the chance to teach some adjunct classes at the same community college where I had previously received my industrial maintenance technician degree. I discovered my passion for teaching and switched to teaching full-time when the chance came up. It was during this time that I had the chance to dig into the robot and programming aspects to the level I had desired. At the writing of this, I am in my seventh year as a full-time instructor. I am a CERT (Certified Education Robot Training) instructor for FANUC programing and vision. I have programmed and taught programming for Mitsubishi, FANUC, Panasonic, NAO, and LEGO NXT robotic systems. I also teach industrial maintenance classes, such as electricity, fluid power, PLC programming and operation, mechanical power transmission, and safety.

My unique background coupled with the fact that I could not find an introductory robotic text that I liked is what led to my writing this textbook. I have worked to bring my classroom conversational style to print in hopes that the book reads better than those other textbooks that seem to forget who the audience is. I am excited to share my knowledge with the next generation of roboticists and help them begin their journey of exploration into the world of robotics!

Acknowledgments

There are some very special people without whom this book would never have come to be, and I would like to thank them. First and foremost, thank you to Nicole Sgueglia and Sharon Chambliss at Cengage publishing for taking a chance on a tech instructor they met through LinkedIn and giving me the opportunity to write this book. I want to thank my lovely wife, Lucia, for supporting me through the writing process and putting up with me during the stressful times. A big thanks to my parents for letting me tear apart all those toys and electronics as a child, which served as a great basis for my later carriers. A special thanks to my dad, Frank Dinwiddie, for sharing with me his knowledge of things mechanical and his ingenuitive mindset.

I would also like to thank the following reviewers:

Paul Bartomioli, USFIRST.org volunteer
Thomas M. Bishop, Roswell High School, Roswell, GA
Allan Cameron, Ph.D., Sí Se Puede Foundation, Chandler, AZ
James G. Hillwig, Jr., Crawford County Career and Technical Center, Meadville, PA
Nick Yates, math and engineering teacher, Patterson High School, Baltimore City, MD
Dr. Janet L. Larson, Millard Public Schools, Omaha, NE

Keith Dinwiddie
2014

dedication

I dedicate this book to James E. Stone, my best friend, who lost his fight to cancer on February 7, 2012, shortly after I began work on this text. Mr. Jim, as I always called him, took me in as a friend, adopted me as the son he never had, and shared with me his passions and a lifetime of knowledge in various things, including industrial maintenance. Mr. Jim had a big hand in shaping my industrial maintenance career and was a great mentor to me. I know he would have loved to have this book sitting on his shelf.

History of Robotics

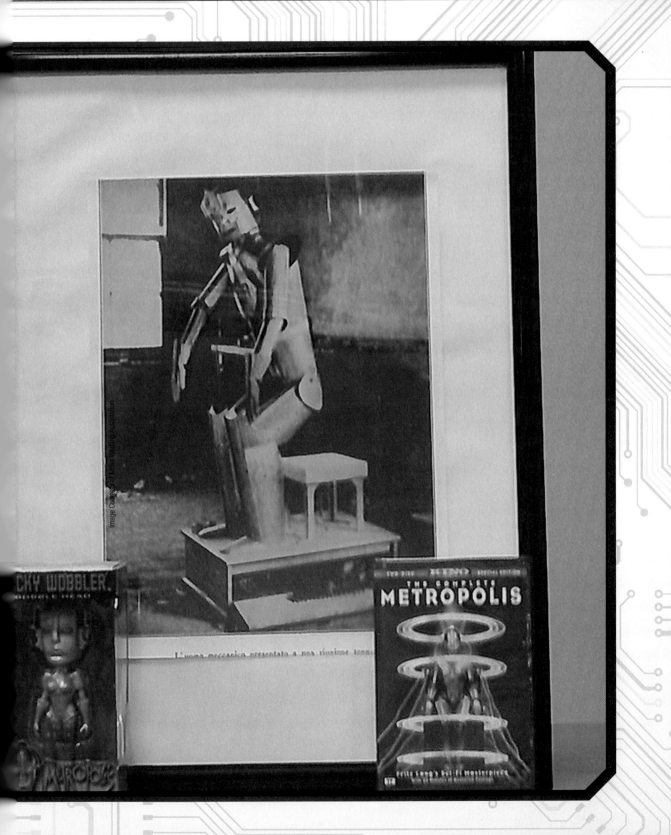

What You Will Learn

- Key events that lead to the invention of the modern robot
- The evolution of the modern robot
- The difference between industrial robots and other robots
- The four Ds of robotics
- Where and why we use robots in the modern world
- The difference between the top down and bottom up development methods for robotics
- What Artificial Intelligence is and how it could impact the robots of tomorrow
- Some of the ways that robotics might be used in the future

Overview

When first learning about a subject, it is often beneficial to study the history involved. When discussing robots, many people envision innovative technology and sophisticated modern machinery—but this is only half of the story. Without the technological and ideological advancements that have occurred over the centuries, we would not have the modern robot we know and love today. In this chapter, we will look at the history leading up to the modern robot as well as future trends for robotics. In the course of our exploration, we will cover the following:

- A timeline of events important to robotics
- Key events in the history of robotics
- What is a robot?
- Why use a robot?
- The top-down versus bottom-up approach
- AI and the future of robotics

Timeline of Events

The following chart details the timeline of many events that have been important in the development of modern robotics.

Key Events in the History of Robotics

Looking at the timeline of events, we can begin to see the centuries of human development and ingenuity that have led to the modern marvel that is the robot. The robots used in our homes, entertainment, and industry are not the result of the last few decades, but are the culmination of centuries of human effort to create something marvelous. While we will examine many of the events presented in the timeline, I encourage you to review them all and research any that catch your interest.

Table 1.1

Year	Person(s) or Company	Event
BCE		(BCE – Before Common Era)
3000		Abacus invented in the Orient (Ament 2006)
420	Archytas of Tarentum	Creates a wooden pigeon that can fly under steam or compressed air power (Timeline of Flight 2010)
285–222	Ktesibios (or Ctesibius) of Alexandria	Credited as the father of pneumatics (Lahanas, Ctesibius of Alexandria 2010)
200		Chinese artisans create an entire mechanical orchestra (Needham and Ronan 1994)
CE		(CE – Common Era)
10–70	Heron of Alexandria	Improves on Ktesibios work, credited with first steam engine design, designs automata such as *Hercules Killing the Dragon*, powered by water (Lahanas n.d.)
1206	Al-Jarzai	Publishes a book on mechanical and automated devices heralded as the most complete collection of such information at the time (Nadarajan 2007)
1495	Leonardo da Vinci	Creates the *Metal-Plated Warrior*, a suit of armor that sits up, opens and closes its arms, moves its head, and opens its visor (Rosheim 2006)
1525	Hans Bullmann	Credited with building the first real androids with human form (Dalakov, Gianello Torrian 2012)
1540	Gianello Torriano	Creates the *Lute Player Lady*, an automation of a lady playing a lute or mandolin (Dalakov, Gianello Torriano 2012)
1543	John Dee	Creates a very lifelike mechanical beetle using his mathematical knowledge, resulting in witchcraft charges (Fell-Smith 1909)
1564	Pare Ambroise	Publishes a design for an artificial hand that uses mechanical muscles with an organic frame (Mo 2007)
1620	William Oughtred	Plots a logarithmic scale along a single, straight 2-foot ruler (slide rule) (O'Connor and Robertson 1996)
1623	Wilhelm Schickard	Invents the "Rechenuhr," a four-function calculator that performs addition, subtraction, multiplication, and division (O'Connor and Robertson 2009)
1642	Blaise Pascal	Builds his version of the calculator, which tis later mass produced (Blaise Pascal 1623–1662, 2002–2012)
1679	Gottfried Wilhelm von Leibniz	Develops and perfects the binary system of arithmetic (O'Connor and Roberston, Gottfried Wilhelm von Leifniz 1998)
1725	Lorenz Rosenegge	Installs a mechanical theatre at the Heilbrunn chateau, consisting of 256 figures; 119 are animated by means of a single water turbine (Animation Notes #1, What Is Animation? 2010)
1738	Jacques de Vaucanson	Invents three very lifelike clockwork automatons, the *Flute Player*, the *Tambourine Player*, and the *Digesting Duck* (Dalakov, Jacques de Vaucanson 2012)
1752	Benjamin Franklin	Conducts his famous kite experiment and proves that lightning is a form of electricity, leading to further experiments and several common terms still in use today (Independence Hall Association 1995)
1760	Friedrich von Knauss	Creates a true writer machine capable of writing predetermined text or manually controlled by a letter board (Dalakov, Friedrich von Knauss 2012)
1771	Richard Arkwright, Jedediah Strutt, and Samuel Need	Set up the first true factory, which is next to the River Derwent in Cromford, Derbyshire (Simkin 1997)
1772	Pierre Jaquet-Droz	Builds the *Writer*, a child figure that can write with spacing and punctuation and is controlled by a programmable mechanical computing device of impressive complexity (Dalakov, Pierre Jaquet-Droz 2012)

Year	Person(s) or Company	Event
1804	Joseph-Marie Jacquard	Invents the Jacquard loom, an automation system for looms utilizing a punch card control system (Dalakov, Joseph-Marie Jacquard 2012)
1810	Friedrich Kaufmann	Creates a mechanical trumpet player controlled by a stepped drum (Timeline of Robotics 1 of 2, 2007)
1824	Hisashige Tanaka	Masters the art of constructing mechanical dolls and travels to Japan as an entertainer (Toshiba-Cho 1995)
1835	Charles Babbage	Designs an analytical engine complete with processor and memory functions that many credit as the world's first general-purpose computer, despite the fact the actual machine was never built (Charles Babbage n.d.)
1839	Sir William Grove	Creates the first fuel cell (Roberge 1999)
1843	Augusta Ada King	Publishes her notes on Babbage's analytical engine, which is considered the first example of a computer program (O'Connor and Robertson, Augusta Ada King, Countess of Lovelace 2002)
1847	Gorge Boole	Publishes his *Mathematical Analysis of Logic*, which he later develops into what we call Boolean algebra, a type of math widely used in the computing field (George Boole 2012)
1868	Zadoc P. Dederick	Patents a steam-powered man for pulling carriages (Buckley 1868, Zadoc P. Dederick, Steam Man 2007)
1873	William Thomson (Lord Kelvin)	Develops a tide predictor that is a special purpose analog computer (Sharlin n.d.)
1882	Nikola Tesla	Creates the concept for the modern AC induction motor (Vujovic 1998)
1887	Thomas Edison	Invents a talking doll that is marketed to the public in 1890 (Buckley 2007)
1889	Herman Hollerith	Patents his punch-card-driven tabulation machine used in conjunction with the 1890 census (Cruz 2011)
1892	Seward Babbitt and Henry Aiken	Patents a crane with a gripper, designed primarily to help move metal billets (Babbit and Aiken 1892)
1895	Nikola Tesla	Designs the world's first AC power plant at Niagara Falls (Vujovic 1998)
1896	Herman Hollerith	Starts the Tabulating Machine Company, which ultimately becomes IBM (Cruz 2011)
1897	Joseph John Thomson	Creates a series of experiments that helps him infer the existence and characteristics of the electron (Chemical Heritage Foundation 2010)
1898	Nikola Tesla	Patents his design for a radio-controlled boat (Vojovic 1998)
1921	Karel Capek	His play *R.U.R.* (Rossum's Universal Robots) is the first documented usage of the word "robot"; this term is based off the Czech word *robota*, which means "drudgery," or "slave-like labor" (Capek 1890–1938)
1927	Fritz Lang	Debuts the term "robot" in his film *Metropolis* and gives publicity to the term (Helm et al. 1927)
1927	Roy Wensley	Builds *Herbert Televox* for Westinghouse Electric and Manufacturing Co., the first of several robots that the Westinghouse company created (Dalakov, *The Robots of Westinghouse* 2012)
1928	Makoto Nishimura	Creates *Gakutensoku*, a large robotic humanoid capable of movement, facial expressions, and writing (Hornyak 2008)
1937	Alan Turing	Publishes a paper on mathematics of fundamental importance to future computer development (O'Connor and Robertson, Alan Mathison Turing 2003)
1938	Joseph Barnett	Creates *Electro the Moto-Man* for Westinghouse Electric and Manufacturing Co., the most successful of the Westinghouse robots, with 26 routines; a vocabulary of 700 words, operated by voice commands; and featured at the 1939 World's Fair (Dalakov, The Robots of Westinghouse 2012)

Year	Person(s) or Company	Event
1941	Harold Roselund	Directs the building of the first industrial robot for the DeVilbiss Company, based on the patented design of Willard L.G. Pollard, Jr. (Bonev 2003)
1942	Isaac Asimov	Publishes the short story *Runaround* and gives the world his three laws of robotics, which many would consider the guideline for future robotic development (Famous People n.d.)
1943	Warren S. McCulloch and Walt Pitts	Publish *A Logical Calculus of the Ideas Immanent in Nervous Activity*; in their work, they try to understand how the brain works and lay the groundwork for possible thinking computer systems (Marsalli n.d.)
1943	Thomas H. Flowers	With his group, develops *Colossus*, credited as the world's first electronic computer (Copeland 2006)
1948–1949	William Grey Walter	Builds two robots based on his theory of the nervous system (LeBouthillier 1999)
1950–1956	Edmund C. Berkeley	Berkeley Enterprises Inc. creates several robotic systems designed to be "Robot Show-Stoppers," including *Simon*, a miniature mechanical brain; *Squee*, an electronic robot squirrel; and *Franken*, a maze-solving robot (Berkeley 1956)
1957	Planet Company	Demonstrates PLANETBOT—a hydraulically powered, polar coordinate arm with five axes—at an international trade fair (Nocks 2007)
1958	Jack Kilby	Creates the first microchip, an invention crucial to the electronics field (Texas Instruments 1995)
1958	FANUC	Ships the first commercial Numerically Controlled (NC) machine to Makino Milling Machine Co., Ltd. (Fanuc 2011)
1960	American Machine and Foundry	Ships the first VERSATRAN, a programmable robotic arm designed by Harry Johnson and Veljko Milenkovic (Nocks 2007)
1961	George Devol	Receives the patent for an arm-type industrial robot that would be produced by the robotics company Unimate and used by GM (Malone 2011)
1963	Rancho Los Amigos Hospital	Researchers create the Rancho arm, which has six joints to give it the flexibility of a human arm and is designed to help the handicapped (Computer History Museum 2006)
1966–1972	Stanford	One of the first true experiments in Artificial Intelligence (AI) takes place at the Stanford Research Institute with the robotic platform Shakey (Nilsson 1984)
1967	AMF	Ships a VERSATRAN robot to Japan for industrial use (Nocks 2007)
1968	Marvin Minsky	Develops a robot arm based on the design of a tentacle, with 12 joints to give it added mobility options (you can see video of it in operation on YouTube under "AI History: Minsky Tentacle Arm") (Computer History Musem 2006)
1968	Kawasaki Robotics	Starts the production of hydraulically powered robots under the Unimation license (Kawasaki Robotics (USA) 2012)
1968	Ralph Mosher	Develops the Walking Truck at General Electric, a vehicle with four legs instead of wheels and controlled by force feedback (Kotler 2005)
1969	Victor Scheinman	Designs the Stanford Arm, an all-electric, computer-controlled robotic arm used to assemble various parts during lab experiments (Wiederhold 2000)
1970–1973	Waseda University	The bioengineering group at Waseda creates WABOT-1, the world's first anthropomorphic robot (Humanoid Robotics Institute n.d.)
1973	Edinburgh University	The Artificial Intelligence Department creates Freddy II, a robot that can assemble a wooden toy car in 16 minutes from a jumble of parts (Tate 2011)
1973	Richard Hohn	Cincinnati Milacron releases the T3 robot arm, designed by Hohn, which is credited as the first commercially available, microcomputer-controlled industrial robot (Control engineering 2009)

Year	Person(s) or Company	Event
1973	KUKA	Starts to develop robotic systems with the IR 600, produced in 1978 (KUKA 1998)
1974	David Silver	Designs the *Silver Arm*, which can assemble small parts using built-in touch and pressure sensors (Computer History Musem 2006)
1974	Leif Jonsson	Takes delivery of ASEA's (now ABB) all-electric, microprocessor-controlled robot for Magnusson, a company in Sweden; the first commercially available robot of this type, it is still in service 40 years later (you can watch a video of this robot on YouTube under "ABB Robotics – Where it all began") (Ciampichini 2012)
1976	Shigeo Hirose	Creates a soft gripper able to conform to the contours of an object (Computer History Musem 2006)
1976	MOTOMAN	Begins as a robotic welding company in Europe (Yaskawa Motoman 2010)
1977	Hans Moravec	Rebuilds the Stanford Cart, adding sensor systems that would eventually allow it to navigate an obstacle-filled room without help in 1979 (Computer History Musem 2006)
1977	Victor Scheinman	Sells his company VICARM to Unimation, which ultimately leads to the development of the Programmable Universal Machine for Assembly (PUMA) robot (Munson 2010)
1977	FANUC	The FANUC cooperation is established in the United States (Fanuc 2011)
1978	Hiroshi Makino	Develops the SCARA (selective compliant articulated robot Arm) at the Yamanashi University, which sells in the United States as the IBM 7535 and holds the honor of being the first industrial Japanese robot (Yamafuji 2008)
1981	Dr. Robert J. Shillman	Leaves MIT to start COGNEX, a vision system company that in 1982 released the world's first industrial optical character recognition (OCR) system DataMan (COGNEX 2012)
1983	Adept Technology	Produces industrial robots, machine vision systems, and other automation equipment in the United States in the same year as the company's creation (Adept Technology 1996)
1984	Takeo Kanade and Haruhiko Asada	Receives the patent for the first direct-drive robotic system, greatly increasing the speed and accuracy of robotics (Kanade and Asada 1984)
1985	Kawasaki	Begins its own global business as its collaboration with Unimation ends (Kawasaki Robotics (USA) 2012)
1986	Kazuo Yamafuji	Develops the Parallel Bicycle Robot at the University of Electro-Communications, which does not see commercial applications until the Segway is developed in 2001 (Yamafuji 2008)
1989	Colin Angle	Develops Genghis, a six-legged autonomous walking robot for MIT (Angle 1989)
1991	Dr. Mark W. Tilden	Inspired by the Genghis and Attila robots, Tilden creates the BEAM (biology electronics aesthetics mechanics) robotic concept (Hrynkiw and Tilden 2002)
1993	Epson	The 1993 version of the Monsieur robot goes into the *Guinness Book of World Records* as the world's smallest robot; it has a volume of 1 cubic centimeter and consists of 98 components (Seiko Epson Corp. 2012)
1994	MOTOMAN	Introduces the world's first robot controller capable of synchronizing two robots (Yaskawa Motoman 2010)
1994	Carnegie Mellon	Scientists explore the Mt. Spurr Volcano with the Dante II robot, ushering in a new field for robotics: remote data collection in harsh environments (Bares n.d.)
1994	AESOP	AESOP (Automated Endoscopy System for Optimal Positioning) is the first robot approved by the FDA for abdominal surgery (Valero et al. 2011)
1996	Chris Campbell and Stuart Wilkinson	A brewing accident leads to Gastrobot, a robot powered by digested sugar (Hapgood 2001)

Year	Person(s) or Company	Event
1997	Honda	Introduces the P2 robot, a bipedal humanoid robot that walks smoothly for over 30 minutes (Yamafuji 2008)
1997	Cynthia Breazeal	Begins work on Kismet at MIT, a robot designed to learn and respond socially (MIT news 2001)
1998	NASA	Launches Deep Space 1, a craft with artificial intelligence used to conduct testing and tasks needed to prove out new technologies (NASA 2001)
1998	Campbell Aird	Has the honor of being fitted with the world's first fully mobile "bionic," or robotic, arm (BBC News 1998)
1998	LEGO	Introduces the MINDSTORMS line, a popular system for robotic education (Mortensen 2012)
1999	Probotics, Inc.	Releases the Cye robot, billed as the first affordable personal robot for home or office use and designed to perform tasks such as carrying items, leading people, and cleaning (Hendrickson 1999)
1999	Intuitive Surgical, Inc.	Introduces the da Vinci medical robotic system, which is approved for laparoscopic surgery in July of 2000 (Intuitive Surgical, Inc. 2010)
2000	Sandro Mussa-Ivaldi	Mussa-Ivaldi and his team connect a robotic system to the brain of a lamprey to glean insight into how to control prosthetic limbs with the mind (BBC News 2000)
2000	Honda	Creates ASIMO (Advanced Step in Innovative Mobility), utilizing the lessons learned with its P2 and P3 humanoid robots (Honda Motor Co., Ltd. 2007)
2000	Waseda University	Creates Waseda Talker No. 1 to reproduce human vocal movement, which serves as the basis for several improved versions to follow (Takanishi Laboratory 2012)
2000	NASA	Produces the first version of the Robonaut, a dexterous robot designed to work with astronauts (Canright 2012)
2004	Dr. Mark W. Tilden	Creates Robosapien for WowWee: a robotic toy that is the culmination of lessons learned through BEAM research and designed with user modifications or hacking in mind (Boyle 2004)
2004	Aaron Edsinger-Gonzales and Jeff Weber	Design Domo, a robotic research platform at MIT created to advance human–robot interaction, utilizing a unique feedback system to make it safer for human interaction (Edsinger-Gonzales and Weber 2004)
2005	Dr. Hod Lipson	Creates a robotic system capable of simple self-replication, or the ability to create an identical robotic system from simple units (Steele 2005)
2005	Ralph Hollis	Creates the Ballbot, a robot that uses a single ball for locomotion, opening a new avenue of robot locomotion (Hollis 2006)
2005	Brian Scassellati	Develops Nico, a robot designed to be aware of and recognize its own parts and movements; this robot has the mentality of a 1-year-old child and a level of self-awareness never before achieved in a machine (Scassellati and Sun 2005)
2006	MOTOMAN	Introduces a dual-arm robot with 13 axes, with all cabling run internally and a human torso appearance (Yaskawa Motoman 2010)
2006	Intuitive Surgical, Inc.	Introduces the da Vinci S System, offering high-definition vision in robotic surgery for the first time (Intuitive Surgical, Inc. 2010)
2006	Cornell University	Josh Bongard, Victor Zykoy, and Hod Lipson create a robot that can walk without any initial data about its shape and design and can correct for damage and continue to function, adapting its movement to the new limitations (Ledford 2006)
2007	NASA	Partners with General Motors to begin work on the Robonaut 2 (Wilson 2010)
2008	ReconRobotics	Introduces the Recon Scout, the world's first robot designed to be thrown and capable of seeing in complete darkness (Klobucar 2008)

Year	Person(s) or Company	Event
2009	Intuitive Surgical, Inc.	Releases the *da Vinci Si System* that allows two surgeons to work collaboratively (Intuitive Surgical, Inc 2010)
2009	MOTOMAN	Releases the *DX100* controller, which can control up to eight robots or 72 axes of movement (Yaskawa Motoman 2010)
2009	Aldebaran	Demonstrates the NAO robot system at the International Robot Exhibition, which sells commercially in 2010 (Salton 2009)
2009	Microsoft	Debuts Project Natal, which is released as the Kinect system in 2010 and later adapted as a sensor package for various robotic platforms (Lowensohn 2011)
2010	Lely	Introduces a robotic milking system for the dairy industry (Saenz 2010)
2012	NASA	Robonaut 2, a collaboration between NASA and General Motors, completes the first human/robot handshake in space (Kauderer 2012)
2012	Tecnalia	Begins work on adapting Kawada Industries' Hiro robot to work alongside European factory workers, with the goal of a robot that works with people instead of inside a cage (Fundazioa 2012)
2012	Rethink Robotics	Releases Baxter, a low-cost humanoid torso-type robot designed to be easily taught tasks, work safely around humans, and require minimal setup straight out of the box (Guizzo and Ackerman 2012)
2012	University of Pennsylvania	Nader Engheta and his group develop the first metatronic circuit, which uses light instead of electricity; when fully developed, this technology could create smaller, faster, and more efficient electronics (Lerner 2012)
2012	SpaceX	Becomes the first privately owned company to launch and dock a robotic capsule (the Dragon) with the International Space Station (Robotics Trends 2012)
Today	YOU	Begin or continue your exploration of the field of robotics to ensure a bright and robust future for robots!

BCE
Before Common Era

abacus
A counting device that uses beads to help the user keep track of numbers and make complex mathematical calculations easier

End-of-Arm-Tooling [EOAT]
The part of the robot that manipulates parts or performs the programed tasks of the robot.

The timeline contains only a few events that occurred Before the Common Era (BCE), but these set the groundwork for the advances that would follow. The abacus, a counting device that uses beads to help the user keep track of numbers and make calculations, made the list because it represents the first attempts of our ancestors to conquer complex math. It is a proven fact that technical fields and math are inseparable and codependent. Many times, math has been advanced by the technical need to figure out a volume, distance, torque, force, etc., which then in turn led to advances in related technical fields.

Two other major events in the BCE section are the invention of a wooden pigeon that could fly under steam or compressed air power and the understanding of pneumatics. Archytas of Tarentum's flying pigeon acted as a spark to get the thinkers of the time to explore how steam and compressed air worked, why it worked, and how could it be used. Ktesibios (or Ctesibius) of Alexandria would add his own discoveries to create the basic principles of pneumatic power as we now know it. His work would breathe life into other inventors' creations as well as give man a powerful force-multiplying tool. Pneumatic power is still widely used to this day and is the primary power source for movement in the world's fastest robots as well as a common force for End-of-Arm-Tooling (EOAT) movement. (EOAT is the part of the robot that manipulates parts or performs tasks; you will learn more about this in a later chapter.)

In about 200 BCE, Chinese artisans created an entire orchestra operated by air power and ropes; this in turn moved cams and gears that gave life to their wooden creations. The skilled operators of these devices would time their actions to create performances in which their creations mimicked human movements. Over two millennia ago, audiences watched these mechanical marvels perform in much the same way we might watch a band or orchestra perform today! We often think that all of man's great technological inventions were created in the last century or two, but that is simply not true. There is a very good chance that other great things happened before the Common Era [CE] for which we simply have no verifiable records, thus becoming lost to the sands of time.

CE
Common Era

Heron of Alexandria and his steam engine design emerged shortly after the start of the Common Era. Images of trains and tractors from the 1800s are often the first things to come to mind when we think of steam engines; however, now we find that the first steam engine was designed somewhere between 10 and 70 CE. This added another power source to the arsenal of the early inventors as well as advanced the field of pneumatics (steam is after all a form of air power). Heron would go on to create automata, which are devices that work under their own power, often designed to mimic people. Heron's *Hercules Killing the Dragon* was an automata powered by water in which the dragon would "breathe" a stream of water representing fire at Hercules and then Hercules would slay the dragon with his club. While this may seem a bit simplistic by today's standards, it was a technological marvel at the time of its creation.

automata
Devices that work under their own power and are often designed to mimic people

Let us now move forward to 1495 and Leonardo da Vinci's *Metal-Plated Warrior*. Many of us know da Vinci for his works of art and engineering, but his moving suit of armor was an engineering feat overlooked for many years by the modern world. The suit of armor was able to sit up, move its arms, move its head, and raise its visor via a system of gears and pulleys situated inside the armor. This invention, created to entertain da Vinci's dinner guests, inspired automata that followed and was a marvel of engineering genius. Modern engineers have copied da Vinci's drawings and diagrams of the inner workings of the suit and confirmed that it did indeed move as reported by creating a replica from the designs that survived.

In the 1500s, inventors created several lifelike automata to wow the masses. From Hans Bullmann's humanoid automata, credited as the first real androids; to Gianello Torriano's *Lute Player Lady*; to John Dee's beetle that was so lifelike he had charges of witchcraft leveled against him, the inventors of the sixteenth century created a great knowledge base for future robotic endeavors. It was also during this time that Pare Ambroise published his idea for an artificial hand using mechanical muscles along with organic components. Ambroise's approach was to create a prosthetic or artificial replacement that closely resembled the lost hand in appearance and function instead of something fashioned out of wood with a hook in it. This was the beginning of the long journey of modern prosthetics. It seems that the 1500s was the time frame in which human desire, knowledge, and acceptance all came together to bolster the engineering and scientific fields necessary for the great creations that would follow.

prosthetic
An artificial replacement for a missing or damaged organic part that often closely resembles the part it is replacing in form and function

If we think of the sixteenth century as the advancement of the machine, we can describe the seventeenth century as the advancement of the controller. During the 1600s, William Oughtred gave us the slide rule, a device for quickly looking up complex calculations that proved to be a tremendous timesaver. The slide rule was in popular use until the mid- to late-twentieth century, when the scientific calculator took over this function. Speaking of calculators, Wilhelm Schickard and Blaise Pascal both created simple function calculators during the 1600s. These calculators could add, subtract, multiply, and divide using complex gearing systems and properly calibrated analog, or number, displays. These devices made mathematical calculations easier, faster, and more accurate, helping early inventors as well as those who needed to do large amounts of calculation. These inventions inspired greater computation systems as well as device control systems in the years that followed. Gottfried Wilhelm von Leibniz rounded out the "century of the controller" with his development and perfection of the binary system in 1679. This is the same system used by digital computers and many modern technologies to process large amounts of data quickly using 1s and 0s.

The eighteenth century brought many improvements of previous technology as well as few innovations of its own. Lorenz Rosenegge created a mechanical theater of automata powered by a single water turbine, an updated version of the Chinese mechanical orchestra from Before the Common Era. Jacques de Vaucanson created clockwork automatons, or self-operated machines, of staggering realism and complexity. His Digesting Duck walked, quacked, ate, and expelled waste, though it did not truly digest the food. Mid-century Benjamin Franklin conducted his famous kite experiment and investigation into electricity. Franklin's discoveries laid the groundwork for our understanding of electricity and all the benefits that the mastery of this power source brings to the modern world. Friedrich von Knauss created a machine capable of writing predetermined sentences as well as working like a typewriter when keyed manually. In 1772, Pierre Jaquet-Droz improved on Knauss's design and created a childlike figure that would write out phrase in the same fashion that a person would. Pierre's creation had a programmable mechanical controller that created the written text as well as the human actions of the childlike figure. His design allowed for changes in behavior and operation without a complete rebuilding of the automaton, which was a big step forward in the field. It was also during this century that Richard Arkwright, Jedediah Strutt, and Samuel Need set up the world's first true factory, planting the seeds for the industrial revolution that would later transform the world.

The nineteenth century brought many important and influential events that shaped our modern world. Right at the turn of the century in 1804, Joseph-Marie Jacquard created his punch card control system for the mechanical loom, which used various holes in a metal or wooden rectangle called a "card" to control the timing and flow of actions. To change the operation of the loom, one simply put in a card with a different arrangement of holes in it. Jacquard designed his system as an add-on to a loom instead of requiring a machine specially built for his device, making it compatible with many of the mechanical looms of his time. The punch card system was widely used for machine control until the late 1970s to early 1980s when the Computer Numerically Controlled (CNC) systems took prevalence, using computer code to control the action of machines instead of holes in cards. Over nearly two centuries, various inventors improved on Jacquard's design, but it took the invention of the computer to replace the punch card system as a go-to source for machine control!

The 1800s saw the birth of another control technology that is still around today. Friedrich Kaufmann created a mechanical trumpet player controlled by a drum with high and flat spots on it that triggered levers and cams during rotation. To alter the operation of this system, one changed the arrangement and spacing of high spots on the drum. This system has also survived the test of time, earning a place in industry alongside Jacquard's punch card system for machine control as well as being one of the early control systems for robots.

Charles Babbage finishes out the first half of the nineteenth century with his design of an analytical engine that is credited as the world's first general-purpose computer (even though he never built it). Augusta Ada King read Babbage's published work on the analytical engine and wrote a paper on how to use his device. Many consider Augusta's paper to be the first example of a computer program. Gorge Boole published his "Mathematical Analysis of Logic," which we know and love as Boolean algebra today. If you are into computer programing or work with

automatons
Self-operated machines, often designed to emulate something living

punch card control system
A control system that uses a card made of a sturdy substance with a series of carefully placed holes that can control the sequence and operation of machines

Computer Numerically Controlled (CNC)
The common descriptor used for machines that use computers and software to control system operation

gantry
A simple two- or three-axis machine/robot designed to pick up parts from one area and place them in another

Artificial Intelligence, I can almost guarantee that you will learn all the ins and outs of this mathematic field.

The second half of the nineteenth century follows the familiar pattern, with many practical applications of older technology along with some new innovations. In 1868, Zadoc P. Dederick created a steam-powered man that moved his arms and legs while pulling carriages, and in 1887, Thomas Edison invented a talking doll for the public. We often think of these types of inventions as part of the modern world, not technology that is well over a century old! In 1889, Herman Hollerith used punch card technology for a tabulation machine and started his own company, which ultimately became IBM, a prime player in the twentieth century computer field. Between 1882 and 1898, Nikola Tesla, the father of modern electricity, conceived the AC induction motor, designed the world's first AC power plant at Niagara Falls, and patented a radio-controlled model boat. This is by no means all that Nikola Tesla gave the modern world, but these were critical developments for the modern robot we have today. In 1892, Seward Babbitt and Henry Aiken patented a crane with a gripper for industry, arguably one of first Gantry type systems. ("Gantry" is the name given to many simple two- or three-axis machines designed to pick up parts from one area and place them in another.) Many of the early robotic material handling systems would mimic the basic design of this crane. Finishing out the list of nineteenth-century inventions, Joseph John Thomson inferred the existence of the electron in 1897. His discovery gave us a crucial understanding of the atom, which laid the groundwork for discovering how electricity works at the atomic level.

The twentieth century bore witness to the birth of the digital age, the modern robot, and many of the technological devices we use and love, but it did not happen all at once. In 1921, Czech writer Karel Capek gave the world the term robot, derived from the Czech term robota, which means "drudgery," or "slave-like labor." Fritz Lang would help popularize "robot" in his 1927 film *Metropolis*, strengthening its place in the twentieth-century dialogue (see Figure 1-1). The word "robot" became widely used in science fiction movies and to describe automaton and automata from the past. The fact that the term "robot" is less than a hundred years old yet used to describe things that predate it has led to confusion and conflict over the term's meaning. We will explore the current definitions of "robot" later in this chapter.

In 1927, Roy Wensley built a humanoid robot known as *Herbert Televox*, which was the first of several robots backed by the Westinghouse Company for promotional purposes. This robot could lift a phone as if answering it, manipulate a few simple switches, and after some modifications utter a couple of sentences. A year later, Makoto Nishimura unveiled *Gakutensoku* as Japan's first attempt at creating a "robot." This

robot
A machine equipped with various data gathering devices, processing equipment, and tools for operational flexibility and interaction with the systems environment, which is capable of carrying out complex actions under either programmed control or direct manual control.

robota
Drudgery, or slave-like labor

Figure 1-1   One of the actors from Karel's play *R.U.R.*; the DVD and bobblehead are from a special anniversary rerelease of *Metropolis*.

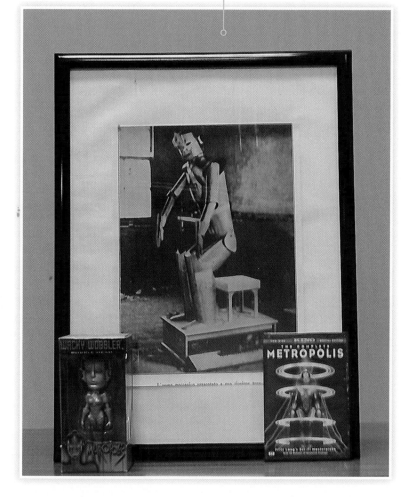

humanoid robot was about 10 feet tall and had elven features meant to be a combination of multiple races. The whole point was to create something lifelike, so Nishimura designed his robot with simulated breathing, various facial expressions, and the ability to write. Joseph Barnett's 1938 *Electro the Moto-Man*, another of the Westinghouse Company robots, could walk around on rollers in its feet, count on its fingers, recognize red and green light, smoke cigarettes, and talk. The operator controlled this impressive system by voice commands sent to a telephone in the robot's chest. The robot's systems converted sound vibrations from the words into signals of light that in turn controlled various relays in the system for action.

In 1941, the DeVilbiss Company built what many consider the first industrial robot, based on the patented parallel robot design of Willard L.G. Pollard Jr., bringing robots into the world of industry. Harold Roselund directed the creation of this robot for spray painting applications to ensure even, consistent coats of paint while minimizing waste. Ultimately, this application for robots found its way into the automotive industry and began what has proven to be a beneficial partnership. The automotive industry is currently one of the top utilizers of robotic technology. Companies such as General Motors and Honda have taken this a step further and have helped with or designed advanced robotic systems.

As Karel Capek gave the world the term "robot" from his play, Isaac Asimov gave the world the three laws of robotics through his science fiction in 1942. Many in the robotics realm accepted these laws as the outline of how robots and humans should work together. The robots in Asimov's writings did not enslave the human race, start global wars, go on killing sprees, or create any other dark story line for the human race; instead, they became stuck in logic loops that created erratic operation due to the poorly thought-out commands of people. In a time when many people were just learning about the wonder of the robot, Asimov's laws provided a path to a bright future where man and machine worked together instead of the dark future some feared. Asimov's three laws are as follows:

- *First Law*: A robot may not injure a human being or, through inaction, allow a human being to come to harm.

- *Second Law*: A robot must obey any orders given to it by human beings, except where such orders would conflict with the First Law.

- *Third law:* A robot must protect its own existence as long as such protection does not conflict with the First or Second Law. (FamousPeople n.d.)

In 1943, Thomas H. Flowers and his group developed the world's first electronic computer, Colossus. The first computers were huge machines that took up enormous amounts of space and required many hours of labor and often days of time to change programing. While Flowers and his group created Colossus, Warren S. McCulloch and Walt Pitts published a paper laying the groundwork for the possibility of Artificial Intelligence (AI): software and/or hardware that is capable of processing data similar to human thought, giving it the ability to deal with questions where there is no clear right answer, or where multiple answers would work as well as ultimately intuitive jumps in problem solving. These two events planted the seeds for the technological world that we know and love, where computers are compact and readily available and robotic systems are capable of completing complex tasks while traversing difficult terrain with little or no help from their human creators.

The latter part of the 1940s and the first half of the 1950s was a period of exploration for the new technologies of the time. From 1948 to 1949, William Grey Walter built two robots designed to emulate his theories on neural systems. These innovative robots of the time could detect light sources and move toward them, detect impact and change directions, and find their way back to recharging

Artificial Intelligence (AI)
Software/hardware that is capable of processing data similar to human thought, giving it the ability to deal with questions where there is no clear right answer, or where multiple answers would work as well as ultimately intuitive jumps in problem solving.

stations. Technology and advancements in the field have improved how robots perform these tasks, but they remain a part of robotic behavior to this day. From 1950 to 1956, Edmund C. Berkeley created robotic Show-Stoppers for the sole purpose of getting people to stop and take notice. While they were more of an advertising gimmick, these "Show-Stoppers" brought robotics into the limelight. These devices ranged from a robotic brain that could take math problems submitted in a punch card format and then give answers via blinking lights, to machines that could play games such as tic-tac-toe, to robots that could solve mazes created by audience members. My personal favorite of Berkeley's robots is *Squee*, named after squirrels, which would scoop up a tennis ball that someone was standing over when directed by a flashlight shone toward the robot. Once it had the tennis ball, or "nut," it would take it back to a control area, illuminated by a light that pulsed on and off 120 times per second, to deposit the "nut" in the "nest."

Robotics experienced two key events in 1958. One was the creation of the first microchip by Jacky Kirby. Microchips are a necessity of the digital age; we use the descendants of Kirby's microchips in modern robotics, electronics, and many aspects of our modern lives. These tiny chips allow for Boolean algebra logic flow, power manipulation and control, data storage, and many other complicated tasks necessary for modern computation and control of devices. The microchip is the invention that allowed computers to become small portable units instead of remaining room-size behemoths. The second event was the FANUC Corporation's shipment of its first Numerically Controlled (NC) machine. NC machines used punch cards (like the Jacquard loom) or magnetic tapes (similar to VHS or cassette tapes) with position and sequence information to control the motions and actions of the machine. This eventually led FANUC into the realm of robotics and, with its trademark yellow robot (see Figure 1-2), to its current position as a top robotic supplier for industry.

Show-Stoppers
Devices built for the purpose of getting people to stop and take notice, typically used to draw attention to products or booths at trade shows

Numerically Controlled (NC)
Numerically controlled machines use punch cards or magnetic tapes with information encoded on them to control the operation of equipment

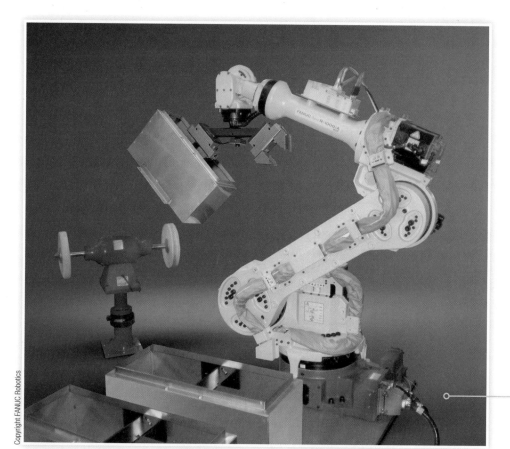

Copyright FANUC Robotics

Figure 1-2   While not an early system, this is an example of the FANUC robots used by industry.

It was also at about this time that industrial robotics really began to expand and take root. In 1957, Planet Corporation exhibited PLANETBOT at an international trade fair; this hydraulically powered robotic arm, with its five axles and polar coordinates, gave many industries their first look at what was considered a useful robot. This robot was not a "Show-Stopper" or limited to painting operations, it had the versatility that industry was looking for. In 1960, American Machine and Foundry shipped its first VERSATRAN programmable robotic arm to an American customer and in 1967, shipped its first robot to Japan. In 1961, George Devol patented his industrial robot arm, which became one of the Unimate robotic models used by General Motors (GM). In 1968, Kawasaki Robotics started producing hydraulically powered robots for Unimation, beginning its rise in the robotics field. The centuries of science, technology, engineering, and math had finally advanced to the point that the industrial robot was a viable technology, and many companies were ready to embrace this new piece of equipment.

While industry was beginning to integrate the available technology, scientists and engineers were working on the next generation of robots. In 1963, Rancho Los Amigos Hospital designed the Rancho arm—a six-jointed arm that mimicked the human arm—with hopes of helping people with disabilities. This configuration is now widely used in industry as it has the ability to copy human movements, freeing operators from mundane tasks that often lead to ergonomic injuries. In 1968, Marvin Minsky developed a robotic arm with 12 joints designed to work like the tentacle of an octopus. While this design is not widely used in modern robotics, it demonstrates that with a little thought and ingenuity robotic systems can copy many of the natural movements of living creatures. At about the same time, Ralph Mosher developed a robot for General Electric known as the *Walking Truck* or *Giant Elephant* (Kotler 2005). This robotic vehicle had four legs instead of wheels and used a force-feedback system that interpreted the motions of the user's arms and legs to drive the robot. In 1969, Victor Scheinman designed the Stanford Arm, an all-electric, computer-controlled robotic arm. He used this arm to assemble parts during various experiments, proving that robots were capable of many tasks that his contemporaries thought impossible. In 1974, Victor Scheinman founded the VICARM Company to promote and sell his robotic arm design.

From 1966 to 1972, Stanford University worked on Shakey, a robotic platform designed for experimentation with Artificial Intelligence, the ability of a machine to think like a person. Many credit this as the first true experimentation with AI and the inspiration for systems that followed. From 1970 to 1973, Waseda University created *WABOT-1*, a robot designed to be as human as possible that earned the title of the world's first anthropomorphic (having human characteristics) robot. While many of the robots previously mentioned had human traits, the complexity of this robot set the bar for what it takes to make an anthropomorphic robot. In 1973, Edinburgh University created *Freddy II*, a robot that used rudimentary AI to assembly a toy car from a jumble of parts. In 1974, David Silver developed the *Silver Arm*, which could assemble parts by touch using built-in pressure sensors. Many of the technologies created through these various experiments found their way into the robotics industry and are a crucial part of the modern robot.

The 1970s was about integrating the new advancements in robotics into industrial robotic systems, some of which are still in service at the writing of this book! Richard Hohn designed the T3 robot for Cincinnati Milacron, released in 1973, that was the first commercially available robot controlled by a microcomputer. The T3 utilized the microchip technology that Jack Kilby gave the world so that the controller of the robot took up less space on the factory floor. The following year,

anthropomorphic
The term used to describe things that move in the same manner as humans would, imparting to the motion the feeling of a living presence

return on investment (ROI)
The common term used to describe the length of time that it takes a piece of equipment to pay for itself

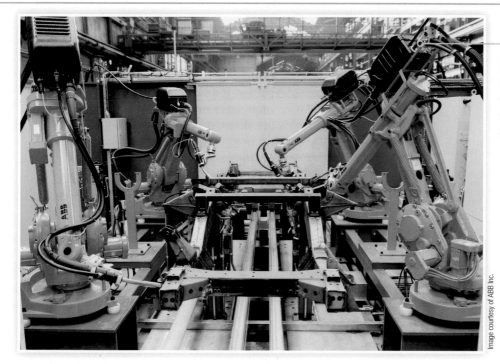

Image courtesy of ABB Inc.

Figure 1-3   While these robots are not the robots sold to the Magnussons Company in Sweden, they are very similar in design.

ASEA (now known as ABB) delivered the first fully electric microprocessor-controlled robot to the Magnussons Company in Sweden. Forty years later, this robot is still in service, handling and polishing parts for Magnussons! This illustrates how, with proper care, a robotic system similar to Figure 1-3 can last for decades, giving industry a very profitable return on investment (ROI). (Chapter 12 covers ROI in detail for those wanting to know more.)

In 1976, MOTOMAN started a robotic welding company in Europe that would survive the harsh competition of the 1980s and emerge as one of the key players in modern robotic systems (see Figure 1-4). The following year, Victor Scheinman sold his company VICARM to Unimation, which ultimately lead to the Programmable Universal Machine for Assembly (PUMA) robot. The PUMA robot is another example of the older systems that are still around; it was the primary reason that the Staubli Group bought Unimation from Westinghouse in 1989 to create a new Staubli robotics division.

We finish out the 1970s with Hiroshi Makino's Selective Compliant Articulated Robot Arm (SCARA), developed in 1978 (see Figure 1-5). This robot holds the honor of being the first Japanese industrial robot and one that proved sound for use in industry. Today, companies like KUKA and MOTOMAN sell this type of robot for use in the precise handling of small components for assembly, fastening, and soldering as well as tasks requiring a large amount of downward force. The SCARA robot has become a favorite among various electronics companies for the rapid placement of components on circuit boards.

In the 1980s, manufacturing started to invest in robotics, as they realized the value of the systems, leading to a greater demand for robots. The increased demand inspired small businesses to try their hand at robots, creating a boom of new robotic manufacturers. Unfortunately, many of these companies would last only a year or two at the most. Some of these companies simply did not have the expertise

Figure 1-4   A current production MOTOMAN robot.

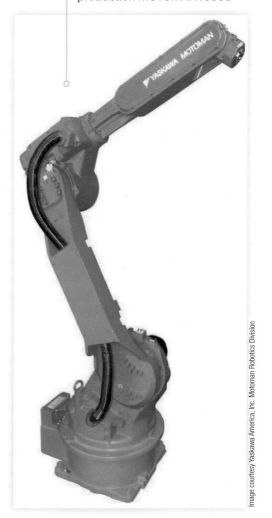

Image courtesy Yaskawa America, Inc. Motoman Robotics Division

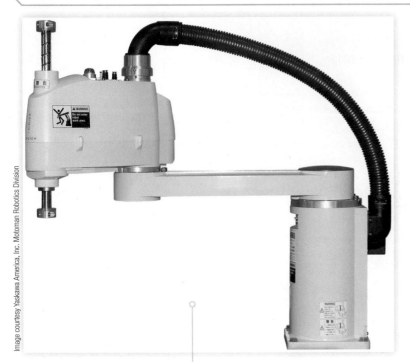

Image courtesy Yaskawa America, Inc. Motoman Robotics Division

Figure 1-5 MOTOMAN's SCARA–type robot. The popularity of this design led several of the robotics companies to invest in and create this type of robot.

character recognition
The ability to read written or printed letters, numbers, and symbols

Biology Electronics Aesthetics Mechanics [BEAM]
The study of robotics involving simple systems without complex controllers

bionic
A robotic system designed to mimic the human part it replaces that is controlled by nerve impulses, usually through the use of transmitters implanted into the user's body

to compete in the field or could not capture the market share they needed. Others were more successful and drew the attention or ire of the larger, established robotics companies. If a successful small company had an impressive product or a fair amount of market share, the larger companies would often try to buy them out or collaborate with them in some way. If this failed, the larger companies would simply work to take market share from the smaller company and force it out of business, thus eliminating their competition. While the names of these small companies are not important for our purposes, it is important to note that you may come across some of these systems still working in industry. As seen from our ABB robot example, robots can survive for decades with proper maintenance, and many manufacturers are reluctant to buy a brand-new robotic system when the one they have is still working.

This turbulent time for robotics also saw some new technology emerge. In 1982, COGNEX released DataMan, the brainchild of Dr. Robert J. Shillman, which was capable of character recognition, or the ability to read written or printed letters, numbers, and symbols. This system allowed manufacturers to track parts, process written data, and give robotic systems the ability to "see." Takeo Kanade and Haruhiko Asada patented their direct drive robotic system in 1984: where the motor and joint connected directly instead of using gears, pulleys, and belts or sprockets and chains to create motion. This resulted in increased speed and accuracy of the robot while reducing maintenance requirements and explains why the majority of modern systems use direct drive.

By the 1990s, robots had earned their place in industry, and the dust had pretty much settled on who would survive the robotic boom of the 80s. While there were many great advances during the 1990s, several deserve special attention. One was the birth of Biology Electronics Aesthetics Mechanics [BEAM] robotics, the brainchild of Dr. Mark W. Tilden. This approach to robots is about making simple systems with very basic control systems if any control system at all is present; rather than following the traditional theory and emulating man, they are patterned after lower life forms. Tilden's idea of BEAM robotics was inspired by Colin Angle's six-legged robot that could move independently, designed for MIT in 1989. This system was simplistic in many ways but could carry out complex operations with a minimal amount of programming and equipment. (We will explore the methodology behind this approach in deeper detail later in this chapter.)

In 1993, Epson created a working robot that was only 1 cubic centimeter in volume. Consisting of 98 different components, Epson's robot won the Guinness World Record for smallest robot and became what many consider an early nanotechnology endeavor. In 1994, MOTOMAN built the world's first controller capable of controlling two robots at once, with the systems working together in a complementary fashion. With an ability to accomplish larger, more complex tasks, this system outmatched its competitors. That same year, scientists at Carnegie Mellon explored the Mt. Spurr volcano with its *Dante II* robot, opening the door for robots to collect data from environments that are dangerous or difficult for their human counterparts. This was also the year that AESOP (Automated Endoscopy System for Optimal Positioning) received approval by the FDA for use in human surgery, opening a completely new realm of robotic operation.

In 1996, Honda introduced the world to what it called the *P2* robot. Resembling a person and able to walk smoothly on two legs for over 30 minutes, this robot was a huge accomplishment at the time. Honda continued to work on this system, improving and refining the technology to make it more lifelike and changing its name to ASIMO (Advanced Step in Innovative Mobility)(see Figure 1-6). (If you have not seen this robot in action, you can find many videos of the various ASIMO robot versions on the Web or YouTube.)

In 1998, doctors fitted Campbell Aird with the first bionic arm prosthetic, or robotic arm designed to mimic a human arm and controlled by nerve impulses. The arm proved the functionality of this technology and served as a template for future prosthetics. Another advancement in medical robotics took place in 1999, when Intuitive Surgical, Inc., introduced the famous *da Vinci* medical robot, which received FDA approval in July 2000.

As one might expect, the new millennium brought with it some great innovations in robotics. In 2000, ASIMO made its first official debut, and NASA created the first version of the Robonaut, a robotic torso designed to perform the functions of an astronaut with a tracked vehicle base for movement (see Figure 1-7). Both of these robots helped take robotics one step closer to a human level of functionality. That same year, Sandro Mussa-Ivaldi successfully integrated a robotic system into the brain of a lamprey, hoping to find insight on how to improve the field of biologically controlled prosthetics. This was also the year that Waseda University unveiled the *Waseda Talker No. 1*, a robot designed to reproduce human vocal movement and sound. Students updated the Waseda Talker yearly in an effort

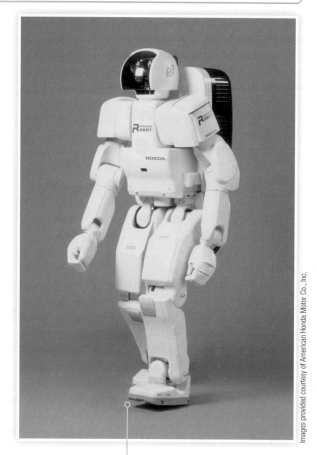

Images provided courtesy of American Honda Motor Co., Inc.

Figure 1-6   Honda's *P3* robot, the second iteration of the robot that is ASIMO today.

NASA

Figure 1-7   The second version of the Robonaut with some of the tools of the trade for working in space.

figure 1-8   Here is my personal collection of WowWee robots; Dr. Tilden's signature Robosapien is in the back (right) and Femisapien is in the back (left). From left to right in the front are the Cyber Spider, Roborover, and Mr. Personality robots.

to perfect the system. This technology could help future robots become more life-like and improve the natural feel of interaction with said systems.

In 2004, Dr. Mark W. Tilden gave the world his Robosapien, the first of several robots that he created with the WowWee Corporation (see Figure 1-8). Tilden based this humanoid robot on his BEAM principles with a benefit unheard of in robotic toys: he designed it to be hacked or modified by the user! Tilden wanted something that anyone interested in BEAM robotics could experiment with instead of the tamper-proof toys of the time.

That same year, 2004, Aaron Edsinger-Gonzales and Jeff Weber created *Domo* for MIT, a robotic research platform with the intent of advancing human–robot interaction. Because robotic systems are so powerful, they are usually fenced off or isolated from the operator in some way. (See Chapter 2 for more on robot safety.) While this is a necessity for safety, it is not always the best way to integrate a robotic system into the industrial process. *Domo* represented one of the first focused efforts to bring the robotic system out of isolation and place it side by side with human workers; it employed a unique feedback system specifically designed with human safety in mind. The seeds planted by this research would begin to bear fruit in 2012, as industry bought into the concept of robots and humans working together instead of isolated from each other.

In 2005, Dr. Hod Lipson created a robotic system that could self-replicate (make copies of itself) when given robotic elements to work with. While these systems were simple in design and required specific elements to build duplicates, they represented the idea of robots creating new robots to complete tasks, repair damage, or share the workload. That same year, Ralph Hollis created *Ballbot*. This system used a single ball for locomotion and opened up a new form of robot motion from point A to point B. This was also the year that Brian Scassellati created *Nico*, a robot capable of recognizing its own body parts and possessing the cognitive ability of a 1-year-old child. While most robots know the position of their parts, they are unable to recognize those same parts when presented with visual data about them. In other words, a robot with a camera system may have images of part of the robot, but it does not process that data as "Oh, that is part of my robotic arm." *Nico's* ability to make these data correlations represented a huge step forward for AI.

In 2006, MOTOMAN gave industry a dual-arm robot that resembled a human torso without a head (see Figure 1-9). The whole point was to mimic the motions of human workers and equip the robot with two arms instead of the standard single arm. Another innovation of this system is that the cabling and air hoses run internally, preventing damage from contact, impact, or friction—a common problem with many systems. At about the same time, Josh Bongard, Victor Zykoy, and Hod Lipson working at Cornell University created a robot that can walk with no initial data about its shape or design! In other words, you can remove motors, add motors, change the robot's orientation, etc.—when powered on, the robot will figure out what it has to work with, how it is orientated, and how to move. This kind of programming and problem solving could be a huge benefit to those working with AI or systems that need to continue to function in spite of any damage received.

Skipping ahead to 2009, MOTOMAN released a robot controller that can manage up to eight robots or 72 different axes of movement, opening up exciting new motion options and raising the bar once more (see Figure 1-10). That same the education world got its first look at the NAO robotic system developed by the French company Aldebaran (see Figure 1-11). This humanoid

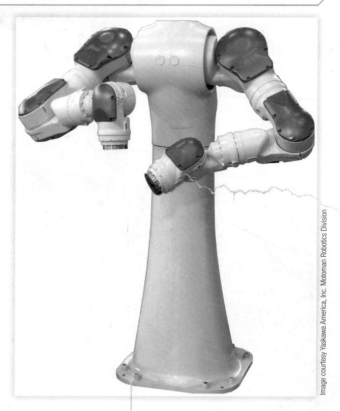

Image courtesy Yaskawa America, Inc. Motoman Robotics Division

Figure 1-9   The two-armed torso-style robot MOTOMAN developed for industry.

Image courtesy Yaskawa America, Inc. Motoman Robotics Division

Figure 1-10   The MOTOMAN DX100 controller, which can handle up to eight robots or a total of 72 axes of movement.

Figure 1-11   The NAO robot that students at Ozarks Technical Community College are working with.

Figure 1-12   The second version of the Robonaut performing the first human robot handshake in space.

robot sports 25 different axes of motion, tactile sensors, microphones, ultrasonic sensors, and cameras along with a whole suite of AI features to give students lots of options for experimentation. Many schools and universities began buying this system after the 2010 commercial release for use in advanced robotic studies and research. (You can find videos of the results on the Web and YouTube.) 2010 also saw the release of the Kinect by Microsoft. This system turned the human body into a game controller. However, robotic enthusiasts quickly figured out ways to hack the Kinect and use it as an advance sensor package for robots. Microsoft later released a free software package to help robotic enthusiasts hook their creations into the Kinect to give the robot information about its world.

Robonaut 2, a collaboration between NASA and GM, completed the first human–robot handshake in space (pictured in Figure 1-12) during 2012. While this may not seem like much, it represents the culmination of sensor, feedback, and safety controls necessary for robots to not only work near people, but directly with them. This handshake may prove to be for robotics what Neil Armstrong's first step on the moon was for man, a crucial first event with many exciting things to follow. In this spirit of human–robot interaction, in 2012, Kawada Industries tasked Tecnalia to adapt its Hiro robots to work with employees in Europe instead of isolate behind safety cages.

Late in 2012, Rethink released the Baxter robot, which works in the in trial environment without the need for a cage! Baxter has a human to tablet that doubles as both interface and head for the robot, force sen 360-degree camera system, and AI software that helps make the robot sa easy to use. Baxter can adjust to changes in its work environment as slowing to safe levels when humans are detected close by.

Another interesting development in 2012 was the work of Nader E and his group at the University of Pennsylvania on a new type of elec dubbed "Metatronic" circuits that manipulate light, not electricity. This nology performs the same function as modern electronics—such as restricti flow, logic management, and amplification—but it uses light instead of electricity! If this technology can be perfected, we can expect benefits such as significant increases in signal processing speeds, smaller components and devices, reduced power consumption and heat generation, and signals unaffected by magnetic fields. This technology could very well prove to be a game changer on par with Jack Kirby's microchip of the twentieth century. Imagine having a device that is the size of a tablet with the computational power of a high-end, top-of-the-line computer and a battery able to handle 24 hours of heavy use! On the industrial side of things, this technology would mean that design engineers could place electronic control devices right next to main power feeds without worrying about ghost signals or corrupted data. This could lead to smaller control cabinets, remove the need for dedicated cable paths, save space as well as materials, and reduce or eliminate the danger false signals present to equipment and people alike.

I hope by now that you understand more about the events that led to the modern robot and why they are important. The list provided in this book by no means includes every event that had an impact on robotics, nor did we fully cover the events that did make the list. There is a wealth of information available, and I encourage you to do some investigation of your own. I will caution you that not all the information out there is legitimate; do not believe everything you find on the Internet. During my own research, I found "facts" that others had included in similar timelines to be false, but I had to dig through reputable resources to do so.

What Is a Robot?

From our exploration of the timeline, we can see that there has been an amazing array of systems and devices that we could label as a robot. Was Archytas of Tarentum's wooden pigeon the world's first robot? Or perhaps it was Heron of Alexandria's water-powered automata *Hercules Killing the Dragon*. Alternatively, was da Vinci's *Metal-Plated Warrior* the world's first robot? Truthfully, the answer depends on how we define "robot."

The term "robot" has been in use for less than a century; thus, it is a new term when compared to many of the words we commonly use. Czech writer Karel Capek created the word "robot" for his play to describe a highly advanced factory machine that becomes human. When we add in the root word this was taken from, "robota," with its meaning related to hard, menial work, it becomes a bit unclear what robot was truly intended to convey. Thus started the debate about "robot"; did Capek mean for "robot" to describe the drudgery side of his transcending machine or the human aspects it takes on? Add to this the liberal use of the term since its creation to describe fictional machines, industrial part movers, machines that emulate human actions, toys, information-gathering devices and programs, and you have the many definitions of the term that we have today.

With all this in mind, let us look at a few of the common definitions of a robot.

- From the OXFORD ENGLISH DICTIONARY ONLINE (by permission of Oxford University Press):
    1. A machine capable of carrying out a complex series of actions automatically, especially one programmable by a computer.
    2. Another term for *Crawler* (in the computing sense).
    3. South African, a set of automatic traffic lights.

- From Merriam-Webster's Collegiate® Dictionary, 11th Edition ©2013 by Merriam-Webster, Inc. (www.Merriam-Webster.com).
    1. A: A machine that looks like a human being and performs various complex acts (as walking or talking) of a human being; also, a similar but fictional machine whose lack of capacity for human emotions is often emphasized; B: an efficient insensitive person who functions automatically.
    2. A device that automatically performs complicated, often-repetitive tasks.
    3. A mechanism guided by automatic controls.

- Robotics Industries Association (RIA), 2012:
    Industrial robot: An automatically controlled, reprogrammable multipurpose manipulator programmable in three or more axes which may be either fixed in place or mobile for use in industrial automation applications.

As you can see, the term "robot" is still being refined—and we are still trying to decide exactly what a robot is. Therefore, before we can decide which classical device was the first robot, we have to settle on which definition to use. For this textbook, we will use one definition for robots in general and one for industrial robots specifically, as the latter is a very specific branch of robotics.

Here is the definition of the general term "robot":

- A machine equipped with various data-gathering devices, processing equipment, and tools for operational flexibility and interaction with the systems environment, which is capable of carrying out complex actions under either programed control or direct manual control.

This definition should be broad enough to encompass the various systems we think of when we talk about robotics, but specific enough to prevent everything from becoming a robot. One of the key features defining the modern robot is "operational flexibility." By this, we mean you can change what the system does, how it does it, and why it does it, usually by changing the programming and a few parts. This capacity is what separates the robot from other machines in the home or industry. Robots are about options, even if we do not use them. A robotic arm may pick and place parts today, spray paint tomorrow, and by the end of the week detect leaks in a pressure vessel with nothing more than some programming changes and different tooling. A battle bot may use a spiked arm and bump sensor in the arena during the first battle and then feature a spinning blade of death and ultrasonic sensor by the next battle.

battle bot
Robots used to compete in various forms of robotic combat

Let us look at an example that may help drive the point home. You purchase two wireless remote-control pickup trucks, one from a department store and one from a hobby shop. Both trucks run on battery power, go forward and in reverse, and are controlled with a wireless remote. The one from the department store works on a specific frequency range, has internal parts that are hard to get to, and has all its control components on a specialized circuit board with no room for expansion. The one from the hobby shop consists of a control module with extra connections, motor, gearbox, radio receiver with multiple frequencies, shock absorbers, detachable wheels, detachable chassis, and power pack, all of which are easy to remove or replace. Given its limitations, it would take massive amounts of modification to change the department store model from a toy into a robot. However, the modification of a few parts and the addition of some sensors could easily turn the hobby store vehicle into a complete robotic system. The real key to the definition of a robot is flexibility—this is the difference between a robot and automated or specialized equipment.

Some robotic systems build in "operational flexibility" during the initial design, thus preventing the need to crack open the case and change components. Take robotic lawnmowers, for instance. These systems automatically take care of the lawn based on internal sensors, programming, and possibly some external devices to set the boundaries of the mow area and to help the mower find the charging station as needed. Many of the better mowers can tell where they have been and avoid covering cut grass and thus wasted time and energy. These robots have built-in sensors that can detect obstacles—such as flowerpots, pets, and people—to insure that there is no damage done during their operation. They have internal power sensors to know when it is time to stop mowing and return to the recharging station as well as the ability to find the charging station. The units have updateable programs or firmware, know when they have mowed the entire yard, can detect rain, have antitheft devices, and run programs that instruct the system on how often and what time of day to mow as determined by the owner. With all these options built in there really is no need to modify the system, though it is possible if so desired. In this type of specialized robotic system, most of the flexibility stems from the ease of programming and updates that are downloaded into the system to make it run more efficiently or to correct any operational issues.

When it comes to the term "robot," the population at large and industry often have different ideas. While we may think of WowWee's Robosapien, *R2-D2* from Star Wars, or another fictional system, people in industry are thinking about machines that can perform tasks related to the manufacturing process with precision and consistency, often at a greater speed than their human counterparts.

The Robotic Industries Association's (RIA) 2012 definition of an industrial robot is one of the best and the one we will be using in this book:

- An automatically controlled, reprogrammable multipurpose manipulator programmable in three or more axes that may be either fixed in place or mobile for use in industrial automation applications.

As we look at this definition, it is clear that we are talking about something useful to industry as opposed to a hobbyist's creation or a home-service robot. Another important distinction is the fact that the industrial robot works primarily under automatic control. This does not mean that there is no way to move or operate the system manually; rather, when it is in production mode, the system performs its duties as directed by the programming without the operator directing every move. We use industrial robots in places where the tasks are dangerous to people, are repetitive in nature, are highly precise, and/or require great force. Often robotic systems used for industrial purposes perform their tasks without direct human supervision for most, if not all, of the work shift. The axes referred to in the definition are the number of directions in which the robot can move. By definition, an industrial robot must be able to move in three different directions, or axes. You will also notice that the RIA definition includes the word "multipurpose"; an industrial system must have built-in operational flexibility to be a robot, which is what separates a robotic system from a three-axes milling machine.

Using our definition of a robot, I believe that the first robots would be the creations of Hans Bullmann, which looked like living people or animals and reportedly could respond to their environment. While none of his creations survived, the descriptions of the systems seem to fit all the criteria; a similar creation by Gianello Torriano, an inventor of Bullmann's time, has survived to prove that the technology did exist then. A slight change of our definition could alter who gets this credit, thus explaining the controversy over the first robot. Using the RIA definition, credit for the first industrial robot goes to the DeVilbiss Company for its spraying machine based on the patented design of Willard L.G. Pollard Jr. This machine had the required axes and range of functionality to make the cut.

The definition of "robot" will likely continue to undergo changes and refinement as we become comfortable with the term and come to a consensus on what we consider to be a robot. If you go into the field of robotics, there is a good chance that you will help make these future refinements.

Now that we have settled on a definition for a general robot and an industrial robot, we can move forward with our exploration of robotics.

Why Use a Robot?

From the hobby robot built for fun and discovery to the highly developed and specialized industrial robot, there are many reasons for using a robot in the modern world.

In the industrial setting, we typically use robots for situations that we call the four Ds of robotics: dull, dirty, difficult, or dangerous. These conditions were a driving force behind the robot's acceptance by both management and workers in industry.

The first D, *dull*, describes tasks that are repetitive in nature and often require little or no thought. The danger to the human worker in this situation is often damage to the body as a result of doing the same thing over and over and over again for weeks, months, or years on end. These injuries tend to involve the joints of the body or the back and can range from sprains to carpal tunnel syndrome to injuries requiring complete joint replacement or back surgery. Another danger is that the worker may become bored and not pay attention to what he or she is doing (sometimes known as "running on autopilot"). While this inattention could lead to part or machine damage, the main concern is that the person will be injured by the machine's moving parts due to carelessness. Such injuries can range from cuts, bruises, and deep lacerations to lost fingers, complete amputations, and death. Given the dangers involved and the fact that these types of tasks rarely change, this is a perfect task for robots. A robot is able to perform thousands, if not millions, of repetitive actions with only minimal maintenance and no loss of focus. This also frees up people to do more important things, such as checking part quality, making machine adjustments, or other tasks better suited to humans.

The second D is *dirty*. These tasks involve processes that produce dust, grease, grime, sludge, or other substances that people would rather avoid. Such tasks may pose a mild health risk (causing allergic reactions or irritation of the skin), but for the most part, these are simply jobs that leave people covered in "something" by the day's end. While this kind of work does not bother some people, many of us would rather avoid the mess. Dirtiness can also affect worker satisfaction, which in turn can affect productivity, product quality, and **turnover rate** (how often workers quit a job or an employer). Now, robots do not care about getting dirty and do not have skin that can be irritated. When robots work in a dirty environment, the only real concern is any damage to its mechanical parts. To prevent this, we can seal robots in plastic or design the robot specifically for a dirty environment. To protect a human often involves suits that are hot to wear, a respirator that requires special training and medical clearance, and other protective equipment that adds to the difficulty of the job—not to mention the potential discomfort for the worker.

The third D is *difficult* (see Figure 1-13). The construction of various machines or certain production processes may require people to bend, twist, and/or move in ways that are difficult for the human body. While the rotational joint of a robot may have 270 degrees of movement, the human elbow is limited to about 90 degrees (I do not know anyone who can rotate his or her wrist 360 degrees). Sometimes the difficulty is a combination of position and weight: the person may be able to move as needed, but the added weight of the part causes strain to joints and muscles. Or a person may encounter physical limitations due to the length of his or her arms or overall build. Instead of searching for a seven-foot-tall contortionist, it makes sense to use a robot; especially since we can specify its reach, payload, and number of axes.

The fourth and final D is *dangerous* (see Figure 1-14). This environment can be excessively hot, contain toxic fumes or radiation, involve working with unguarded machinery, or involve conditions that pose a high risk of injury or illness for humans. This is a perfect job for a robot because we can either repair or

turnover rate
How often workers quit a job or working for an employer

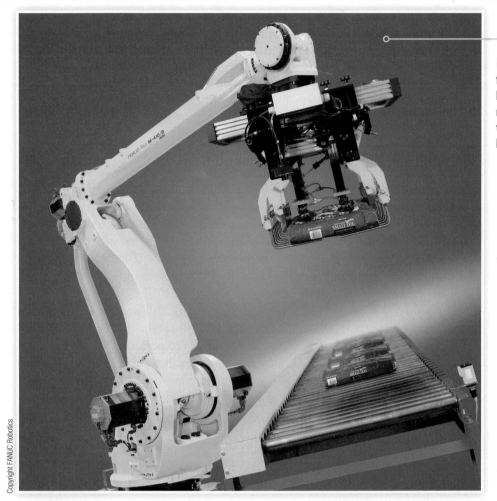

Copyright FANUC Robotics

Figure 1-13 It is difficult for people to handle multiple bags of dog food at once, not to mention the impact that has on the human body, but this robot handles the task with ease.

Image Courtesy of SCHUNK

Figure 1-14 Here we have a robot picking up a red-hot piece of metal, a dangerous task for which it is well suited. In the background, notice how chipped and worn the paint is on the robot as well as the debris on the tooling. This is dangerous and dirty all in one!

replace a damaged machine; however, a person may require months or years of medical assistance and even then never return to his or her former quality of life. There is no profit margin in industry that is worth risking the worker, especially when a robot could do the job just as well (if not better).

precision
The exact or accurate performance of a task within given quality guidelines

consistency
The ability to produce the same results or quality each time

repeatability
The ability to perform the same motions within a set tolerance

fringe benefits
Health and dental insurance, retirement, life insurance, Social Security, and other items that the company pays part or all of the cost on behalf of an employee

Figure 1-15 A robot painting parts, taking advantage of its precision and consistency.

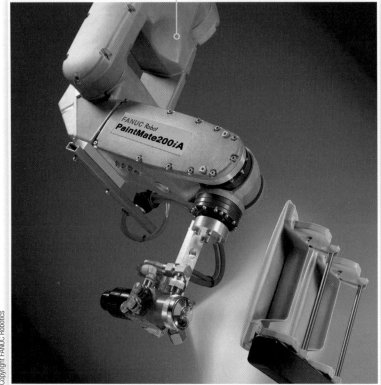

A great example of this is the robotic systems used to clean up the Chernobyl nuclear facility. These robots work in a radioactive environment that would be harmful if not fatal to humans, cleaning up the radioactive waste for proper disposal. When they have completed the job or no longer function, they go to the same facility as the rest of the nuclear waste for storage until the radiation levels are safe once more.

Beyond the four Ds of robotics, another good reason for using robots is precision: the performance of tasks accurately or exactly within given quality guidelines. With human workers, the differences in vision, body construction, and background all contribute to varying levels of possible precision. In industry, this variance can translate into increased production times or parts that do not function properly; this is where the robot comes into play. Robots have the capacity to perform tasks repeatedly with levels of precision that are difficult if not nearly impossible for humans to match. This is largely due to fact that many robots have tolerances ranging from 0.35mm to 0.06mm or 0.014in to 0.002in! This makes a robot especially useful in the production of electronics, aerospace components, and other precisely manufactured parts.

The precision of robots lends consistency and repeatability that is difficult for their human counterparts to match. Consistency is the ability to produce the same results or quality each time; repeatability is the ability to perform the same motions within a set tolerance. There are multitudes of factors that affect a human worker's performance: illness, injuries, fatigue, distraction, emotions, job satisfaction, temperature, and ability, to name a few. Any of these can affect a worker's production level and the quality of the parts he or she is producing. Because industrial robots do not have emotions and are not organic systems, their work does not suffer from these conditions. With a robot, you have to worry about the program controlling the system, the mechanical and electrical components of the system, and the validity of the signals received and transmitted by the robot.. As long as these are in good order, the robot will perform in the same way no matter if it has been working for 1 hour or 48 hours, whether it is Monday morning or Friday afternoon, whether the company is hiring or going through layoffs, or any other condition that might affect human workers. This consistency translates into a cost savings in materials used, higher part quality, and manufacturing times that help reduce the overall part cost (see Figure 1-15).

Cost savings is a factor that has helped to drive the integration of robotics into industry and another "why" for the industrial robot. In industry, workers are generally paid a set per-hour or per-piece wage along with what is known as fringe benefits, which include health and dental insurance, retirement, life insurance, Social Security, and anything else that the company pays part or all of the cost of for the employee. With a robot, you cover the initial cost of the system, the cost of parts and maintenance, and the cost of electricity or fuel to run the system. In many cases, a robot can pay off the initial costs in two years or less and from that point on, they operate at the cost of replacement parts, consumed energy, and maintenance—which for some systems can be as low as $0.42 per hour! This is in addition to any time or material savings that come about due to the robot's precision and speed. For instance, the welding of a car frame by human workers used to take four to six hours, but now a team of robotic systems does the same task in 90 minutes. In addition to the time saved, the robot's precision allows smaller welds to have the tensile strength of larger human welds; this saves the company money in materials used and fewer reject parts. This also frees up human workers to do important tasks such as quality inspections, process modifications, and other tasks that require judgment-based decisions, which are difficult if not almost impossible to automate.

Once we move beyond the world of industry, we find that robots have found their way into many aspects of modern life. For the average consumer, we have service systems that help with the chores of the modern world, systems that enhance the quality of life for those who are disabled, and systems that exist purely for entertainment. To help those in dangerous professions, we have systems that risk their parts so that humans do not have to risk their lives, systems that gather information on dangerous situations, and systems that can carry the fight to the enemy. In academic circles, we have systems used for research as well as educating those who are learning about robotics. In the business realm, robots deliver mail, wow customers, or allow workers to interact with the office from afar. We will explore some of the various ways robots touch our lives in the remainder of this section, but by no means will we cover all the robots out there and the ways they are used.

We will begin by examining robots for entertainment, as this is how many first discover the fascinating world of robots. From the "Show-Stoppers" of old to WowWee's Robosapien, robotic creations have fascinated the masses and drawn both our attention and money. Today, you can buy robotic pets that mimic real animals but do not die if neglected, or robots that help entertain young children while teaching important skills such as shape recognition, sharing, reading, and other important life skills. There are sophisticated bears and dolls that can act as a robotic friend, or systems that can play war games with child-safe weapons for those who want more action. For those who like hands-on projects, there are robotic platforms such as Vex robotics, Lego Mindstorms NXT, OWI robotic arms, various quadcopters (with built-in stabilization and flight-assisting functions), and many more (see Figure 1-16). For the hard-core hobbyist, there is a wide range of

Figure 1-16
The Robosapien alongside a LEGO NXT system built as Alpha Rex.

figure 1-17 This is a system I built from scratch using an Arduino processor, a couple of servos, a motor driver chip, some old CDs heat formed, and spare parts from a spider toy that I had.

telepresence
The interaction with others from a remote location using technology

components and tutorials out there to create a robotic marvel from scratch, for whatever purposes you may desire (see Figure 1-17). As you can see, there are many entertainment options when it comes to robots.

Of course, when talking about entertainment, we cannot leave out the various animatronic systems used at theme parks or the large robotic systems designed to wow and entertain the masses. These systems create magical tours through fairy tales as well as bring our nightmares to life in order to give us a good scare. Hollywood has used these systems for years to make movies more realistic and to bring various creations to life (see Figure 1-18). Many companies use these systems to interact with potential customers and give their sales pitch at trade shows or conventions. When salespersons are competing with the chaos of large crowds and a huge number of competitors for your attention, robots provide the gimmick needed to win some of your precious time and to deliver their pitch. When it comes to the entertainment side of robotics, the "why" is often the same answer one would give for watching a YouTube video or a movie: because we enjoy it. As long as we remain fascinated by these scientific marvels and enjoy having them in our lives, we can expect to see new systems and innovations arise in an effort to capture a bigger market share of our time, interest, and money.

Telepresence, where a person interacts with others from a remote location using technology, is another fascinating and growing field of robotics. The advanced systems in use have two main categories of equipment: the input side and the output side. Inputs include cameras, microphones, and any sensors necessary for navigation. Output includes video display, speakers, and in many cases a method of moving the system around. The inputs gather data from the location of the telepresence system and transmit these to the user, giving them vital information for interaction; the outputs allow the user to create a sense of presence even though he or she may be miles or continents away. Business was the first area to buy into this technology

figure 1-18 The *ABB* robot handling a *Terminator* robot for a shot in the *Terminator Salvation* series.

Image courtesy of ABB Inc.

even though the early systems were little more than video chat setups. Today, high-tech interface and robotic systems allow workers to live where they chose and work from home over a high-speed Internet connection while still being a vital part of a company. As this technology has matured and gained acceptance, its presence has expanded into new areas of society. One use gaining popularity is in senior centers, where robotic units allow friends and loved ones to visit easily and more often than might otherwise be possible. This technology also allows children with severe allergies or other limiting health conditions the chance to interact with their peers and classmates. Before this technology, these children often had the choice of being isolated in a controlled environment or literally risking their lives to interact with others. As the acceptance for this technology grows, we can expect to see expanded use and a greater variety in the systems offered.

Research is one area that has heavily invested in the robot. We use robotic platforms to test new design ideas, technologies, and programing methods or algorithms. (Algorithms are systematic procedures for solving a problem or accomplishing a task.) Research robots go places and do things that are dangerous or difficult for their human counterparts. Why risk exposure to toxic gases and other such dangers inside a volcano when we can send in a robot? Is the area radioactive? Send in a robot! What lives at the bottom of the deepest part of the ocean? Send in the robot sub and find out.

algorithm
A systematic procedure for solving a problem or accomplishing a task

Space exploration is one realm of science and research that has latched onto the robot. In 1981, the Canadarm arm helped astronauts move heavy payloads to and from the space shuttle. Deep Space 1, launched in 1998, helped prove the capabilities of AI software and new technologies, ushering the space program into a new era. The Robonaut began in 2000 with a dream to create a humanoid robot to work with humans or take the risks for them and a partnership with NASA's Johnson Space Center, General Motors, and Oceaneering (see Figure 1-19).

Figure 1-19 Here you can see the Robonaut 2 showing off the strength and capabilities of the system.

Through lessons learned, this evolved into a second version in 2007 and set a milestone for robotics in 2012 by shaking hands with an astronaut in space. That same year, the Mars rover Curiosity made its safe descent to the Martian soil, sending back vital information for potential manned visits. These are but a few of the various robotic systems that have enhanced our understanding and interaction with the universe that lies outside of Earth's atmosphere over the years. In the future, robotic systems may build human habitats on the Moon and Mars before the astronauts leave Earth or perhaps make cargo runs between mineral rich asteroids and the earth. Robots may harvest parts from dead satellites to fix working ones or create completely new systems. The possibilities are endless, with the only limits being human imagination and desire.

In the mid-1990s, robotics took an active role in the surgical field of medicine; the number of systems and operations possible has grown steadily ever since. From the AESOP system (Automated Endoscopy System for Optimal Positioning) introduced in 1994 to the famous da Vinci medical system, robots have helped doctors perform delicate surgeries with increased precision and a substantial decrease in the size of incisions. When a doctor performs surgery, he or she must make incisions large enough to use the tools of their trade, allow their fingers or hands entrance with maneuvering room as well as giving them line of sight, all of which translates into larger incisions and longer recovery times for the patient. Robots pack things such as cutting tools, vision, forceps, and other necessary tools into a much smaller package, thus reducing the size of incisions and the amount of trauma to the patient. In fact, the damage reduction to the patient is great enough that surgeries that once required a lengthy hospital stay have become outpatient surgeries!

There are benefits to these systems beyond the time saved in recovery and damage to the patient. Doctors control these medical robots at a terminal that shows what the robot's camera sees while allowing them to control the motion of the system and any of the surgical tools involved. They still have a team that assists with various aspects of the surgery, but the doctor is away from the patient, allowing for greater concentration on his task as well as concentration on his tasks as he focuses on the screen and controls without having to move around the assisting staff tending the patient. This setup has also allowed doctors to work together in areas where two sets of human hands could not fit, opening up a whole new range of options for surgery. Another idea under exploration is surgery where the doctor controlling the robot is not in the room with the patient, but located in a separate room with the controller. This setup saves floor space in the operating room, prevents exposure of the controller to biological hazards, and further reduces the distractions that a doctor might deal with in the operating room. There is still an operation team with the patient, only the doctor is in another location. As this use of the robotic surgical systems is explored and perfected, we could see a time where specialists will be able to perform lifesaving surgeries from miles away, in different states or even countries!

Another field in the medical realm where robotics has taken hold is prosthetics. Between the dangers of modern life, our love of outdoor/extreme sports, and the seemingly ever-present dangers that our military men and women face, many individuals have had their quality of life greatly reduced by amputations and severe spinal injuries. These injuries once meant a life of learning to do without, compensating, and in some cases very little possibility for living a normal life again. The field of robotics is rapidly changing that. Today we have a range of mechanical options to replace missing limbs, systems that allow those in wheelchairs to walk once more, and even systems to help those paralyzed from the neck down.

Pare Ambroise introduced the concept of combining flesh and mechanical parts in his 1564 paper detailing an artificial hand made up of organic and mechanical components; however, it would be 1998 before the world saw a working version of this technology proved by Campbell Aird's fitting of the first bionic arm. One of the advancements that made this possible is the biosensor, which senses signals from the muscular or nervous system and allows the user to control the prosthetic by willing the muscles to move as they once did. The biosensor picks up the electrical activity associated with these actions through wires incorporated into the surface of the patient's skin as well as needle-sized electrodes implanted directly into the muscles. Along with this improvement in control, we have improved the quality of prosthetic arms with the use of tendon drives that closely mimic organic tendons in strength and function; microcontrollers that allow for natural motion, smaller size, less weight, and individual finger control; and advancements in the skin of the prosthetics that can detect both pressure and temperature. When it comes to prosthetic legs, there are segmented feet, special joints, designs for specific purposes, and Bluetooth communication between pairs of prosthetic legs, all of which help the user to have a more natural walking or running motion. Gone are the days of peg legs and metal hooks: amputees now have a variety of options to help them recapture their lost motion and quality of life.

For those who are paraplegic, or have lost all mobility in their lower limbs, there are some new and exciting robotic options available. There are several versions of robotic chairs that allow the user to traverse steps and rough terrain and adjust their height to reach items or interact with other people. These systems take the motorized wheelchair of old and give it a robotic boost in capability and options. A new and very exciting development is the use of robotic exoskeletons that allow paraplegics to walk once more! Exoskeletons are robotic systems that strap onto and around the user's body to enhance strength, endurance, and, in some cases, mobility. One of the companies leading the charge in this new technology is ReWalk. This system consists of a robotic exoskeleton that fits around the user, a controller that coordinates the system's movements, and a set of specialized crutches that help the wearer keep his or her balance as well as time the movements of the system via tilt sensors that detect the position of the body. The ReWalk system enables users to sit, stand, walk, and even climb stairs and walk outdoors. The ReWalk system hit the market in 2012 in Europe; at the writing of this book, it is awaiting FDA approval in the United States.

While biosensors and tilt sensors can help many of those with disabilities, quadriplegics require a more advanced control method in order to interact with assisting devices. Quadriplegics are people who are paralyzed in the torso and limbs. Since they have very limited conscious control of their muscles, signals to prosthetic or assisting devices have to come either from voice commands, facial muscle control, or directly from thought. Sandro Mussa-Ivaldi and his team led the charge in 2000 with experiments that involved hooking electronic components to the brains of a lamp ray in an effort to increase our understanding of thought control for prosthetics. The device BrainGate built on this in 2004 when it received FDA authorization for human testing. The goal of the BrainGate project was for patients to control computer devices with nothing more than thought. In May 2012, this device allowed a woman paralyzed by a stroke to control a robotic arm and get herself a drink of water for the first time in years! Her thoughts, routed through the BrainGate device, controlled a robotic arm that picked up the bottle and brought it to her mouth so that she could drink. ReWalk is also

biosensor
A device that picks up the electrical activity associated with movement thru wires incorporated into the surface of the patient's skin as well as needle-sized electrodes implanted directly into the muscles

paraplegic
A person who has lost all mobility in his or her lower limbs

exoskeletons
Robotic systems that strap onto and around the user's body to enhance strength, endurance, and in some cases mobility

quadriplegic
A person who is paralyzed in the torso and limbs (in other words, from the neck down)

exploring ways that a robotic exoskeleton can return mobility back to those who are quadriplegic, but they have a ways to go. The hope is that someday these lines of research will give normalcy of life back to those currently trapped in their own bodies due to illness or injury.

We also find robots in jobs that fit into the four Ds of robotics outside of industry. We have robots that mow yards, vacuum floors, clean out rain gutters, or take on tasks that are a risk to human life. The robot has found work in the dull and dirty realm of jobs in homes, taking over the simple tasks that we would rather not do. While the type of systems available and the list of chores robots can perform are somewhat limited currently, this field is beginning to expand. With limited technology, the early robotic sweepers cost about $300 or more, and the first robotic lawnmowers were well over $2,000. Over time, these devices have evolved; as they have become simpler to use, they have also slowly come down in price. Once there was only a robotic floor vacuum, now there are robotic sweepers, moping robots, pool cleaners, and research platforms. Where we once had lawn equipment that could only cover a limited area and required some form of device to prevent runaway systems, we now have robots that can figure out where they have been, avoid obstacles, and mow on schedule. As long as the public continues to look for robots to do chores for them, manufacturers will continue to expand and improve the systems available.

We also find robots helping our soldiers and law enforcement officers perform some of the more dangerous aspects of their jobs. One such job is dealing with explosive devices. Before the robot, this meant destroying a device from afar or sending a person into harm's way. Today, a robot can be sent in to take the risk. There is even research into robotic systems that can search out dangerous explosives and neutralize them without direct human control.

Another perilous area where robots are getting involved is the gathering of information and handling threats posed by people. Whether it is a SWAT team dealing with a hostage situation or the military trying to neutralize an enemy target, information about the situation is crucial for planning. Before the robot, this intelligence gathering often involved people putting their life on the line and perhaps making the situation worse. Today, we use robotic systems to help gather the information and, in some cases, neutralize threats. Aerial drones gather information on the location of equipment and people while a human pilot is in a safe location miles or continents away. Some of the larger drones can carry bombs or missiles to destroy enemy targets without risking the lives of our soldiers. We can toss robots into buildings to gather information while the operator is a safe distance away. We have tracked land robots that can be equipped with cameras, sensors, and an array of offensive weapons, depending on the task. These systems can find criminals or enemies, transmit surrender instructions, and deal with those who are foolish enough not to comply. What is the best part about these systems? If they receive damage, we can replace a few parts and they are as good as new; or, if there is too much damage, we salvage the usable parts from the system and get a whole new robot. When a person is damaged, he or she has a minor injury at best, with the worst-case scenario involving the loss of life.

Whether it is the four Ds of robotics, simple human desire, or performance requirements, the robot has worked its way into many aspects of the modern world. While the specifics of why we use a robot are usually linked to an application, the broad answers remain the same: the robot is either performing a task we want it to do, a task we as humans do not want to do, a task that is dangerous, or a task that requires precision and/or force that we cannot match. It is because of this that the robot has earned its place in the modern world and will continue to be a part of our world in the future.

The Top-Down versus Bottom-Up Approach

How do we go about creating a robot? Just like any other invention of man, there has to be a driving ideal, a concept, a dream that we hope to achieve in order to begin the process of bringing something new into the world. Need is often the driving force behind the inventions of man, thus bringing about the famous quote by Vulgaria in 1519, "Necessity is the mother of invention" (Titelman 2001). In many of the historical cases we have looked at, the need, desire, or idea was to create something that could emulate people in some way. Some robots were designed to play music and entertain, some were designed to do things that resembled thinking, some were designed to wow the masses, while others were copies of natural creatures or actions. When we look to some of the more modern robots, we see that the need was to get people out of harm's way, to free people from tasks they did not want to do, and to simply entertain. During the evolution of the robot, two distinct schools of thought have emerged: the top-down and the bottom-up approaches, both of which have a different view on how we should build and advance robots.

The top-down approach is where we start with something very complex, such as people or independent thought, and then try to create something as close to that as possible with available technology or innovative new designs. If you look through the history of robotics, you can see this trend. From da Vinci's suit of armor to the instrument-playing automata of old, from the first robotic movers in industry to the Robonaut, we see the inventors' desire to create something lifelike, something that could perform the same tasks as people, something emulating a complex living system. This approach to robotics can lead to much frustration on the part of those creating the systems, but it has also led to great advances in robotics and technology alike.

The bottom-up approach is where the inventor starts with a very simplistic input and control system and then sees what happens. Mark Tilden pioneered this method of robotics with his introduction of the BEAM (Biology Electronics Aesthetics Mechanics) method. Inspired by Colin Angle's *Genghis* robot, a simple six-leg walker, Tilden began to explore making very simple, insect-like robots. His basic reasoning was that insects are very simple creatures with small brains; however, they can still interact with the world, so why not try to copy this in robotics? At a time when the robots of the world tended to be complex systems, BEAM robotics was going back to the basics and looking at how simple robotic systems could evolve into machines that are more complex.

We also see the three laws of robotics in effect for both systems. The top-down method has generally tried to follow Asimov's three laws of robotics:

- *First Law:* A robot may not injure a human being or, through inaction, allow a human being to come to harm.

- *Second Law:* A robot must obey any orders given to it by human beings, except where such orders would conflict with the First Law.

- *Third Law:* A robot must protect its own existence as long as such protection does not conflict with the First or Second Law. (FamousPeople n.d.)

top-down approach
The robot evolution approach in which one starts with something very complex, such as people or independent thought, and then tries to create something as close to that as possible with the technology available or innovative new designs

bottom-up approach
The robot evolution approach in which the inventor starts with a very simplistic input and control system and then sees what happens

Since the bottom-up method is all about working with simple systems, Mark Tilden created his own laws for BEAM robotics:

1. A robot must protect its existence at all costs.

2. A robot must obtain and maintain access to a power source.

3. A robot must continually search for better power sources. (Hrynkiw and Tilden 2002)

As you can see, Tilden's simple set of laws match his simpler approach to robotics in general. Asimov's first law is about the protection of human life; Tilden's first law is about protecting the robot. This illustrates the difference between the two approaches: the top-down method works to create grand machines that rival or surpass their human creators; the bottom-up approach is about creating simple robots and seeing what happens. Asimov's second law is about being of use to people; Tilden's second rule is about the robot's function. Here, we can see that the driving force behind the top-down approach is the need to be useful to people; the bottom up focuses on its continued existence. When we get to Asimov's third law, we finally see something about the robot protecting its continued operation; Tilden's third law is about improving the robot's situation. The differences between the sets of laws give us great insight into the mind-set behind those creating the systems. With the top-down method, inventors design the robot to accomplish a function as safely as possible, with the preservation of the system secondary to other concerns. With the bottom-up method, inventors try to keep the system running and functional to see what happens.

At this point, you might be asking yourself, why bother with bottom-up systems? If the robot only exists to exist, what good is it? Well, one of the startling discoveries of this branch of robotics is the emulation of animal behavior without a program or human direction of the robot. Tilden and his research team built various models and configurations of BEAM robots based on Tilden's three basic laws, using things such as Walkman tape players and other scavenged parts that they had at hand. These BEAM robots had some means to collect power, some means to store that power, some method to sense out the power source, and some method to move to that power source with no set programing or complex controller added. They released these robots into a synthetic habitat and something wondrous happened. Over time, the robots developed patterns of operation and found territories or areas of the environment that they favored. There were even robotic fights over these territories with the better-designed BEAM bot winning the resources of the area and the loser having to find a different area to gather its power. Here, we have these simple basic systems emulating the insect world with no complex programming or controllers! This intriguing result kept the bottom-up method in the robotic research mix and sparked more than one roboticist's interest in BEAM robotics.

territories
Areas of an environment that are favored

Many hope that as research continues into the bottom-up method, we will learn more about creating machines capable of making decisions or Artificial Intelligence. The thought is to take these basic insect-like robots and slowly improve them until they can emulate more intelligent life forms. While the top-down method is working to create a thinking machine through advanced programing and technology, the bottom-up method is trying to evolve something greater than the sum of its parts. Each evolution of the robot involves taking what worked well in a previous unit and trying some new twist, or creating a completely new line based on lessons learned and an untried configuration.

Another thing to consider when comparing the two methods is that the bottom-up method is young compared to the top-down method. Inventors and innovators have been using the top-down method for centuries, while the bottom-up method has only received cursory exploration until the past couple of decades. As BEAM robotics and the bottom-up method have a chance to mature, we will likely see more innovations and discoveries from this field. Already many hobby robot experimenters have seen the benefits of BEAM research in the form of the highly modifiable Robosapien robot designed by Mark Tilden. The Robosapien represents a marrying of the two approaches in robotics as it has many internal systems based on Tilden's BEAM research as well as a complex controller managing the robot's operation. Another unique feature of the Robosapien is the fact that Tilden designed the system in such a way that owners can actually get into the robot and experiment with its operation, as shown in Figure 1-20. Here again, we can see the bottom-up method at work: these systems were designed to be modified or "evolved" into something different and hopefully greater than when they came out of the box.

Like most things in life, both the top-down and bottom-up methods have value and their place in robotics. Both methods have advanced what we know about robotics, allowing for the creation of systems greater than their predecessors. As the bottom-up method gains maturity, it may well give us the insight we need to create the advanced systems that the top-down method has ever strived and struggled to create. The future may see a blending of the two methods that leads to a completely new approach to robotic creation, similar to what we see in the Robosapien. Only time will tell.

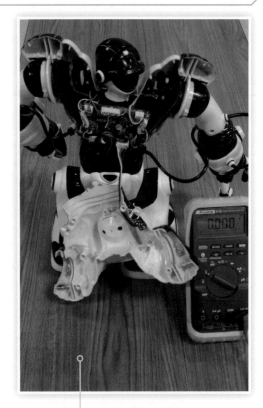

Figure 1-20   The inner workings of the Robosapien, accessible with the removal of a few screws.

AI and the Future of Robots

With this varied history the question that comes to mind might be, "What's next?" This section is about some of the likely answers to that question.

One of the broad areas likely to see continued growth is Artificial Intelligence (AI), which is the term used to describe a software or system with the capacity to make human-type decisions and solve problems intuitively or creatively. This software or system would be able to handle a real-life situation for which there are several ways to proceed with no clear-cut best answer or address problems for which it has no specific programming. The desired result of AI research is to create a system that could respond to the conditions of its environment quickly and effectively without the operator having to reprogram the system. Ultimately, this research could lead to systems in industry that could prevent machine crashes, increase productivity, and possibly even save lives. In our everyday lives, this could lead to robotic companions for a variety of circumstances and interaction levels such as robotic childcare or caregivers for those with medical needs or disabilities. One day we could even see robots of the type that Isaac Asimov envisioned in his writings, helping man to the betterment of all.

Let us explore an example of a problem that AI could help with. For our example, we have a robotic system on a wheeled base taking a package from the warehouse to the main office; it has the directive to complete this task as fast as possible with no specific route designated. The first problem that the control system runs into is the fact that there are three quick routes to the office; all of which are the same distance and should take the same amount of time (in theory). At this point, the system would need to examine such things as number of turns, are there any inclines, would there be anything to delay the robot (such as cross traffic), and any other factor that might affect travel time. This is where the AI would come in. It would assign values to the conditions and then add that to the calculations of time for each route. Using this new data, our robot analyzes the three routes and picks the one that represents the least travel time. Our robot is off and running when it hits a new problem. In the robot's programming, there is a directive that people always have the right of way, no matter what. A line of people ready to leave work for the day blocks the robot's current path. The programming dictates that the system must wait for the people to move before it can progress. However, its internal clock is showing that it has already waited for 30 seconds, while its sensors show that the people are not moving. Now if you or I had to deal with this situation, we would likely just excuse our way through the line and go on, but the robot in question has nothing in its program detailing this option. Without some way of making creative decisions, the robot must either wait for the people to move or backtrack and take another path, adding the time already spent plus backtracking to the total time of the trip. Here is another place where AI could come up with a solution outside of that clearly outlined in the programming. One answer might be to utilize the space between the people and a wall to find where the line ends, or it might ask for a clear path if it has speech capabilities. Another solution might be to create a new path from the current location while avoiding the human blockage.

The whole point of AI is to create robotic systems that have a capacity to deal with the chaos of life. When a robot is in a tightly regulated environment, there is little need for AI or advanced problem solving: a well-written program can deal with whatever arises. However, life rarely provides tightly regulated environments. Even in industry, where conditions should be fairly consistent, there is the human factor to deal with as well as unexpected machine faults. As AI technology advances, robots will be able to work beside their human counterparts instead of isolated in cages and other safety areas as well as deal with the chaos of unexpected events. AI can also make it easier for humans to direct robots in the tasks that they perform. In fact, robots like Baxter are already using this type of technology to correct for part misalignment and ease the process of teaching the robot what to do.

Artificial Intelligence is not a new field in robotics, and we have already made some advancement in this field. We have systems that can take fuzzy logic problems, which are situations where there is more than one right answer or no clear-cut answer, and come up with a plan of action. We have robots that can figure out how to deal with obstacles in their path, and systems that can map out their environment with little or no human aid. We have systems that can figure out how to move with no programmed knowledge about its shape or how it should try to move. We have robots that have the self-awareness of small children, with a similar decision-making ability. We also have video games that can learn the gaming style of the player and react accordingly to make the gaming experience more challenging. Programmers create AI programs using high-level mathematical formulas that allow the system to revise the way it sorts and weighs data with the ability to improve decisions based on the collected data and, in some cases, modify the core program itself. As those working in robotics continue to refine these mathematical formulas and programs, we can expect robotic systems to become "smarter" or more lifelike in the way they respond to the problems of the world.

Another field that holds a lot of room for advancement is robotic vehicles. The Defense Advanced Research Projects Agency (DARPA) holds yearly competitions for long-distance driverless vehicles. These vehicles have to traverse a fair amount of difficult driving conditions that simulate many of the problems human drivers have to deal with. During the first competition in 2004, none of the entrants

fuzzy logic
Situations where there is more than one right answer or no clear-cut difference between several plans of action, and a computing system must choose one

finished the course. However, just one year later, five vehicles successfully completed the course. Since then, DARPA has been increasing the difficulty of the course while the entrants have been fine-tuning their vehicles or creating completely new systems to compete. One of the benefits of this competition is robotic troop and material transports that are in trial use on several military instillations. On the civilian side, we now have cars that can parallel park without driver assistance and systems to help avoid collisions. While fully robotic vehicles are not currently commercially available or approved to drive on the open road, you may see a day when robotic driving becomes a standard option for new vehicles.

Exoskeletons, mentioned earlier in this chapter, is another field where we currently focus a large amount of military and civilian robotic research. The military is hoping to find a way to augment soldiers' capabilities and enhance their performance in the field with this technology, while many of the civilian projects are about giving mobility to the disabled. Both sides have had success with the technology, but it is still in the early stages of development. ReWalk is working to use this technology to give mobility to those who cannot move their arms or legs. A company in Japan is working to develop a version of the technology that will augment the average person's strength up to 10 times what he or she can normally carry while reducing fatigue from activities such as walking or carrying loads. The military is working with contractors on whole-body systems to help soldiers move heavy equipment as well as systems for the common soldier that will support the weight of their pack and gear, preventing fatigue and extending their effective range. As exoskeleton technology matures, we may see a day where many of us will use robotic assisting suits in our jobs or everyday lives to increase our abilities and provide capabilities that would otherwise be beyond the limits of our physical body.

When we look to the future of robotics in the world of entertainment, the sky is truly the limit. In the entertainment field of robotics today, one can find toys, kits for experimenters, instructions on how to build systems from parts, and animatronic show-stoppers. Robots have found their way into trade shows, conventions, and the hearts of audiences as they share information, show off capabilities, or simply entertain the masses. From systems like "Titan the robot," which has a human operator inside, to repurposed ABB robots that have been on stage with Bon Jovi, we have witnessed that people enjoy seeing this technology in action. The field of robotic entertainment will continue to grow with the direction and focus determined by human interest.

Of course, no discussion of the future of robotics would be complete without a look at the industrial side of things. Industrial robotics is still a prime driver of innovations in the robotics field and, as such, the industrial robot will continue to advance. Terrific examples of the future of the industrial robot include units like the Robonaut, created by a partnership between NASA and General Motors (GM); ASIMO, which is the work of the Honda Corporation (see Figure 1-22); and Motoman's double-armed robot that resembles a headless torso (see Figure 1-21). All these systems have a very human feel to them in the way they look and, perhaps most importantly, how they operate.

Image courtesy of Yaskawa America, Inc. Motoman Robotics Division

Figure 1-21
MOTOMAN's dual-arm robot working on parts from both sides, similar to how a human might complete the task.

Figure 1-22 ASIMO in action, running across the stage during a demonstration of his capabilities.

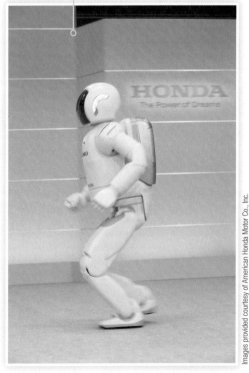

Images provided courtesy of American Honda Motor Co., Inc.

The driving goal behind these systems and others like them is to create robots that can mimic the flexibility of their human counterparts while working alongside human workers. Add in AI technology, and you have an industrial robot that is easy to program, able to respond to unexpected situations, and safe to work alongside their human operators.

These are only a few of the possible advancements that may be in store for the field of robotics; this is by no means a complete list of the developments to come. As technology continues to advance, as we refine current systems, as we discover new ways to utilize current technology, and as our interests and desires change, so too will robotics. Those who have a passion for robotics and strive to find new and exciting ways to use these systems will write the future of robotics. People like you, the reader of this textbook, will be the ones to decide how the future of robotics unfolds.

review

Through the course of this chapter you have learned about important events leading up to the modern robot, what a robot is, why we use robots, the theories behind the development of robotics. You have caught a glimpse of what the future of robotics could hold. This is a lot of information to take in; so you may want to refer back to this chapter from time to time as you gain knowledge about robots and a deeper understanding of it all. During our explorations in this chapter, we hit on the following topics:

- **A timeline of events important to robotics.** Here we have a timeline of the events that influenced and created the future where the robot we know and love was born.

- **Key events in the history of robotics.** We looked at some of the key events from the timeline worthy of deeper exploration.

- **What is a robot?** We explored the various definitions of a robot, settled on the two used for this book, and explored the systems that fall under these definitions.

- **Why use a robot?** We looked at the reasons we use robots both inside and outside of industry.

- **The top-down versus bottom-up approach.** This section was all about the two main concepts behind robot creation and how they compare.

- **AI and the future of robotics.** We took a look at the current trends in robotics to predict some of the areas that will continue to expand in the future.

key terms

Abacus	Bionic	Fringe benefits	Repeatability
Algorithm	Biosensor	Fuzzy logic	Robot
Anthropomorphic	Bottom-up approach	Gantry	Robota
Artificial Intelligence (AI)	Common Era (CE)	Numerically Controlled	Return on Investment
Automata	Character recognition	(NC)	(ROI)
Automatons	Computer Numerically	Paraplegic	Show-Stoppers
Battle bot	Controlled (CNC)	Precision	Telepresence
Before the Common Era (BCE)	Consistency	Prosthetic	Territories
Biology Electronics Aesthetics	End-of-Arm tooling (EOAT)	Punch card control system	Top-down approach
Mechanics (BEAM)	Exoskeletons	Quadriplegic	Turnover rate

review questions

1. Who do we credit as the father of pneumatics, when did he make his discoveries, and why was this important?

2. What was the *Metal-Plated Warrior*, and who created it?

3. To whom do we give credit for creating the first real androids in human form?

4. When did Benjamin Franklin conduct his famous kite experiment, and why was it important?

5. What was the Jacquard loom? Who invented it, and when?

6. What are three things that Nikola Tesla is famous for?

7. What are Isaac Asimov's three laws of robotics?

8. What was the first true experimental platform for Artificial Intelligence?

9. What was the first commercially available micro-computer controlled industrial robot?

10. When did BEAM robotics come to be, and who created this approach?

11. When did work begin on the Robonaut 2 and what historical event was the Robonaut 2 part of in 2012?

12. What is the general definition of a robot used in this textbook?

13. What is an industrial robot as defined by RIA?

14. Who created the first industrial robot?

15. List the four Ds of robotics and give an example of each as it relates to industry.

16. List three ways robotics has helped with research.

17. What benefit to a patient does a robotic surgical system offer?

18. What are some of the mundane tasks that we use robots for around the home?

19. Compare the top-down and bottom-up approaches to robotic creation.

20. What happened with the bottom-up systems when put in a synthetic environment?

21. What does AI stand for, and what is it?

22. DARPA has a yearly robotic contest. What does DARPA stand for, what kind of robotic contest does it hold, and what are some of the outcomes from these contests?

23. What are some of the reasons that the military is looking into exoskeletons for soldiers?

24. What is the main benefit of using a robot to perform a dangerous job like reconnaissance in a hostage situation or dealing with explosive devices?

Safety

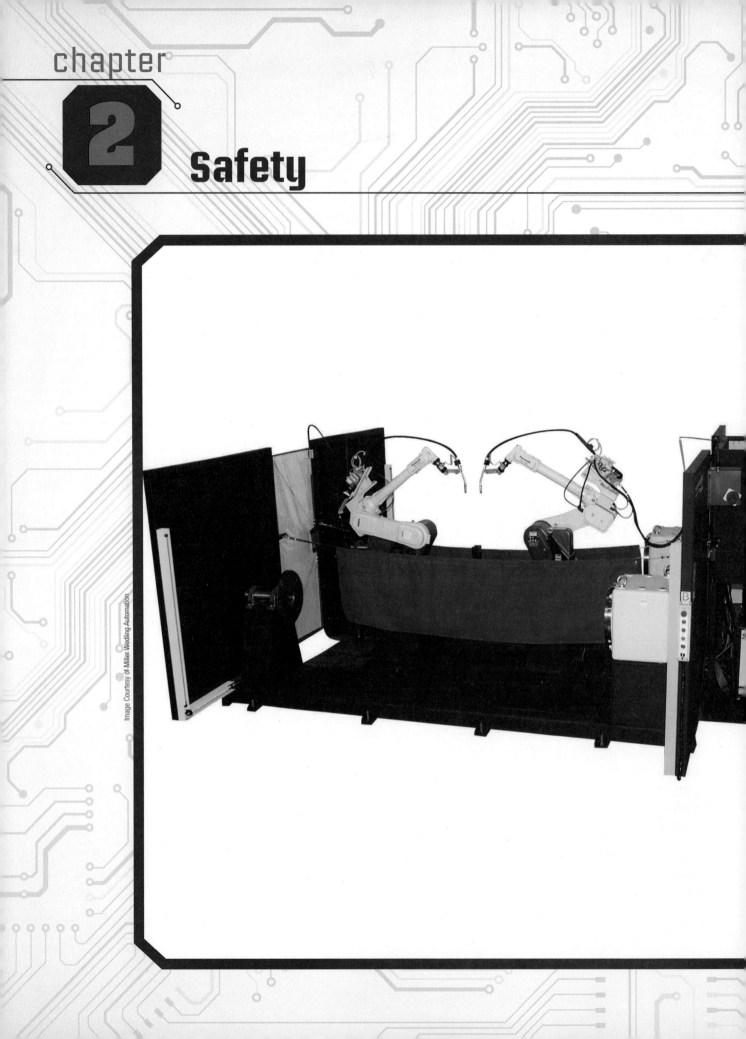

Image Courtesy of Miller Wedling Automation

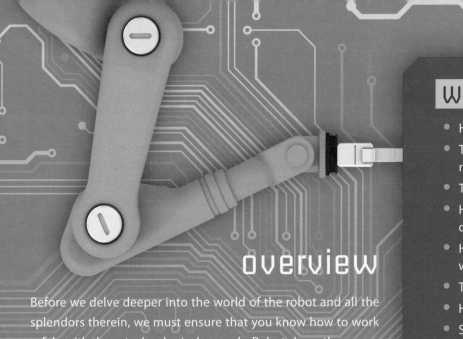

What You Will Learn

- How to work safely with robots
- The three conditions that will stop a robot
- The three zones around a robot
- How to insure the safety of people in the danger zone
- How some of the common safety sensors work
- The dangers of electricity
- How to deal with emergencies
- Some basic first aid procedures

overview

Before we delve deeper into the world of the robot and all the splendors therein, we must ensure that you know how to work safely with these technological marvels. Robots have the power to move heavy materials, perform machining functions, fuse metal, deposit various substances, and move much faster than humans. The same functions that make a robot valuable are also the greatest dangers of working with a robot. From the person who works around a robot, to those responsible for maintenance, to the creators of robots, safety needs to be a primary concern at all times. To help familiarize you with the safety requirements of robots, we will cover the following topics in this chapter:

- Robots Require Respect!
- Danger zones
- Guarding

- Safety devices
- Electricity and you
- Handling emergencies

Robots Require Respect!

When it comes to safety, think about the three Rs of robotics: "Robots Require Respect." This is another catchy phrase (like the four Ds of robotics) to help you remember how to work safely with robots. We have created robots to carry out amazing tasks in both industry and our everyday lives. These systems are capable of moving parts (both heavy and light), welding metal, machining, providing surveillance, defusing bombs, delivering military strikes, entertaining us, and executing many other tasks, all of which involve some level of inherent danger. When we become comfortable with a robotic system, we often start to ignore these dangers and put ourselves in harm's way.

A robot performs its actions via programming, direct control, or some combination of the two. That is it. They do not have feelings, they do not have moods, they do not have intuitive thought, they do not think as we do, and they do not have their own agenda. While it is true that we are working on AI programs and have given robots the ability to deal with complex situations where there may not be a clear-cut right answer, they are still only performing

food for thought

2-1 The dangers of being in the robot's work envelope

During my time in industry, several accidents occurred involving robotic systems in which people learned hard lessons. The following is an incident that I feel truly exemplifies the danger of disrespecting the robot.

This situation involved an employee on the assembly side of the plant and a robotic system designed to pick up compressors for central air units and place them inside a welding machine. The compressors weighed about a hundred pounds, give or take 15 to 20 pounds depending on the model. The assembly line at the time of the incident had a sensor with some type of unidentified intermittent problem. This was a known problem, and operators had observed other employees activating the sensor manually to get the line flowing once more. Thus, we have the setting for our tale of woe.

On this day, the assembly line stopped as it had many times before, prompting the star of the story into action. He climbed up on the conveyor in an effort to trigger the faulty sensor and start thing up again. He had seen others do this and thought that he could save production time by just "taking care of it." Unfortunately, this individual made two critical errors. One, he was physically within the work envelope—the area that the robot can reach, specifically the place where the robot picked up compressors. Two, he triggered the right combination of sensors to signal to the robot that a part was ready for pickup. Once the robot received the proper signals, it began its program to pick up a compressor and load it into the welding machine. However, instead of a compressor, the robot's grippers closed on the employee's leg. The robot did not care that it had a leg instead of a compressor in its grippers, so it went about its programmed mission.

Imagine, if you will, that you are working on fixing a problem with the line when suddenly a robot grabs your leg and starts trying to load you into another machine. You have no E-stop to stop the robot. There is no one at the controls to save you. You cannot reason with it, plead with it, threaten it, or bribe it. You are now at the mercy of the three conditions that will stop a robot!

At this point, you are likely wondering if the system stopped, the employee got free, or if he was shoved into the welder. What saved this person's life was that his **femur** (the large bone in the thigh) broke. The whipping action of the robot, yanking him off the conveyor and swinging him around to feed into the welder, proved too much for the bone, and it broke. This in turn allowed the employee's leg to slip from the gripper under acceleration, and he flew free, to end up out of harm's way inside the robot's cage. The robot did not alarm out. The robot did not stop. There was no mechanical failure. What saved this individual was that the injury made his leg pliable enough to slip free of the robot's grip.

The individual made a full recovery, but was in the hospital for several weeks. His tale serves as a harsh reminder of the consequences for one who does not respect the robot. If his leg had not broken, there is a good chance that this accident would have cost him his life.

---

work envelope
The area that a robot can reach during operation

as programmed. Because of this functionality, there are only three conditions that stop a robot:

1. **The program/driven action has been completed**. Remember, the program or direct control is what we use to control robots, so once the robot has run its program or we have stopped sending action signals, it simply stops and waits for the next command. A sensor or other system may initiate the next command/program, which explains why robots sometimes seem to start up for no reason, but this is the robot working under program control.

2. **There is an alarm condition**. Almost all modern robotic systems, industrial or otherwise, have some sort of alarm system. In many cases, this system monitors such things as safety sensors, E-stops, load on the motors, vision systems, and other available devices that give the robot information about the world around it and its internal systems. The alarm function stops the robot in an effort to prevent or minimize harm to people, other equipment, and the robot.

3. **There is some type of mechanical failure**. Remember, robots are mechanical systems and are susceptible to breaking down, just like any other machine. Motors fail, bolts break, air hoses rupture, wiring shorts out, connections work loose, just to name a few conditions; any of these situations could cause a robot to stop. In a worst-case scenario, the robot would keep operating but perform its tasks erratically or unpredictably.

When humans disrespect the robot and put themselves in harm's way, they often learn about the three conditions that stop a robot the hard way. Now, if the robot in question is a low-powered system, say a Robosapien or iRobot floor cleaner, then there is likely to be little damage (if any) to a person. However, what would happen if we up the power of the system to, say, a lawn-mowing robot or something that a bomb disposal unit might use? These systems have enough power to cause damage to a person in the right spot at the wrong time. If we go up another step and look at the world of industrial-grade robots, no matter what the use, then the danger to humans becomes severe to deadly. If you have the misfortune to be in the danger zone when a robot starts up or you miscalculated the robot's next action, there is no amount of pleading you can do, no bribe you can offer, and no reasoning with the system to make it stop if you are in harm's way. The only way the robot will stop is when one of the three conditions mentioned previously occurs. If you do not respect the robot, then you too might learn this hard lesson.

Danger Zones

You may now be wondering if it is ever safe to work around a robot. The answer is an absolute yes. Every day, we use thousands of robots safely and effectively in many facets of the modern world. One of the first things that we do to create a safe working environment is to determine the various zones around a robot. Each zone has a risk level and requires a certain level of awareness based on the risk. For our purposes, we will be covering three main zones in this section: the safe zone, the cautionary zone, and the danger zone.

Safe Zone

The safe zone is where a person can pass near a robot without having to worry about making contact with the system. This area is outside of the reach of the robotic system as well as beyond the area that the robot can affect. The distance from the robot to the safe zone depends on the type of robot, the maximum force of the system, and what it is doing. As you have probably guessed, the more powerful the robotic system, the farther away you will need to be in order to be in the safe zone.

> **safe zone**
> The area where a person can pass near a robot without having to worry about making contact with the system

Let us look at an example to help shed some light on what a safe zone truly is.

For this example, the unit in question is a lawn-mowing robot. This robot works with a buried perimeter wire that limits the area in which the system can work and defines its work envelope or the area that the robot can reach during operation. For the sake of the example, the perimeter wire is around the front yard, limiting the system's work envelope to the front yard, mow every Sunday afternoon. Now for the final piece of information: you are babysitting a 3-year-old and want the child to be able to enjoy a sunny Sunday afternoon safely.

A quick scan of the setup for this example reveals the potential for trouble. There is a robotic system with a programmed job to do and a potentially inquisitive 3-year-old whose safety you are responsible for, both of which are in motion on Sunday afternoon. Obviously, you do not want the child playing around the lawn-mowing robot, as this could be a recipe for disaster. So instead, you place the child in the backyard, an area outside of the robot's work envelope with the added bonus of the house to block anything the robot might hit and send flying during operation. In other words, you place the child in an area where he or she will be safe from any action that the robot takes.

Cautionary Zone

The **cautionary zone** is the area where one is close to the robot but still outside of the work envelope. While the robot cannot reach you in this area, there could be danger during its operation from things such as chips, sparks, thrown parts, high-pressure leaks, crashes, overspray, or intense light from welding. Since this is often the area where operators perform their tasks, they must understand the potential hazards involved to work safely. This is not an area for just anyone.

A good example of a cautionary zone is the operator station for an industrial robot that performs **pick-and-place operation** (picking up items from one area and placing them in another). This type of system will often pick up a raw part from a conveyor or container, remove a finished part from a machine, place the raw part into the machine for processing, and then deposit the finished part on a conveyor or similar mechanism and wait to start the process all over again. The operator of these systems is usually responsible for such tasks as loading raw parts onto a conveyor, checking the dimensions and quality of finished parts, making corrections to the process as needed, and any other work needed to complete that portion of the production process. The requirements of this job often place the operator in close proximity to not only the robot, but also other production equipment. Workers spend millions of hours in these areas each year without injury or incident, though accidents do happen on occasion. Because of the potential for incident, this area requires the worker to be aware of all of its dangers as well as how to handle any situations that might arise. Because this is the normal space for the operator, the cautionary zone will contain stop buttons, emergency stops (E-stops), the controller for the robot, the teach pendant, and other ways to stop or control the system as needed.

Danger Zone

The **danger zone** is the work envelope, the area the robot can reach and where all the robotic action takes place. The various axes of the robot and the design of the system define the work envelope, so, once again, you need to be familiar with the robot to know its danger zone. Each robot will have its own danger zone. You must exercise extreme caution in this area because it has the highest potential for injury or death. When you are in the danger zone, you must watch out for the robot, any tooling used by the robot, and any **pinch points**, which are places where the robot could trap you against something solid.

In industrial settings, we have to protect the danger zone in some way so that when people enter this area, the system either slows to a safe velocity with extra sensitivity for impacts or stops its automatic operation. A popular method for protection is to place metallic fencing around the robot, creating a cage that keeps people out of the danger zone while providing one or two entrances for necessary repair, cleanup, or other normal job requirements. These entrances have sensors in place that stop automatic operations or in some other way render the robot safe for humans to be near when the doors are opened (see Figure 2-1).

When we put ourselves in the danger zone, we should always have some way to shut the robot down or know how to stop the system if needed. If you enter the danger

---

cautionary zone
The area where one is close to a robot, but still outside of the work envelope

pick-and-place operation
The common description used for robotic operations where parts are picked up from one location and then placed or loaded into another

danger zone
The area that a robot can reach and where all the robot's tasks take place

pinch point
Any place where a robot could trap a person against something solid

Occupational Safety and Health Administration (OSHA)
The federal agency charged with the enforcement of safety and health legislation in the United States

teach pendant
A handheld device, usually attached to a fairly long cord, that allows people to edit or create programs and control various operations of a robot

Figure 2-1   This is a trainer that my students use. This system has a completely encapsulating cage that creates two zones, a danger zone inside the Plexiglas and a safe zone outside the safety glass (as long as the door remains closed). If the door opens, it creates a cautionary zone directly in front of the robot. If you look in the upper right corner of the picture above the door handle, you can see the red door switch that senses when the door is open and keeps the robot from running in automatic mode.

Figure 2-2   A close up of the teach pendant for the FANUC trainer. Note the big red E-stop on the upper right corner of the teach pendant as well as the E-stop on the controller directly below the teach pendant.

zone with no way of stopping the robot, you are asking for trouble. In industry, it is an Occupational Safety and Health Administration (OSHA) requirement that anyone entering this area be in possession of the teach pendant. (OSHA writes and enforces safety regulations for industry.) The teach pendant is a handheld device, usually attached to a fairly long cord, that allows people to edit or create programs and control various operations of the robot. It is equipped with an E-stop, or Emergency Stop, should the need arise (see Figure 2-2). Let us explore two examples that demonstrate all three areas.

We will use a bomb disposal robot as example 1. The bomb technician controls the system with an operator station. The robot's task is to dispose of a

E-stop
An emergency stop used to shut down most of the powered operations of the machine and stop all motion as quick as possible

2-2   OSHA

The Occupational Safety and Health Administration (OSHA) officially formed in April 1971 as the result of an act signed into law by President Richard Nixon on December 29, 1970. OSHA's sole mission is to assure a safe and healthful workplace for every working person. The agency does this by inspecting work sites and factories to ensure that they are following the guidelines that OSHA has written or mandated by reference. Many of the rules that OSHA enforces come from such regulations as the National Electrical Code (NEC) or entities such as the National Fire Protection Association (NFPA).

OSHA enforces these rules by going to factories and work sites to conduct inspections where personnel look for violations or things that do not conform to the rules. When OSHA agents find a violation, the company in question is fined and then given a certain amount of time to fix the problem (or face steeper fines and penalties). OSHA averages about 40,000 to 50,000 of these inspections each year; many of which are in response to written complaints from workers.

If you would like to know more about OSHA or how to report safety violations, check out **http://www.OSHA.gov**.

suspected improvised explosive device at a shopping mall. Police have cleared out the mall and set up a perimeter complete with "do not cross" tape enforced by officers who ensure that everyone stays beyond this area. The area outside of the red tape is the safe zone. Next, the bomb technician enters with his truck full of equipment, his protective suit, and the robot. He suits up and prepares the robot to take care of the suspected bomb. The area where he sets up is the cautionary area: it is closer to the danger zone than the area beyond the red tape, but away from the explosive device, which is where the robot will be working. The bomb technician is wearing safety equipment to reduce the potential for injury from flying debris, indicating that there is some potential danger where he is located. Once set up, he sends the robot over to the device to begin the disarming process. The area the robot can reach, interact with, and affect is the danger zone. In this instance, the danger zone changes geographically as the robot moves into position. It also changes in size, depending on the task the robot is performing. When the robot is in the process of approaching the device, the danger zone is only the area that the robot can reach. Once it reaches the explosive and begins to work, the danger zone increases to include the potential blast area. In some unique cases, like this one, the danger zone is larger than just the work envelope of the robot. This is why it is necessary to understand all of the dangers involved when classifying the various zones.

For example 2, we will return to our industrial pick-and-place robot. This robot works with several machines and moves parts around in a defined area enclosed by a metal mesh cage. The operator has a workstation where he or she monitors operations, checks parts, and makes adjustments as needed. There is a clearly marked main aisle near the operator, but it is 15 feet away from the robot's work envelope. The aisle is for anyone to use and is considered a safe path for people or a safe zone. The operator station is within the cautionary zone, where there is some potential for injury; however, it is safe for people to be as long as they have the proper training and safety equipment. Inside the cage, as seen in Figure 2-3, one is exposed to all the dangers of the system; thus, anywhere in this area is considered to be the danger zone. Anyone who enters this area must have access to an emergency stop to comply with OSHA regulations. Remember, we meet the E-stop requirement by taking the teach pendant with us, but this also gives us the ability to control the robot and prevent bad things from happening in the first place (see Figure 2-4).

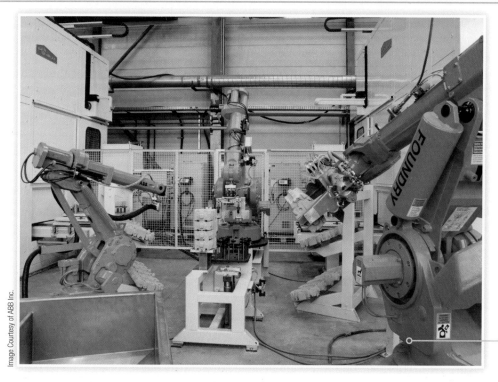

Image Courtesy of ABB Inc.

Figure 2-3 Three robots handle parts inside their caged-off area. Notice how much floor space is considered in the danger zone as denoted by the white metal cage behind the robots.

Image Courtesy of ABB Inc.

Figure 2-4 An operator in the danger zone working on the program with the teach pendant in hand.

These examples help demonstrate that each robotic system is unique and thus requires an individual assessment of the risks it possesses and the whereabouts of its boundaries zones. Remember, the more powerful the system and the more dangerous its functions, the greater the potential for injury or death. Failure to respect the requirements of the cautionary zone or danger zone is a quick way to experience firsthand the horror of being at the nonexistent mercy of a robot.

Guards

Devices or enclosures designed to protect us from the dangers of a system

Expanded metal guarding

Metal that is perforated and then stretched to create diamond-shaped holes with eighth-inch pieces of metal around it used to enclose equipment while providing a clear line of site

Guarding

In our discussion of danger zones, we briefly talked about guarding robotic systems. Guards, for our purposes, are devices designed to protect us from the dangers of a system. The two types we will explore are guards installed directly on the equipment and those placed a set distance from the device. No matter the placement, a guard's main purpose is to keep people safe; only rarely do they improve the operation of the equipment. In fact, there are many situations where guards will limit the operation of the equipment and/or make it more difficult to work with.

Guards that protect us from moving parts such as chains, pulleys, belts, or gears mount directly to the robot and are part of the system. Usually this guarding is supplied by the manufacturer in the form of the outer structure of the robot (see Figure 2-5). This guarding will often be made of a sturdy plastic or light metal that will allow for removal during repairs and preventative maintenance. Depending on how the guarding fits together, it may be necessary to move the robot into specific positions to remove certain pieces. If you are having trouble removing a piece of guarding, look for hidden screws or other pieces that may be holding it in place. Trying to force a piece of guarding off is a good way to break it and possibly damage parts underneath. Always repair or replace damaged guarding to ensure proper operation of the system and the safety of those who work around it.

Guarding that encloses the work area or envelope of a robot, as mentioned earlier, can be made of various materials, though expanded metal or metal mesh is the most common. Expanded metal guarding is metal that is perforated and then stretched to create diamond-shaped holes with eighth-inch pieces of metal around it (see Figure 2-6); metal mesh consists of thick wire welded and/or woven together to create a strong barrier that is easy to see through (see Figure 2-7). We weld these into metallic frames, usually angled iron pieces, which make up the panels of the robot cage. This creates a robust guarding system that is easy to see through, but strong enough to resist thrown parts, robot impacts, and people falling into or leaning on it. Add a few sensing devices, which we discuss next, and you have an OSHA-approved system to insuring the safety of workers in the cautionary zone.

Image Courtesy of SCHUNK

Figure 2-5   This robot has a fairly complex shell that allows the various parts to move without obstruction but keeps anyone on the outside from being at risk.

Metal cages are not the only way we can guard the work area of the robot. A system that is fast gaining popularity is a camera-based system mounted on the ceiling above the robot that detects when people pass into the danger zone, triggering the robot to respond accordingly. These systems often have a projector that defines the monitored area with clearly visible white lines, so that operators can see the danger zone. With a camera system, it is easy to adjust where the danger zone is defined and increase the area protected as needed, preventing the cost and down time associated with moving and fabricating metal cages.

The Baxter robot performs danger zone guarding by using a 360-degree camera system in its head that detects when humans are in the danger zone. When it senses someone within range, the robot slows to what is considered a safe movement speed by the RIA, OSHA, and ISO while also monitoring the sensitive collision detection system that will stop all movement of the robot if an impact is detected. This combination of safety features allows Baxter to work outside of the cage that restricts so many robots. As this style of robot interaction

Figure 2-6 An example of expanded metal guarding rotating parts.

metal mesh
The common name for guarding that uses either welded or expanded metal strips with small openings, allowing visual checks, to protect people from something potentially hazardous

Image Courtesy of ABB Inc.

Figure 2-7 Behind the ABB robot is the metal mesh cage used to protect people from the danger zone.

becomes more popular, there is a good chance that you will encounter other industrial systems with similar protection methods.

We have not covered every existing form of guarding, but you should have a good idea of what guarding is and why we use it. The Baxter robot and Tecnalia's work with the Hiro robot illustrate that industry desires a change from placing the robot behind the fence to having it work side by side with people. Industry has been a strong driving force behind many of the changes in robotics, so it is likely that this desire to get the robot out of the cage will usher in new and exciting ways to protect people in the danger zone. It is hard to predict what the future of robotic guarding holds, but we can expect it to continue to change and expand as we refine the current technology and invent new systems to ensure user safety.

Safety Devices

The modern robot uses a multitude of various sensors and devices to ensure that the humans who work around or happen into the danger zone are as safe as possible. Without these devices, many of the tasks we perform with and around robots would have a greatly increased risk of injury or death for those involved. We want to use robots to enhance our lives, not as a detriment. This section of the chapter examines the devices that make the robot safe for humans in or near the danger zone.

Figure 2-8 shows a common device used to ensure the safety of those who work around robots as well as directing robot operations known as the proximity switch (or "prox" switch). This device generates an electromagnetic field and senses the presence of various materials by changes in this field instead of physical contact. We use a prox switch in many applications, such as sensing when parts are present or when machinery is in position as well as tracking items on conveyor lines. When it comes to safety, we tend to use the prox switch to ensure that something is in a specific position before an operation takes place. This ranges from part placement, machinery doors being open or closed, devices to grip parts being open or closed, ensuring that the doors to the guarding around the danger zone are closed, or any other condition that is crucial to proper operation and safety.

A close relative of the proximity switch is the limit switch, which senses the presence or absence of a material by contact with a movable arm attached to the end of the unit (see Figure 2-9). We use limit switches in much the same way we use prox switches, but the main difference is that the limit switch will actually make contact with whatever it monitors. Because it makes physical contact instead of depending on a sensing field, we can extend the range of a limit switch by simply adding a longer arm for contact. These switches commonly ensure that guarding doors are closed, prevent the robot from traveling past a certain point during movement, prevent impact, and in general provide physical information about the world around the robot.

Many times, we use proximity and limit switches as a safety interlock: a system where all the safety switches must be closed or "made"

proximity switch
A device that generates an electromagnetic field and senses the presence of various materials by changes in this field, without physical contact with the item sensed

limit switch
A switch that senses the presence or absence of a material by contact with a movable arm attached to the end of the unit

safety interlock
A system where all the safety switches have to be closed or made for the equipment to run in automatic mode

Figure 2-8
Various proximity switches found in industry.

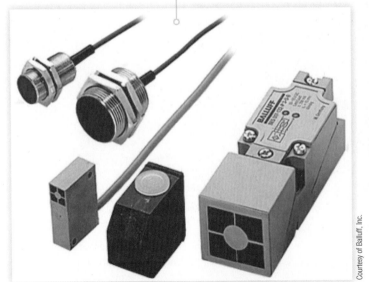

Courtesy of Balluff, Inc.

for the equipment to run in automatic mode. If at any point during the operation of the equipment these switches open or lose connection, the system automatically drops into a manual or alarmed condition, with many systems doing both. By tying these sensors to the entrances for a caged robot, any opening of the cage entrance automatically puts the robot in a state that is safer for people. If we have covers that are easy to remove, we can use the same trick to make sure that they are in place before the robot runs in automatic. The downside to using interlocks on the door to a robot cage is that someone could potentially open the door, step in, close the door, and then use the teach pendant to reset any alarms and start the robot once more. To get around this, we use presence sensors as well; these sensors detect when a person is inside the danger zone and are tied into the system to prevent automatic operation. The camera system mentioned earlier as well as pressure mats or light curtains (both of which are discussed below) are examples of presence sensors that we could use to add another layer of protection.

A pressure sensor detects the presence or absence of a predetermined or set level of force. For safety purposes, these devices are used in mats that can detect the weight of the operator and respond accordingly. These devices prevent the operator from entering the danger zone and closing the cage behind him or her, creating a condition where the robot could run in automatic and thus expose the operator to risk of injury or death.

Many robots can also tell when a motor has encountered something unexpectedly by sensing the increased demand for power. As a motor encounters resistance, it naturally starts to use more energy to try to overcome this force. When this happens, the part of the robot that drives the motors recognizes the additional power draw and can shut down the robot while sending an alarm to the teach pendant, letting the operator know that something is wrong. In the past, robots primarily used this method to detect major collisions; however, now it is part of the safety system for robots like Baxter, freeing some of the newer systems from the traditional cage. With the older technology, it took a fair amount of resistance to make a large enough difference in the power draw for the system to respond. With the new programming and sensing technology, the robot can look for major impacts and the power draw that they create when no one is in the danger zone and minor fluctuations in power consumption when people are near the robot, adding another layer of safety.

We also have a wide range of safety devices based on interrupted or reflected light that work similar to a photo eye. The photo eye sensor emits an infrared beam that strikes a shiny surface or reflector, which reflects the light back to a receiver in the unit (see Figure 2-10). These devices often have a couple of contacts: one that allows voltage through when the beam is detected by the receiver, and one that allows voltage through when no beam is detected, giving the user a wide set of options. The downside to the photo eye is that it

Figure 2-9
An example of a limit switch.

presence sensors
Sensors that detect when a person is inside the danger zone that are tied into the system to prevent automatic operation

pressure sensor
A device that detects the presence or absence of a predetermined amount of force

photo eye sensor
A device that emits an infrared beam that is reflected back by a shiny surface or a standard reflector to a receiver in the unit to detect the presence or absence of materials

Figure 2-10   A standard photo eye and its reflector.

laser photo eye
A sensor that uses a concentrated beam of light known as a laser to sense the presence or absence of objects, often over large distances, by the reflection of the laser light to a receiver

light curtain
A sensor that houses the emitter and receiver separately to create an infrared sensing barrier, often used to sense people in or entering a dangerous area

is a short-range sensor. To get around this, we can either separate the part that sends the beam (the emitter) and the part that receives the beam (the receiver), or we can change from infrared light to laser light. The laser photo eye works like the infrared photo eye, except that the light emitted is a concentrated beam, allowing greater distances of travel before the returned signal becomes too diffused for sensing purposes. When we house the emitter and receiver separately, we often call the sensor a light curtain. A light curtain may have a few or a large number of transmitters and receivers, allowing them to cover and protect a large area (see Figure 2-11). Light curtains protect workers in areas where parts are loaded and unloaded, detect objects, and create invisible sensing walls across large areas of space. The downside to the light curtain is the need for precision alignment of the emitter array and the receiver array. A very small amount of error in this alignment results in the infrared light traveling to one side or the other of the receiver and thus never creating a complete sensing path, which the machine interprets as something being in the danger zone.

Stop buttons and Emergency Stops (E-stops) are another crucial part of the safety circuit. The stop or pause button enables us to stop the system during

Image Courtesy of Miller Welding Automation

Image Courtesy of Miller Welding Automation

Figure 2-11 The two yellow bars on either side of the open space are light curtains. In this case, when the robot is not running, the operator can freely enter the work area and change out parts (or do whatever else is needed); however, if the operator crosses the invisible barrier while the system is in auto mode, the robot immediately stops operation.

normal operation and gives us control over when the system runs. In many modern systems, "stop" or "pause" are options on the teach pendant. Sometimes we need a faster, more decisive response—for these situations, we have the E-stop. An E-stop is wired into multiple systems on the robot and thus provides a more immediate and system-wide response than the stop button (see Figure 2-12). When we press an E-stop, the robot will halt its motion as quickly as possible and will often shut down many of the power and drive systems in an effort to prevent or minimize damage and injury. This is one of the ways we can tell the robot to stop, cease and desist, freeze, quit, or do no more (however you prefer to think of it). The stop tells the robot to stop, but generally lets it complete its current motion or slow to a stop. The E-stop is like yelling "Freeze!" Industrial robots that have teach pendants (which we will talk

Figure 2-12 This is a FANUC control panel, with the big red E-stop on the lower right-hand side. It also has an on, off, cycle start, and fault reset button.

about in detail later) have an E-stop installed on this device. Remember, you are required by OSHA regulations to take the teach pendant with you so that you will have access to an E-stop anytime you enter the danger zone.

We will dig deeper into the operation of many of these safety devices in Chapter 6 as well as describe how other sensor systems come into play. In the meantime, I encourage you to do your own research on the subject of safety devices and see where it leads.

Electricity and You

No matter what power source a robot uses, you must pay it the proper respect. Hydraulic power can generate bone-crushing force as well as release hot fluid at high pressure. Pneumatic power can generate enough noise to damage the ears, blow chips and dust into your eyes, or whip around busted hoses with all the mayhem and danger that that entails. Electricity can stop your heart, paralyze your muscles, stop your breathing, and literally burn you inside and out! We will explore the dangers of hydraulic and pneumatic power later in this book; however, I want you to be especially aware early on of the dangers of electricity. Since there is a high probability that any robot experiments you perform will involve electrical systems, it is my sincere hope that you will pay attention to this section and work carefully around all electrical systems.

Electricity is something we live with every day and is a crucial part of the modern world. We use it to heat and cool our homes, bring light to the darkness, cook and preserve our food, and power many of our modern entertainments. Whenever the power goes out, we are reminded how much we rely on electricity for many facets of our lives. Just as we depend heavily on electricity, so too is it a crucial part of the modern robot. We find electricity at work in the control systems, drives, sensors, and peripheral systems of the robot, thus making it very beneficial to know how to work with electricity safely.

Before we get too deeply into the safety side of electricity, we need to define a few terms:

Voltage– This force drives electricity through the system; the greater the voltage, the greater the driving force. We measure this force in volts (V).

Amperage– Amperage is a measurement of how many electrons, or how much electricity, is flowing through a system. We measure this in amps (A).

Resistance– This is a measurement of how much force is working against, or resisting, the flow of electrons. We measure this in ohms, represented by the Greek letter Ω (omega).

Electricity– This is a flow of electrons from a point with more electrons to a place with fewer electrons. We often control this flow and harness its power to do meaningful work or tasks.

In typical electrical systems, we control the flow of electrons using insulated wire and proper connections to route this force through specific systems and perform the desired work. Under normal circumstances, electricity is a safe and reliable power source. However, when we become careless or something goes wrong, we can become part of the **circuit** or path that electrons flow through. This is where using electricity becomes especially dangerous. When a person gets an electric **shock** or becomes part of the circuit, the electricity enters his or her body at the point of contact with the electrical system, passes through the body, and then exits at a **grounded point** or point somehow connected to Earth. The three main factors that determine the severity of an electrical shock are:

Voltage
A force that drives electricity through a system

Amperage
A measurement of how many electrons, or how much electricity, is flowing through a system

Resistance
A measurement of how much force is working against the flow of electrons

Electricity
A flow of electrons from a point with more electrons to a place with fewer electrons

circuit
A path that electrons flow through

shock
When a person becomes a part of the electrical circuit and has electrons flowing through his or her body

grounded point
A point somehow connected to the earth

1. The amount of current that passes through the body.

2. The path that the electricity takes through the body.

3. The duration of the shock.

The amount of current that passes through the body determines how much damage the affected area suffers. Remember, current is the number of electrons passing through the system. The amount of current will depend on the voltage that is providing the driving force and the resistance of the material it is passing through. The formula that represents this relationship is known as Ohm's law. Named after German physicist Georg Simon Ohm (1787–1854), it states that the current through a conductor between two points is directly proportional to the potential difference across the two points, or $I = V/R$, which is that the current is equal to voltage divided by resistance. V is the amount of voltage present to drive the electrons or potential difference, I is the current or number of electrons flowing, and R is the resistance in ohms to the flow of the electrons in a given system. We can also manipulate this formula to find voltage or resistance using algebraic rules, giving us $V = I \times R$ or $R = V/I$.

Example 1

How much current could flow in a system with 100 ohms of resistance powered by 120 volts?

$$I = V/R$$
$$I = 120/100 = 1.2 \text{ amps (or 1.2 A)}$$

Example 2

How many volts would it take to push 3 amps through 100 ohms of resistance?

$$E = I \times R$$
$$E = 3 \times 100 \text{ or 300 volts (or 300 V)}$$

Example 3

What is the resistance of a system that allows 4 amps to pass through when powered by 120 volts?

$$R = V/I$$
$$R = 120/4 = 30 \text{ ohms (or 30 } \Omega)$$

The more current that passes through a person who is shocked, the worse his or her injury will be. If you look at Table 2-1, you can see that 0.003 to 0.010 amps of current can be a painful shock. This is on par with events similar to shocking yourself on a doorknob after walking across a room. However, if the shock increases to 0.100 to 0.200 amps, there is a high possibility of ventricular fibrillation, a condition where the heart quivers instead of actually pumping blood. We are talking about only a portion of an amp being able to stop your heart from working! Moreover, 2 to 4 amps of current can stop your heart altogether, damage your organs, and possibly cause irreversible damage to your body. Many industrial robotic systems use voltages from 220V to 480V and have amperages ranging from 30A to 100A. Most 110V plugs, like the ones in the classroom, have a 15A or 20A breaker that protects the system. This means that any of these systems have more than enough current available to do severe and irreparable damage to you, should you become part of the circuit.

ventricular fibrillation
A condition where the heart quivers instead of actually pumping blood

Table 2-1   Electrical Effects on the Human Body

Current in Milliamperes	Effect on the Body
1–3	Ranges from unnoticed to mild sensation
3–10	Painful shock
10–30	Muscle contractions and breathing difficulty begins, with loss of muscle control possible
30–100	Severe shock with high possibility of respiratory paralysis
100–200	Ventricular fibrillation highly possible
200–300	Severe burns and breathing stops
2,000–4,000 (2 amps–4 amps)	Heart stops beating, internal organ damage occurs, irreversible bodily damage possible

Note: 1 milliamp is equal to .001 amps.

The second main factor in the severity of a shock is **the path that the electricity takes through the body**. If current or amperage enters your body at the tip of your finger and then exits through the center of the palm of the same hand, your heart and other internal organs would be safe. Burns to your finger or hand may occur, you may even lose a finger and part of your hand to the event, but you will more than likely survive. If that same shock comes in through your finger or hand, travels through your body, and then out your foot (which is a common path), your lungs, heart, and any other organs through which the current passes are in danger of electrical burns and failure, which is life threatening.

The last factor in the severity of a shock is **the duration of the shock**. It does not take long for damage to occur, as 0.1 amps passing through the heart for one-third of a second can cause ventricular fibrillation, but the longer the current passes through you, the more damage it will do. A worst-case scenario is where a person becomes part of the circuit, loses muscular control so that he or she cannot get free, and there is not enough total current flowing through the fusing devices to open the circuit. In this situation, the damage continues until a muscle spasm

food for thought

2-3 The danger of wearing metal around electricity

A caution I would like to add here is the fact that metal is a great conductor of electricity, so any metal that you have in contact with your skin, such as glasses frames, necklaces, rings, piercings, etc., provides a great initial path for a shock. Having metal in direct contact with or through the skin greatly reduces the skin resistance factors from Table 2-2.

Imagine for a moment the horrific effects of having your metal frame glasses or facial piercing bringing electricity into your body, making the initial point of entry your head. Either case creates the potential for electricity to pass through the brain, damaging the control center for the body. While this is not a common occurrence, this path is more dangerous than through the entire body, as shown by Table 2-2.

If you have piercings or such that you cannot remove, you will need to take extra precautions when working around electricity. You may need to cover them with some type of insulating material or make sure to keep that part of your body far away from anything that has electricity flowing through it. Failure to do so could lead to an unwanted electrical lobotomy.

Table 2-2   Material Resistance and Amperage Flow Chart

Material	Resistance in ohms	Current When 120V AC Is Applied
Dry wood, 1-Inch thick	200,000–200,000,000	0.0006 to 0.0000006 amps
Wet wood, 1-Inch thick	2,000–100,000	0.06 to 0.0012 amps
1,000 foot of 10 AWG copper wire	1	120 amps
Dry skin	100,000–600,000	0.0012 to 0.0002 amps
Damp skin	As low as 1,000	Up to 0.12 amps
Wet skin	As low as 150	Up to 0.8 amps
Hand to foot (inside body)	400 to 600	0.3 to 0.2 amps
Ear to ear (inside body)	100	1.2 amps

of the victim breaks the connection, the damage to the body destroys the conductive path, someone frees the person, or the amperage is finally great enough that protection devices open the circuit. Unless someone is nearby or responds immediately, this scenario means almost certain death.

If you look at Table 2-2, you can see some common resistance values and the current that can pass through them when exposed to 120V. For instance, the 1,000 feet of copper wire will have 120 amps passing through it ($E = IR$ or $120V = 120$ amps $\times$ 1 ohm). If we look at skin, you can see that dry skin will barely allow any current to pass through; however, damp skin can let enough through to be dangerous. Damp skin would be lightly sweaty skin, washed but not fully dried hands, etc. Wet skin is a huge risk—this is anytime the skin has a noticeable layer of sweat or water on it. From the table, we can see one of the worst-case scenarios is when current passes from ear to ear, with 1.2 amps doing damage directly to the brain.

No matter the voltage and amperage level present, you should always treat electricity with respect. Many times, it is when we are in a hurry or no longer consider the dangers of working with electricity that tragedy strikes. In the next section, we will discuss how to deal with someone getting shocked or electrocuted.

Handling Emergencies

Unfortunately, the modern world we live in tends to create circumstances that we label as emergencies. An **emergency** is a set of circumstances or a situation that requires immediate action and often involves events (or potential events) that have caused injury to people and/or severe damage to property. I am sure that most of you reading this book have experienced some kind of emergency in your life and can relate to how intense these situations can be. Often we have to react as fast as possible to these situations and have little time to sit back and think about what to do. With this in mind, we will look at some good general rules for dealing with emergencies and then drill down into some specifics for various situations.

emergency
A set of circumstances or a situation that requires immediate action and often involves events (or potential events) that have caused injury to people and/or severe damage to property

General Rules

General Rule 1: Remain Calm
If you allow your emotions to get the better of you, especially fear, the odds of dealing effectively with any situation are dramatically reduced. Fear clouds judgment and blocks logical thought patterns. Fear can prevent a person from thinking,

speaking, and/or acting. Fear is also contagious. If a person is injured and senses your fear, it will only add to that individual's own negative reactions. For instance, which of the following statements do you think would be more beneficial? "I have training in first aid and I am here to help you" or "You're bleeding a lot! What should I do?!" For most, the second statement will only make the situation worse and focus the victim's mind on the negative aspects of the situation.

General Rule 2: Assess the Situation

Just because someone has been hurt and needs help does *not* mean that the situation is safe for you to help them. Many times, rescuers must first take care of the dangers of a situation before they can worry about helping the injured. Blindly rushing in is a good way to become another victim or worse, lose your life for the mistakes of another. Unfortunately, history is full of those injured or killed why trying to help others. This is another part of being calm; you have to think before you act so that you act wisely, and not impulsively, irrationally, or emotionally. What if you are the only one there who could call for medical help, and instead you become victim number two? Who will call for help then?

General Rule 3: Perform to the Level of Your Training

When you are responding to an emergency, you need to make sure that you are making things better, or at the very least, no worse. If you are unsure what to do, you need to contact someone who has a higher level of training who can advise you. There is a wealth of resources out there such as 911, the Centers for Disease Control's (CDC) poison hotline, first responders, doctors, and many more to whom you could turn for help with the situation. You may also have someone in your facility or nearby that can help you deal with the situation. *Do not* try something you saw in a movie once or heard about randomly on the Internet! Often this will only make the situation worse and may put your life in danger as well.

General Rule 4: After It Is Over, Talk It Out

Emergencies are high stress, high emotion situations. Even if no one is hurt, there is a good chance that your adrenaline will be flowing, you may have been scared or upset, your heart may be pounding, and you may find that it has troubled you on some deeper level. These are common reactions and nothing to be worried about, but you should talk with someone about the situation. The where, what, how, and who of the emergency will determine who is best to talk with, but you should take the time to talk about it with whoever is appropriate, especially after an emergency involving severe injury or death. For lesser events, friends and family are a good group to turn to. For worse events, you may want to talk with those who were involved, or with a counselor. For accidents in the workplace, a protocol often details the individuals with whom you follow up to ensure that victim information remains private while providing this support.

Specifics of Emergency Response

Bleeding

There are multitudes of ways that we can cut, damage, or break the skin, all of which can result in bleeding. Because of this, we shall start here with our deeper look at emergencies. For minor cuts and abrasions, the procedure is simple. Clean the wound, apply some kind of antiseptic ointment, and cover with a nonstick bandage. For serious bleeding, the first step is to stop the loss of blood. To do this, we take a clean bandage, cloth, or gauze and apply firm pressure directly to the wound. This will likely hurt, but almost all bleeding can be stopped in 5 to 15 minutes with this method. If the material you are using on the wound becomes soaked with blood, you will need to apply another on top of the first *without* removing the original. Removing the original covering has the potential to rip open any clotting that has occurred and allow the wound to flow freely again. When possible, you can also elevate the injured portion of the body above the heart, as this will help to reduce the pressure on the wound and accelerate clotting, which in turn stops the bleeding.

When direct pressure does not stop the bleeding, a tourniquet may be required. A tourniquet is a tightened band that restricts arterial blood flow to arm or leg wounds to stop severe bleeding. Because the tourniquet stops blood flow below the place where it is located, it can also damage the tissue of the area it affects. Tourniquets may lead to severe tissue damage and even amputation. They can save lives, but are a last resort when all else has failed. They should only be applied by those with the proper training. If you would like to know more about tourniquets, I highly encourage you to take a first aid course offered through a reputable source.

tourniquet
A tightened band that restricts arterial blood flow to wounds on the arms or legs to stop severe bleeding

Burns

Burns are another common injury and thus worthy of a deeper look. Burns are classified on a scale of degrees from minor (first-degree burn) to severe (fourth-degree burn). A first-degree burn is easily recognizable by the reddening of the skin with a dry appearance; it is painful and heals in about a week. A second-degree burn is more severe and often includes blistering of the skin along with reddening or a whitish appearance; it can look dry or wet; some sensation of the area may be lost; and it can take up to three weeks to heal, with the most severe requiring medical assistance. A third-degree burn is a very severe burn; it is often blackened or ash white in appearance; the skin may be leathery; open wounds are often present; it does not hurt, as all the nerves are destroyed; it takes months to heal; and requires immediate medical attention. A fourth-degree burn extends past the skin into the muscle and bone of the victim and has the same characteristic of a third-degree burn.

When dealing with burns that do not have open wounds, such as first-degree and minor second-degree, you will want to submerge the area in cool water for 10 to 15 minutes or until the pain subsides and then wrap with a dry, nonstick, sterile bandage. Do not pop blisters should they appear and seek medical help if the area burnt is a sensitive area or pain persists. *Do not* place burns that have open wounds—such as severe second-degree, third-degree, or fourth-degree burns—in water and do not try to remove any clothing that may be stuck in these burns. Cover with a cool, moist, sterile, nonstick bandage or cloth and seek immediate medical help. Infection is the number one enemy when dealing with burns, so it is paramount to let the medical professionals deal with cases of broken skin burns.

Blunt-force Trauma

Care of blunt-force trauma (or impact that does not penetrate the skin) depends on the level of the injury. The severity of the injury usually depends on the size of the object that hits the person, how much force is behind the impact, and where the impact is on the person's body. For minor impact injuries, you may want to apply an ice pack and monitor for continued swelling, discomfort that does not fade, or other signs of serious injury. For major impact injuries, apply an ice pack and seek medical help, as there is an increased chance of internal injury that is not outwardly visible. In the case of broken bones, immobilize the limb as best you can with a splint device from your first aid kit. If you do not have a splint, you can use materials such as wood, rolled up magazines, or anything rigid placed on either side of the broken bone and held there by a cloth wrap of some kind. The whole

blunt-force trauma
An impact that does not penetrate the skin

point is to immobilize the injured limb for transport to medical help. If there is a fear of head trauma or internal injuries, keep the victim calm and get him or her to medical help as quickly as possible.

Shock

The last situation I want to focus on is shock, or electrocution. You learned earlier what can happen when electricity passes through a person's body, but we did not review how to deal with this situation. If the person is still being shocked, you will need to either cut the power to circuit or use a nonconductive item (like a wooden broom handle or dry rope) to get the person out of contact with the circuit. DO NOT touch the person! If you do, you will become a part of the circuit and get shocked as well. If the person is already clear of the circuit, make sure that there is no chance that you will be shocked before you check the person. Once the victim is clear of the circuit and it is safe for you to do so, check for a response in the victim. If the victim responds, you will want to get him or her checked out by a medical professional as soon as possible. If the victim does not respond, call for help and have someone call 911. If you know CPR, perform the steps and respond accordingly, otherwise try to find someone who is CPR trained, as often a severe electrical shock will stop the heart or cause ventricular fibrillation.

Of course this is not the entire list of possible situations you might face in an emergency, nor do we have the time to cover them all. If you remember the general rules from this section and keep your wits about you, there is a good chance that you can deal with most of the emergencies that you may encounter. If you are worried about your ability to perform first aid or want to get a deeper knowledge of any of these topics, then I would encourage you to seek out additional training. Some employers have programs where employees can volunteer for training to deal with emergencies, and that might be an avenue for you to learn more. The more training you have on these topics, the better you will be prepared to deal with any emergencies you may encounter.

review

By now, you should have a deeper understanding of what it takes to work safely with robotics as well as how to handle any emergencies that might arise. The robotics environment is constantly changing as we adopt new devices and new standards to ensure that we can work safely with robots. As you work with robots along your chosen path, you will learn more about the specifics of how safety applies to your field.

Here is a quick review of the topics we covered in this chapter:

- **Robots Require Respect!** We discussed what happens when you no longer fear the robot as well as how to work safely with robots.

- **Danger zones.** This section outlined the various zones around a robot as well as who could be in those areas.

- **Guarding.** This section talked about how we keep people out of the danger zone.

- **Safety devices.** We discussed some of the devices used to ensure people's safety around the robot.

- **Electricity and you.** This section outlined the effects that electricity can have on the body and how to work safely with electricity.

- **Handling emergencies.** We talked about emergencies in general and then drilled down into first aid for some injuries that you might encounter.

key terms

Amperage	Guards	Photo eye sensor	Shock
Blunt-force trauma	Grounded point	Pick-and-place operation	Teach pendant
Cautionary zone	Laser photo eye	Pinch point	Tourniquet
Circuit	Light curtain	Presence sensor	Ventricular fibrillation
Danger zone	Limit switch	Pressure sensor	Voltage
Electricity	Metal mesh	Proximity switch	Work envelope
Emergency	Occupational Safety and	Resistance	
E-stop	Health Administration	Safe zone	
Expanded metal guarding	(OSHA)	Safety interlock	

review questions

1. What are the three Rs of robotics?

2. What are the three conditions that stop a robot?

3. What are the safe zone, cautionary zone, and danger zone as they relate to a robotic system?

4. What is a common way that we keep people out of a robot's danger zone?

5. Who made it a requirement that any time you enter a robot's work envelope you take the teach pendant with you, and what is one of the main benefits of having the teach pendant?

6. What are some of the tasks we use proximity switches for?

7. What do we use pressure sensors for when it comes to robotic safety?

8. What is the difference between a safety interlock and a presence sensor?

9. Describe what happens when a person is shocked.

10. What are the three factors that determine the severity of a shock?

11. What is the formula for Ohm's law?

12. What is ventricular fibrillation and at what amperage passing through the body does this become a high possibility?

13. When it comes to industrial robotic systems, what are the common ranges for voltage and amperage?

14. What are the general rules for dealing with emergencies?

15. How do we stop serious bleeding?

16. How do we treat minor burns?

17. How do we treat severe burns?

18. What do we do for broken bones?

19. When a person is being electrocuted, what is the first thing you must do?

3

Components of the Robot

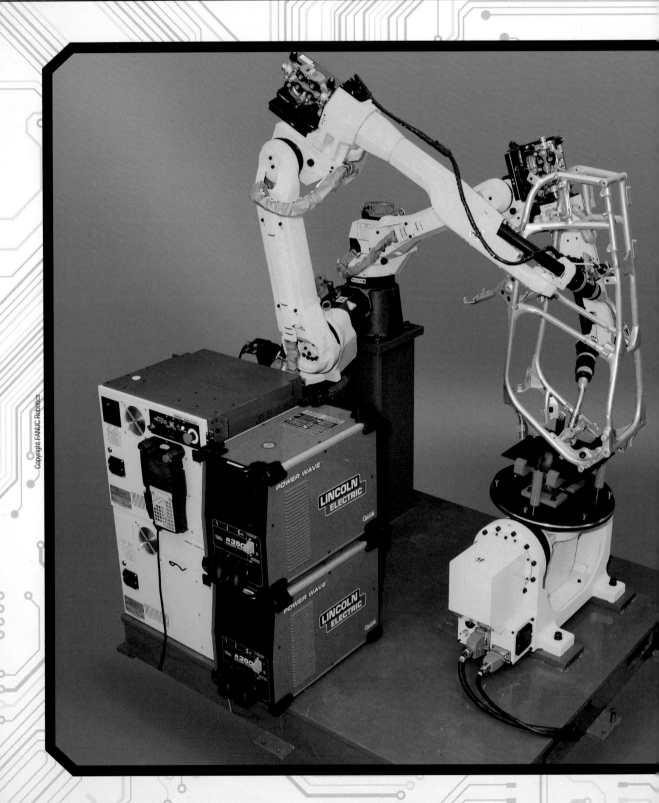

Copyright FANUC Robotics.

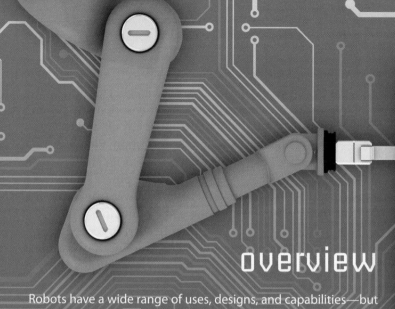

overview

Robots have a wide range of uses, designs, and capabilities—but in the end, a robot is a collection of subsystems that make up the whole. In this chapter, we will look at each of these subsystems and their role in making the robot a functional unit. As a general rule, the more complex the robot, the more components and systems it uses. The better you understand the components and subsystems of the robot, the easier it will be for you to work with the robot regardless if you are operating, designing, repairing, or setting the system up for a new task. To help you understand the systems, this chapter will cover the following topics:

- Power supply
- Controller/logic function
- Teach pendant/interface
- Manipulator, Degrees of Freedom (DOF), and axis numbering
- Base types

Power Supply

No matter the type, complexity, or function of the robot, it must have a source of energy to do work. The power supply used in a robot often depends on what the robot does, its working environment, and what is readily available. Potentially, any power source we can harness could power a robot, but the main sources used for modern robotics are electric, hydraulic, and pneumatic power. As we look closer at each power source, we will discuss the pros and cons of each.

Electric Power

Electricity, as previously discussed, is the flow of electrons from a place of excess electrons to a place of electron deficit. We route these electrons through connected systems of components, known as circuits, to perform some type of work. In an electrical system, voltage is a measurement of the potential difference or imbalance

Electricity
The flow of electrons from a place of excess to a place of deficit that we route through components to do work

voltage
A measurement of the potential difference or imbalance of electrons between two points and the force that will cause electrons to flow

amperes/amps
A measurement of electrons passing a given point in one second

Resistance
The opposition to the flow of electrons in the circuit and the reason why electrical systems generate heat during normal operation

of electrons between two points and the force that will cause electrons to flow. We measure the flow of electrons in amperes or amps: One amp is equal to 6.25×10^{18} electrons passing a point in one second, and this flow does the work in the circuit. Resistance is the opposition to the flow of electrons in the circuit and the reason why electrical systems generate heat during normal operation. Electrons can either flow in one direction, which we call direct current (DC) or it can flow back and forth in a circuit, which we call alternating current (AC). The power supplied by the electric company to homes and businesses is AC, while power from batteries and solar panels is DC. The type of robot in question and its functions will determine how much amperage it requires, the voltage level needed to force enough current through the system, and whether the voltage is AC or DC. Small robots that move light payloads require fewer amps than their industrial cousins moving heavy payloads.

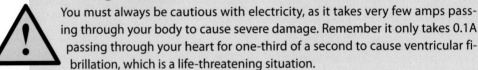

safety note

You must always be cautious with electricity, as it takes very few amps passing through your body to cause severe damage. Remember it only takes 0.1A passing through your heart for one-third of a second to cause ventricular fibrillation, which is a life-threatening situation.

direct current (DC)
Voltage where the electrons flow in only one direction

alternating current (AC)
Voltage where the electrons flow back and forth in the circuit

polarity
The condition found in DC circuits where each component has a defined positive and negative terminal that must be maintained in the circuit for proper operation

conventional current flow theory
Electron flow from the positive terminal to the negative terminal

DC power is electrons that flow at a constant voltage level in one direction only. Due to the way this power flows, every component in DC systems will have a set polarity or positive and negative terminal. It is very important to observe the polarity of components because failure to do so will reverse the flow of electrons through the part, causing erratic operation, or block power flow. For many parts, this will result in permanent damage. This is why when you put the batteries in backward, some things do not work, while others run backward or erratically. If you are replacing a part or building your own robot, you need to pay careful attention to the polarity of the system for proper operation of DC components.

Originally, scientists believed that electrons flowed from the positive terminal of a battery, through the circuit, and back to the negative terminal. We call this the conventional current flow theory. Experiments later proved that the electrons actually flow from the negative terminal, through the circuit, and back to the positive terminal. We call this the electron flow theory. When you are working with DC circuits, it really does not matter which way you think of the system as long as you maintain the proper component to circuit polarity. Many components will have the positive or negative terminal marked, making this task easier. Take a moment to look over Figure 3-1, the circuit example, paying close attention to the polarity of each of the components.

For small robots, such as the Robosapien, LEGO NXT, and NAO robots, a battery pack is often the source of electrical power for the system. A battery pack generates a set amount of amp-hours (Ah), which is the number of amps it can generate over a length of time. For example, a 10 amp-hour battery could deliver 1 amp for 10 hours, or 2 amps for 5 hours. In both cases, we still get the full amp-hours out of the battery, but in the second instance, the system would only run for 5 hours, or half the time of the first system. To get around this problem, you can add another battery or power source in parallel to the first. To do this, you would make sure

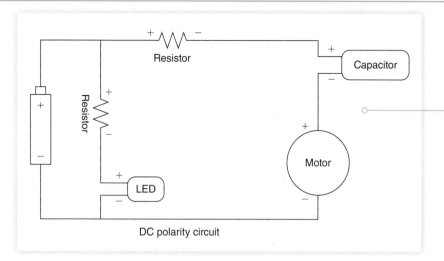

Resistor

Resistor

Capacitor

Motor

LED

DC polarity circuit

Figure 3-1   This example shows the polarity of each component in the circuit. Notice that the polarity is component-specific.

electron flow theory
Electron flow from the negative terminal to the positive terminal

amp-hours (Ah)
A measure of the number of amps a power supply can deliver over a specific time

that both of the positive leads of the batteries are hooked together, and then do the same with the negative (see Figure 3-2). It is important when setting up this kind of power supply to make sure that you use identical batteries, otherwise the battery with more voltage and amperage will spend some of its energy trying to charge the lower voltage and amperage battery. At best, this is wasted energy; at worst, this could cause battery rupture. When we connect batteries this way, we add the amp-hours of each battery to determine the total amp-hours. See example 1.

Example 1

If we connect three batteries in parallel that have 5 amp-hours of energy each, what would be the total amp-hours available?

Total amp-hours = battery 1 + battery 2 + battery 3
(or however many batteries are joined in parallel)
Total amp-hours = 5 + 5 + 5 = 15 amp-hours or 15 Ah

Another issue that comes up with battery use is getting the right voltage level. For instance, AAA, AA, C, and D sized batteries all provide 1.5 volts of DC

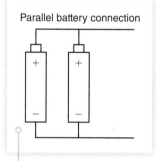

Parallel battery connection

Figure 3-2a
Here are two batteries in parallel, sharing the amperage requirements of the circuit and extending the amp-hours.

Figure 3-2b   A couple of examples of battery-powered robots.

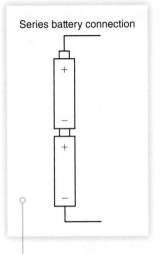

Series battery connection

Figure 3-3

Here is an example of two batteries in series.

electricity, but commonly power systems that require a higher voltage. So how do we get around the 1.5V issue, you might ask? The answer is to connect the batteries in series. When we connect batteries in series, instead of increasing the number of amp-hours, we increase the voltage. We do this by hooking the positive of one battery to the negative of another and then powering the system off the open ends of the proper number of batteries (see Figure 3-3). If you accidentally put one of the batteries in backward where either two positive or two negative terminals are touching, the battery that is backward actually subtracts from the total voltage of the string, so be careful! See examples 2 and 3.

Example 2

What would be the total voltage supplied by three AA batteries connected in series?

Total voltage = battery voltage 1 + battery voltage 2 + battery voltage 3 (or however many batteries are joined in series)
Total voltage = 1.5V + 1.5V + 1.5V = 4.5V

(Note: Remember that AA batteries have a voltage of 1.5V each.)

Example 3

What would be the total voltage supplied by three AA batteries connected in series with the last battery placed in backward or negative to negative?

Total voltage = battery voltage 1 + battery voltage 2 − battery voltage 3 (Remember, we subtract the last battery because it is placed in backward and thus subtracts from the total voltage of the series group.)
Total voltage = 1.5V + 1.5V − 1.5V = 1.5V

By placing one battery backward, we effectively have only 1.5V instead of the normal 4.5V this system should generate.

Many of the high-demand DC systems will use a bank of batteries that consists of parallel sets of batteries in series to meet the voltage requirements and extend the run time (see Figure 3-4). In smaller systems, this will usually consist of sets of AA, C, or D sized batteries, while larger robot systems will often use multiple 12V deep cycle batteries. The difference between a deep cycle battery and a car battery is that the deep cycle battery can be drained fully and recharged many times, where car batteries do not handle full discharge well. In addition, a deep cycle battery will deliver lower amperage over a long period of time, while car batteries deliver a large amount of amperage to start a car over a short period of time. See example 4.

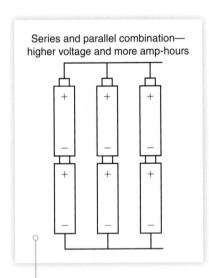

Series and parallel combination—higher voltage and more amp-hours

Figure 3-4

This battery packs three sets of series batteries wired in parallel to increase the amp-hours while maintaining the proper voltage level.

Example 4

What would be the total voltage and amp-hours of a system that has five parallel groups or cells consisting of three 12V batteries connected in series, when each battery has 100 amp-hours of electricity?

Total voltage = battery voltage 1 + battery voltage 2 +

battery voltage 3

Total amp-hours = battery 1 + battery 2 + battery 3
Total voltage = 12V + 12V + 12V = 36V
Total amp-hours = 100Ah + 100Ah + 100Ah + 100Ah +

100Ah = 500Ah

Even though there are 15 total batteries in this system, there are no more than three in series at any given point, so we only use three batteries to figure the total voltage. The three batteries in series all give electrons at the same time to increase the voltage level and thus act as 1 amp-hour amount, so we use the number of groups for calculating amp-hours, not the total number of batteries. One weak or misconnected battery in this scenario could affect the whole system; this is why we try to use the same type and quality of batteries when we create a power source of this nature.

AC power does not have a set polarity because it is constantly changing the direction it flows through the circuit, which means that we do not have to worry about the positive or negative polarity of AC components. Another difference between AC power and DC power is that the voltage or intensity is constantly changing. When we compare a graph of DC to AC electricity, the DC is a straight line showing constant voltage level, whereas the AC is a sine wave showing changing voltage levels (see Figure 3-5). The AC power starts at zero, rises to a positive value, drops back to zero, falls to a negative value, and then rises to zero once more. One complete wave from zero to positive to zero to negative and back to zero is called a cycle. In America, we have 60-hertz (Hz) power, or 60 of the sine wave cycles per second. Because of the nature of AC power, we measure it in Root Mean Square (RMS), which is a mathematical average of the sine wave. Therefore, when you measure 110V at an outlet, you are measuring an average of the peaks and valleys of the sine wave. This is the effective amount of electromotive force (EMF) or voltage for the system.

AC power is a good choice for any robot that will be stationary in an environment that has power readily available, making it a favorite among industry. Using supplied AC power negates the concerns about amp-hours and also removes the space and weight requirements of batteries. With AC, we do not have to worry about hooking batteries in parallel or series; however, we do have to make sure that we know the voltage needed and whether it is single- or three-phase power. Let us take a moment to explore the difference between single-phase and three-phase AC.

Single-phase AC is AC power that has one sine wave provided to the system via a single hot wire and returned on a neutral wire. The power in a standard 110V wall outlet is single-phase AC. At this point, you may be thinking, if there is only one sine wave delivered by one hot wire, why are there two or three prongs on the plugs that I use? While it is true that only one wire is providing power to the system, the other two wires are no less important. The second wire attached to the larger of the two prongs, known as the neutral wire, is the wire that provides a

cycle
One complete sine wave from zero to positive to zero to negative and back to zero of AC power

hertz (Hz)
Sine wave cycles per second

Root Mean Square (RMS)
A mathematical average of the sine wave

electromotive force (EMF)
Another name for voltage that is a measure of the potential difference between two points that causes electron flow in circuits

Single-phase AC
AC power that has a single sine wave provided to the system via one hot wire and returned on one neutral wire

neutral wire
The wire that provides a return path for the electrons and allows for a complete circuit

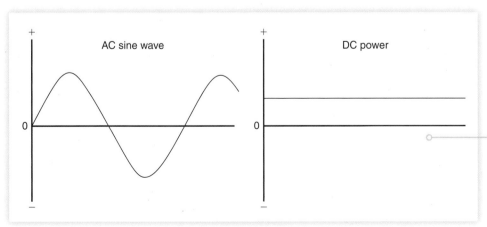

Figure 3-5 You can see the alterations of AC power on the left and the steady intensity of the DC power on the right.

3-1   Single-phase vs. Three-phase

Whenever I am talking about the difference between three-phase and single-phase AC in my classes, I like to use the following example.

Imagine that your vehicle has run out of gas. You have to push it from the bottom of a small hill where you are to the gas station at the top. As you push the vehicle up the hill, you have to take small breaks to catch your breath. These breaks result in the vehicle stopping its forward motion. While you do not have to start over at the bottom of the hill, you do have to put extra energy into getting the vehicle moving once more. This is similar to single-phase AC.

Now, we will use the same example, but this time, there will be three people pushing the vehicle up the hill. You start up the hill when a passerby stops to help. He keeps the vehicle moving forward as you take a moment to catch your breath. At about the time that the person helping needs a break, a third person comes along and adds his muscle to the task. At times, there is only one person really keeping the vehicle moving forward; other times two of the three are pushing the vehicle, at least partially. The key is that the vehicle is always moving forward under force, and thus the forces of gravity and friction do not get the chance to fully stop the vehicle's forward movement. The result is that the vehicle gets to the gas station, but none of the three people are as tired as a single person completing the task. This is similar to three-phase AC power.

For those of you who would like to know more about electricity, I encourage you to do your own research. There is a wealth of information out there on this fascinating subject.

ground

The wire that provides a low resistance path for electrons to flow when the electrons escape the system due to insulation or component failure

Three-phase AC

AC power that has three sine waves that are 120 degrees apart electrically

return path for the electrons and what allows for a complete circuit. Without this wire to complete the circuit, the items we plug in would not work. The third prong on a plug, the round one, is for ground. The ground wire provides a low resistance path for electrons to flow when the electrons escape the system due to insulation or component failure. The primary reason we ground circuits is to protect people from electrocution, with protecting equipment and fire prevention following close behind.

Three-phase AC is AC power that has three sine waves that are 120 degrees apart electrically. This is the primary power source for most industrial facilities due to the great amount of work it can perform and the fact that it is very efficient. With single-phase AC, there are points during the cycle where no voltage flows in the system and thus no work is done. Granted, this only happens for a fraction of a second, but all those fractions of a second add up to create an inefficiency. Three-phase AC avoids this because one of the three phases is always supplying power, so there is no loss of momentum or wastefulness due to lack of electron flow. (See Food for Thought 3-1 for an example that may help to clear up the difference between single-phase and three-phase AC.) With three-phase AC, there will be three hot wires, each supplying one of the three different sine waves. There will also be a ground wire, just as with our single-phase AC circuit, which provides the same function as previously mentioned. What may or may not be present is a neutral wire. With three-phase power, because they are 120 degrees apart electrically, one or two of the hot legs can act as a return or neutral wire. This is why many three-phase systems do not have a neutral. A neutral provides additional safety in three-phase systems, but becomes an added cost in parts, wire, construction, and time. Figure 3-6 compares the difference in sine waves between single-phase AC and three-phase.

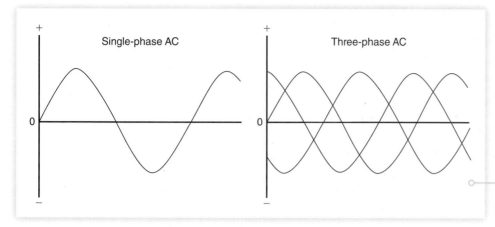

Figure 3-6   As you compare the single-phase on the left and the three-phase on the right, notice how the three-phase always provides electricity to the system, unlike the single-phase, which has zero points or no voltage flowing at times.

Hydraulic Power

Hydraulic power involves the use of a noncompressible liquid given velocity and then piped somewhere to do work. When people first began exploring this power medium, the fluid was water, and often the velocity or speed of movement came from flowing streams or water falling from a higher level to a lower one. We used this power to run mills and direct water where needed; however, it was Blaise Pascal's experiments in 1653 that really began the exploration of hydraulic power. He discovered that pressure applied to a confined fluid is transmitted equally and undiminished in all directions and the resultant forces act upon the surfaces in a container at right angles. Pascal's discovery sparked considerable growth in the field of hydraulics and lead to some exciting new devices such as hydraulic presses, all powered with water-based hydraulic systems.

> **Hydraulic power**
> The use of a noncompressible liquid given velocity and then piped somewhere to do work

For those of you reading this who have some experience with hydraulics, you may be asking yourself: Why do we use oil in the modern world instead of the water of old, especially given the cost? The answer is that water has some drawbacks when used as the fluid for hydraulics. One of the major problems is corrosion or rusting of the ferrous metal parts. If you have ever seen a piece of metal left outside that has rusted, you have an idea of what water was doing to the internal systems of the machine when it was used for hydraulic power. Another problem is that water freezes at 32 degrees Fahrenheit. This meant either keeping the equipment warm somehow or draining it and not using it during the cold portions of the year. Another major issue with using water is the fact that it grows bacteria over time if not cleaned in some manner. Bacteria growth in the system clogs openings and valves, poses health risks for human workers, and creates putrid smells. Because of these drawbacks, when electricity came on the scene in the early 1800s, hydraulic power took a backseat in the power source scene until the early 1900s.

In 1906, the U.S.S. Virginia used oil in the ship's hydraulic system to raise and rotate its guns. This simple change of fluid revitalized hydraulic power and ushered in another period of growth and expansion for the field. With this change not only was the corrosion problem solved, but the system also became self-lubricating! Oil also has a much lower freezing point than water and is not a good habitat for bacteria to grow in, so the other major problems fell to the wayside as well. As engineers experimented with using oil in hydraulics, they discovered that they could generate higher pressures in the system due to the fluid's properties and thus do greater amounts of work.

The final discovery needed to usher in the modern hydraulic system came during World War II. Due to the use of rubber in so much of the equipment needed

Figure 3-7 Here is an example of a hydraulic power supply. You can see the pump in the center, a pressure gauge and filter on the left, and various points to access and return the hydraulic power on the right.

for the war effort, the United States could not get enough natural rubber to meet our needs. We had to find a synthetic alternative. While it was shaky going at first, we eventually mastered synthetic rubber and created new rubber items that were stronger and more durable than their natural counterparts, especially in the field of seals. These new, stronger seals allowed us to increase the pressure of the hydraulic system even more; thus, we have the modern system that give robots the ability to generate tons of force effortlessly (see Figure 3-7).

We use modern hydraulic power when there is a need to generate great amounts of force with the precision control that is common with noncompressible fluids. This allows robots to pick up car bodies and move them with ease, stopping precisely and holding position whenever and wherever needed. While some of these robots tend to be slower than their electric counterparts, they remain a true workhorse in industry. The tradeoff for this extra power is the addition of the hydraulic system to the robot and all the upkeep that goes with it. The oil needs to be filtered and monitored, with periodical filter and oil changes coupled with the removal of any deposits in the hydraulic tank—all based on the type of oil used and how many hours the system runs. While routine tasks are part of maintaining any hydraulic system, they do represent extra work required for hydraulic robots versus their electrical counterparts. The other downside is that eventually almost all hydraulic systems leak. Leaks can range from a nuisance leak that is just a bit messy and easily fixed to the catastrophic failure, which can result in gallons of hydraulic oil sprayed all over the equipment and anything in the immediate area. The most dangerous leak is the pinhole leak, which can turn the oil into a flammable mist and generate high-pressure streams that can shear through metal and human tissue alike (see Food for Thought 3-2 for a worst-case scenario).

When dealing with hydraulic leaks, there are a few key points to remember:

- Hydraulic oil from a running system may be hot enough to burn skin, so try to avoid direct contact until you are sure it is safe.

- If a hydraulic oil puddle is uncontrolled, it can cover a much larger area than you might think. There are special barrier and damming devices that can help to control large spills.

- Leaks that generate an oil mist can be a fire hazard. Most hydraulic oils are stable and take a lot of heat to ignite, except when dispersed as a mist in the air.

food for thought

3-2   The Danger of a Pinhole Leak

During my time in the military, I worked on Huey helicopters. It was during that time that one of my sergeants told me this story.

The Chinook helicopter is a workhorse for the military, with two main rotors that are capable of lifting large and heavy cargos. This helicopter, like many aircraft, depends heavily on its hydraulic system to run various things such as the cargo hatch in the rear, flight controls, and other various powered systems. Because so many systems in the helicopter are dependent on hydraulics, any leaks are a matter for concern.

The Chinook in this story had a hydraulic leak as noticed by reduced fluid in the hydraulic reservoir, but no one could find it. After each flight, the level in the tank would be lower, prompting the mechanics to look all over the helicopter for the telltale signs of a leak. A team of four looked for puddles, loose fittings, oily lines, anything that might show where the system was losing fluid, but they had no luck. This went on for several flights without any clue as to where the fluid was leaking from or where it was going to. The leak was not bad enough to ground the bird, but it was worrisome all the same, as eventually small leaks become big leaks.

Finally, the crew decided to start up the Chinook and its hydraulic system while sitting on the runway and see if maybe they could find the leak that way. While the maintenance crew was making their inspections, one member was checking the outside of the helicopter for any signs of the leak and running his hand along the fuselage. Unfortunately, he was successful in finding it. The leak was a pinhole leak that was against the outside skin of the helicopter, venting hydraulic oil to the atmosphere. Because the leak was so small and the system pressure was so high, it cut right through the outer hull of the Chinook. The leak also cut four fingers from the soldier's hand as he ran it over the leak during the inspection. It happened so fast that the soldier did not even feel it at first! The last time my sergeant saw this soldier, he was on a medevac flight to a hospital in hopes of reattaching his fingers.

My sergeant told us this story so that we would realize how powerful hydraulic power can be and to make sure we would pay attention to what we were doing. While this story is a bit graphic, it has stuck with me over the years and reminded me to be cautious when dealing with hydraulic systems.

- Cleaning up hydraulic oil usually requires the use of some type of absorbing medium.

- You do not want to let hydraulic oil go down any drains connected to the sewer system; it is a contaminant and will wreak havoc at water treatment facilities or on any natural waterways it might reach.

- When finished with the cleanup, make sure you dispose of any oil and oil-soaked materials properly. (See the Safety Data Sheet [SDS] or a supervisor for more information on proper disposal. SDS sheets tell you about chemicals, their dangers, how to handle them properly, and how to dispose of them.)

Pneumatic Power

Pneumatic power is very similar to hydraulic power, with the main difference being the use of compressible gas instead of noncompressible liquid to transmit power. Since the time of Archytas of Tarentum and his wooden pigeon that worked via steam or compressed air in 420 BCE, man has used pneumatic power to operate various mechanical creations. Heron of Alexandria used steam to run the first historical engine right at the beginning of the Common Era. Moreover, Robert Boyle did for pneumatics what Blaise Pascal did for hydraulics. During Boyle's experiments from 1627 to 1691, he discovered that at a constant temperature, the

Pneumatic power
Fluid power that uses air to generate force

pressure of a gas varies inversely with its volume. In other words, if a gas expands it cools, but if it is compressed it will heat up. Boyle also discovered that we need air to transmit sound and built one of the first vacuum pumps.

We have used pneumatics over the years to run motors, power industrial equipment, tunnel through mountains, reduce friction, and even power devices in the home. The primary differences between pneumatic power and hydraulic power are the use of a gas instead of a liquid in pneumatics, and that gas is compressible. This means that if you stop an air cylinder in a position other than fully extended or retracted, the system can flex or move from where you stopped it. You can see the difference for yourself by using an empty plastic soda or water bottle. Fill the bottle as full of water as you can and carefully replace the cap, making sure to get it good and tight. Now try squeezing the bottle and see what happens. If you have filled the bottle fully with water, there should only be the slightest of flexing as the force equalizes in all parts of the bottle before you run into the water's resistance to compression. Now dump out all the water, close the cap tight, and try again. This time you should be able to compress the sides of the bottle much further than before. This is because gases consist of molecules with lots of space between them, so we can force them into smaller areas; this is something you cannot do with the high-density molecules of a liquid.

One of the benefits of pneumatic power is that we can vent the used air back to the atmosphere when we are done with it. With electricity, we have to return the electrons along some path to make it work, and with hydraulics we return the fluid to the tank; however, with pneumatics, we simply vent, or release, the used air at a convenient point. When we vent the used air, we need to be careful of two things: one, it is often a very loud process; and two, we need to watch for small particles that can become airborne due to the pressure of the escaping air. To combat both of these conditions, we use something called a muffler (see Figure 3-8). The muffler slows air as it passes through it, reducing the sound and spreading the air that was flowing in one, focused direction into a circular pattern. Even with the use of a muffler, you should use hearing and eye protection when working with pneumatic systems.

muffler
A device used with pneumatic power to reduce the velocity and noise generated by venting used air

Figure 3-8   On the left side of the picture, you can see the two mufflers: the gold cone is the muffler for the pressure regulation unit pictured and the silver cylinder is a muffler for use where needed. In the center, there is a pressure gauge; we can control the air pressure to the system using the yellow knob on top. The brass connections on the right are points where air is supplied to the system, known as a manifold.

Since many manufacturing facilities had an abundance of air power, pneumatic power was the driving force for many of the early robots. The early systems also used a simpler control system for positioning that consisted of a rotating drum with pegs on it, contacts or actuators for valves, and hard stops to limit the robot's travel. The pegs on the drum would rotate and make various combinations of contacts or actuators to control how the robot moved, with the timing controlled by the speed that the drum rotated and peg placement. The cylinders would extend until the robot hit a hard stop, preventing further motion. Because the cylinder was still under full pressure, this all but eliminated the issues encountered with mid-stroke positioning for air systems. These early pneumatic robots were robust, fast, and easy to work on, but there was a down side. The robot was often noisy due to the impact of the system with the hard stops, which in turn caused extra wear and tear on some of the robots parts. Because of their operation and the noise that went along with it, these early robots picked up the nickname "Bang Bang robots" due to the noise created as they hit one hard stop after another.

Today, due to refinements in the way we control and balance pneumatic power, some of the fastest robots in the world are air powered. These systems move so quickly that their motion seems a blur, and parts or materials seem to be there one moment and then simply gone in the next as the robot completes its tasks. Pneumatics is also a favored power source for End-of-Arm Tooling. Many gripers, suction cups, drills, dispensers, and other devices use air as the primary driver for their operation. One thing is certain, whether pneumatic power runs the whole robot or just various tooling, you can count on pneumatics being a part of the robotic world for the foreseeable future.

With each type of power having benefits and flaws, you may be asking yourself "Which one is best?" or "How do designers pick the type to use?" We can best answer that with three questions:

1. What do we want or need the robot to do?

2. What do we have available?

3. Where will the robot work or operate?

For instance, we likely would not want to use pneumatics for a robot that must lift huge loads and maintain position if stopped at any point. Hydraulically powered robots are not a good choice for cleaning your home, because any leaked fluid might stain or damage the flooring and furniture. In a highly explosive environment, I would avoid using an electric system, as there are often sparks involved with the operation. If pneumatic or electric power would work fine, I might see which I had more of or which was cheaper to use. As you filter robotic applications through these three questions, I believe that you will find it easier to determine which power source is best.

Of the three types we have examined, electric power seems to be the current winner (see Figure 3-9). Electricity is plentiful and cheap, we can store it in batteries, we can produce it from many different sources, and we have a firm understanding of how to control the flow of electrons to do meaningful work. In recent years, manufacturers have developed electric servomotors that give the hydraulic workhorses a run for their money in sheer force while avoiding all the mess and risks of leaks. The once fast pneumatic-only systems are now hybrid systems that join pneumatic and electric power to create systems with blinding speed and precision. This is why learning about electricity is necessary for those who wish to build, maintain, or design robots.

Figure 3-9 Here is a FANUC electrical system moving a car body around, a job that in years past was reserved for hydraulically powered robots.

Copyright FANUC Robotics

controller
The brains of the operation and the part of the robot responsible for executing actions in a specific order and timing or under specified conditions

Controller/Logic Function

Now that you have some idea of the forces used to drive robots, we shall focus on the control side of things for a while. The simple truth is that without some way to control the actions of the robot, the timing of those actions, and the sequence, the robot is an expensive paperweight taking up space. The **controller** for the robot is the brains of the operation and the part of the robot responsible for executing actions in a specific order under specified conditions (see Figures 3-10 and 3-11). The controller takes whatever sensor input is available for the robot, makes decisions based on a system of logic filters and commands called a **program**, and then activates various outputs as instructed by the program. When we want to modify the operation of a robot, it is often simply a matter of changing the program in the controller. Radical changes to the robot's operation, such as adding a new sensor or changing a gripper for a welding gun, often involve adding or changing wiring in the controller, possibly a firmware update, and a program change.

Controllers come in all shapes and sizes and are a reflection of the system they control (see Figure 3-12). Many BEAM systems contain no control chip or program, as these robots use built-in reactions and tendencies to interact with their world. The rest of the robotics world uses some form of processor, ranging from a simple Arduino, Texas Instruments, or other processor chip, to small breakout computers like Raspberry Pi and smartphones, to laptops, tablets, and other full-size and capability computers, to specialized computing systems designed for the hazards and rigor of the industrial world. Each of these controllers holds a place in the robotic world because it is a good fit for the system's needs. For a telepresence robot, you may opt to use a tablet for the brain to take advantage of the speakers, camera,

Figure 3-10 This is an example of a controller for an ABB robot.

Image Courtesy of ABB Inc.

and Wi-Fi functionality of the device. Many hobbyists, like me, use Arduino or Raspberry Pi to breathe robotic life into a group of motors, sensors, and other assorted parts (see Figure 3-13). Industry is full of hot and dirty environments with large amounts of three-phase power in use; this is the reason they require specialized controllers that can survive in these harsh environments.

Just as most robots are useless without a controller, the same might be said of using the wrong controller. If the controller is unable to function properly under the conditions the robot works in, there is a good chance that the system will fail, or worse, operate unpredictably. Imagine an industrial robot that is loading parts one minute and then throwing them across the plant the next. Or perhaps your new quadcopter ignores your commands because the onboard controller malfunctions, and heads full-tilt at your house, or at you! The robot controller needs to be tough enough to handle the various work conditions associated with its location, have enough processing power to keep up with all the subsystems of the robot, and have enough input and output capability to

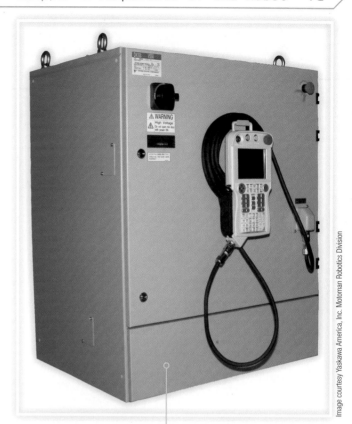

Image courtesy Yaskawa America, Inc. Motoman Robotics Division

Figure 3-11 This controller is for a MOTOMAN robot.

program
A system of sorting and direction function created by the user to define the operation of equipment based on the various monitored input conditions

Figure 3-12 This is the control brick used with the LEGO Mindstorms NXT robotic kits.

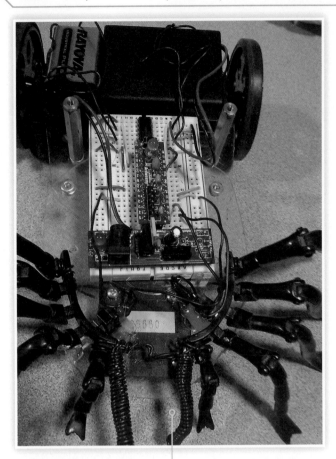

Figure 3-13   Here is a look under the hood at my spider bot with its Arduino controller.

relay logic
Control system that uses a device known as a relay to create various logic-sorting situations, which would in turn control the operation of the system

Relays
Device that uses a small control voltage to generate a magnetic field and make or break connections between field devices using contacts

normally open (NO)
Contacts that do not allow electricity through when in their normal or de-energized state

gather the desired information and control all the functions of the system. This means you may have to get a special controller for harsh environments. If you plan to control two robots with one controller, you will need an advanced controller with lots of inputs, outputs, and computational power. If you need long-range control, then your controller needs the ability to broadcast and receive signals over the anticipated distance plus a safety margin. And so on.

Older robots used something called relay logic to control the operation of the system. Relay logic uses devices known as relays to create various logic-sorting situations, which would in turn control the operation of the system. Relays use a small control voltage to make or break connections between field devices. Relays have contacts that are normally open (NO) which do not allow power through when the relay is de-energized, and normally closed (NC) contacts that do pass power when the relay is de-energized. When we energize the relay or apply power to the coil, all the contacts change state, causing the open contacts to close and the closed contacts to open. By using specific combinations of relays, we create the logic filters (which we will discuss in Chapter 9) to determine the behavior of equipment. The downside to this type of control is that it takes up a large amount of space for all the relays, there is added cost in parts and wire, and any change in functionality requires rewiring the machine. Because of these negatives, relay logic is a dying control type for modern equipment. However, you may encounter this control system still in use with older industrial equipment.

As stated earlier, controllers come in all shapes and sizes, depending on the robot they control. When working with a robot, make sure you understand the capabilities and limits of the controller you are working with. Before you make any modifications to the robot or its systems, make sure the controller can handle the computational portion of the changes as well as having any needed inputs and/or outputs available. Radical changes, such as turning a pick-and-place robot into a welding robot, may require a new controller to handle all the extra control and computational requirements or, at the very least, an upgrade in software. Having a firm understanding of the controller and its operation is the best way to avoid costly mistakes down the road.

Teach Pendant/Interface

To accompany the controller, we need some way to communicate any changes in programming or operations that we desire. In most industrial robots, we use the teach pendant to accomplish this; it allows the operator to view alarms, make manual movements, stop the robot, change/write programs, start new programs, and carry out any of the other day-to-day tasks required to run robots. In industrial settings, everyone who is in the robot's danger zone *must* have the teach pendant with them to help ensure their safety. We refer to the teach pendant as a human interface device because we use it to control or interface with the robots operation. Often, we use the teach pendant to write completely new programs without the need for any other software or computer interaction with the system, saving time and aggravation in the programing process. (We will talk more about this in Chapter 9 on programming.)

normally closed [NC]
Contacts that allow electricity through when in their normal or de-energized state

---

safety note

⚠ Remember that it is an OSHA requirement that operators take the teach pendant with them anytime they are in the danger zone, so that they always have access to an E-stop. The last thing you want is for the robot to start up with you nearby and no way to stop the system.

Teach pendants vary in style, operation, size, and complexity from manufacturer to manufacturer and sometimes from model to model of the same brand of robot (see Figure 3-14). However, even though they come in a wide variety of configurations, there are some standard features that you can expect to find on most pendants. First, those for industrial use will have an E-stop and dead man's switch. The E-stop is an emergency stop to shut down most of the systems of a machine. The dead man's switch, usually a trigger- or bumper-type switch on the back of the pendant, is required to move the robot manually. If you release the dead man's switch or press down too hard, the robot stops moving (see Figure 3-15). This is because when something goes wrong, say the robot hits you upside the head, your reaction is often to either let go or grip the switch too tight—this is why these actions stop all movement. The other common items you can expect from a teach pendant is a

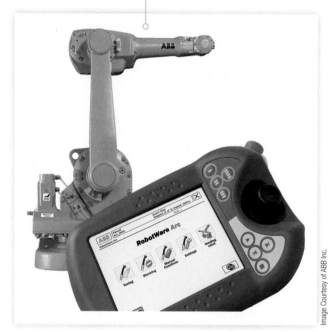

Figure 3-14 Here is a common teach pendant for the ABB robot. Notice the red E-stop in the upper right corner.

Image Courtesy of ABB Inc.

teach pendant
A device that allows the operator to view alarms, make manual movements, stop the robot, change or start a new program, and carry out any of the other day-to-day tasks required of those who run robots

Figure 3-15 The two yellow triggers shown are the dead man's switches for the Panasonic welding robot teach pendant. It does not matter which one is depressed, but at least one has to be depressed at all times for manual movement of the robot.

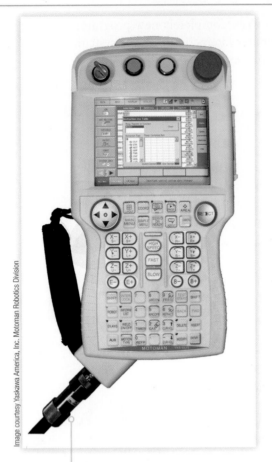

Image courtesy Yaskawa America, Inc. Motoman Robotics Division

Figure 3-16 This is the DX teach pendant for use with MOTOMAN systems.

Copyright FANUC Robotics

Figure 3-17 Here is the teach pendant for the FANUC systems that my students use in class.

dead man's switch

A switch, often found on the back of teach pendants, that must be held for the system to run in manual and stops all movement when released or pressed too hard

Manipulators

Available in all shapes and sizes, what the robot uses to interact with and affect the world around it by activating and positioning the End-of-Arm tooling

display to give you information, some manner or button combination to move the robot, and some way to record robot position for programming purposes. What you push to move the robot, the other options that may be available, and the data that you can find on the pendant are up to the manufacturer. Generally speaking, there is a good chance that you can control and modify most if not all of the desired functions of the robot from the teach pendant. Take a look at the images provided in this section to get an idea of the various teach pendant configurations that you may encounter (see Figures 3-16 and 3-17).

Some of the new industrial systems, like the Baxter robot, are getting away from the traditional teach pendant and using things such as special modes and the sensors of the robot coupled with AI to learn new tasks. Some systems utilize special software, a wired or wireless data connection, and a computer to make changes to its operation as well as to monitor changes in the system. In the hobby world, it is becoming common to use cell phones, radio controllers, or even Xbox/PlayStation controllers to direct robots and change behavior on the fly. No matter what system you are working with, you need some way to communicate or interface with the controller to make the robot perform as desired. After all, it is the ability to change the system's function that separates robots from toys or machines.

Manipulator, Degrees of Freedom (DOF), and Axis Numbering

It is all fine and good for the robot to have a brain, the controller, and brawn, the power supply, but without a way to interact with the system's environment, the options are limited. This is where the manipulator comes in. Manipulators come in all shapes and sizes and are what the robot uses to interact with and affect the world around it (see Figures 3-18 and 3-19). For industrial systems, these are often robotic arms or overhead systems with a series of rods that move the tooling around. Sometimes these consist of complex systems that closely resemble human or animal anatomies. Other systems are unique in function and purpose, making it difficult to easily identify and classify the parts of the robot. For the sake of clarity, we will use arm-type examples for this section, as they are the easiest to understand when one is first learning about robotics. As your knowledge and experience grow, you should be able to translate what you learn here to the diverse and complex systems you encounter.

Manipulators are systems of movable parts. We call each part of the robot that has controlled movement an axis. Each axis of the robot gives the robot a Degree of Freedom (DOF), or one more way that the robot can move. The more DOFs a robot has, the more complex and organic the movements. For instance, many gantry-style robots in industry have only three axes. They can move up and down, which is referred to as their Z axis; back and forth, which is called their Y axis; and rotate the End-of-Arm tooling, which is often called the A axis. As you can imagine, the motions of this system are very square and limited. Most industrial robot arms

Figure 3-18   Here is an example of a robot arm produced by MOTOMAN.

Image courtesy Yaskawa America, Inc. Motoman Robotics Division

axis
Each part of the robot that has controlled movement

Degree of Freedom (DOF)
Each axis of the robot gives the robot one more way that the robot can move

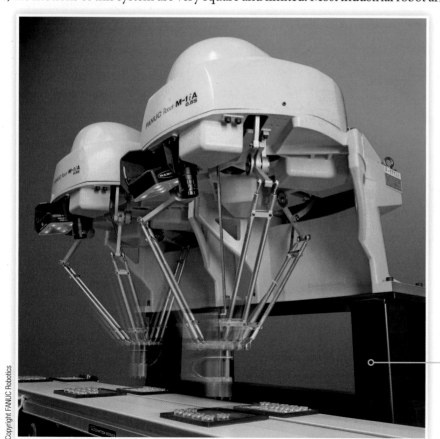

Copyright FANUC Robotics

Figure 3-19   Pictured are two FANUC Delta robots, a configuration gaining popularity in industry.

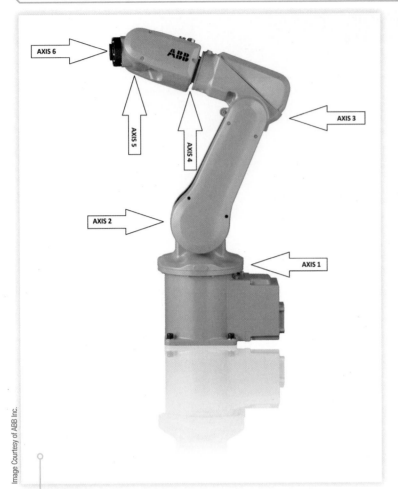

AXIS 6

AXIS 5

AXIS 4

AXIS 3

AXIS 2

AXIS 1

Image Courtesy of ABB Inc.

figure 3-20   Pay close attention to how the axis numbering starts at the base and works toward the wrist where tooling attaches in this robot axis numbering example.

have five or six axes, with six being the more common. A six-axes robotic arm (six DOF) is capable of mimicking most human motions, giving the system a great deal of flexibility.

When it comes to determining which axes are which, it is common to start at the base and number outward. The **base** of a robot is where the robot is mounted or bolted down for stability with nonmobile types, or the platform on which a manipulator mounts for mobile types. From the base, we travel toward the End-of-Arm tooling and number each axis as we go, starting with one. Figure 3-20 of the ABB robot shows the process of axis numbering. When the manipulator is on a mobile base, we commonly assign this axis's number last, as an additional or optional axis to the system. If the robot in our example of axes numbering were on a mobile base, we would refer to the base as axis 7. Axes numbering is of primary importance when troubleshooting robot malfunctions. The alarm will often tell you which axis is at fault, but this information is relatively useless if you are unable to determine which physical axis of the robot faulted out.

The axes of the robot are broken into two main groups: the major axes and the minor axes. The **major axes** are responsible for getting whatever tooling we are using into the general area it needs to be (see Figure 3-21), while the **minor axes** are responsible for the orientation and positioning of that tooling. If we compare the

base
The portion of the robot that we attach to either the solid mounting surface or mobile unit

Based on: Miller Welding Automation

figure 3-21   This picture illustrates the type of motion achieved with the major axes of an arm-type robot.

robot to the human body, the major axes would be like your torso and arm, while the minor axes would be like your wrist, with your hand representing the tooling. From the axes numbering example, the major axes would be 1–3 while the minor axes would be 4–6. We often define the minor axis as pitch (axis 4), which is the up and down orientation of the wrist; yaw (axis 5), which is the side to side orientation of the wrist; and roll (axis 6), which is the rotation of the wrist. When these two systems combine, you get the modern six-axes robot that uses the major axes to copy our body motions and the minor axes to copy our wrist motions while performing tasks.

While on the subject of axes, we need to take a moment to look at external axes. External axes are axes of motion that often move parts, position tooling for quick changes, or in some other way help with the tasks of the robot. They are not a part of the manipulator and are not a part of the major or minor axes of the robot. These axes are under the control of the robot controller and often figure into the motion of the robot, but are separate from the manipulator. Many robot companies offer these as part of packaged robotic systems that are ready for the customer to use right off the truck. Since external axes are not a part of the manipulator, we do not count them in our DOF, but they do present added motion options and thus increase the overall flexibility of the system.

While the six-axes arm-type manipulator is a favorite of industry, it is not uncommon to find systems with differing numbers of axes. Motoman makes a robot that has a base that looks like a human torso with two arms attached in such a way that they look like human arms, giving the robot more than double the standard six axes (see Figure 3-22). The NAO robot at the time of the writing of this text comes standard with 25 DOF for the humanoid model (see Figure 3-23). Marvin Minsky's tentacle robot developed in 1968 had 12 joints or DOF. On the other side, we have simple gantry-style robots with three or four axes that are used in many industrial applications. The PLANETBOT developed by the Planet Company in 1957 had five axes of hydraulically powered motion. No matter what configuration you encounter in the field, if you remember the rule of axis numbering, you should be able to determine how many axes the robot has and the number of each axis. If all else fails, put the robot in manual mode and move axis 1 to see which portion of the robot moves. Continue with this method until you have identified all the axes of the robot.

Base Types

We briefly mentioned bases in the previous section of the chapter, but here we will dig deeper into some of the various configurations you will encounter in the field. We will not be able to cover all the possible bases in the world of robotics, but we can cover the commons ones and their uses. The two broad categories for bases are solid mount and mobile.

Figure 3-22 This is one of MOTOMAN's double-armed robots.

major axes
The axes of the robot that get the tooling and minor axes of the robot into the general area that work is performed, usually the first through third axes

minor axes
The axes of the robot that position and orient the tooling of the robot, usually the fourth through sixth axes

External axes
Axis or axes of motion that are not a part of the main robot often used to move parts, position tooling for quick change, or in some other way help with the tasks of the robot

Figure 3-23 You can see the NAO robot showing off what 25 DOF can accomplish in this picture.

Solid Mount Bases

solid-mount bases
A nonmobile base to which a robot is attached with bolts and other fasteners and from which it works

For many of the systems used in industry, solid-mount bases are the preferred way to go. These involve mounting the robot firmly to the floor or other structures using bolts and fastening systems (see Figure 3-24). A solid mount allows the system to maintain a very specific coordinate base to work from, which in turn allows industry to make full use of the precision of robotic systems. Many times in industry, any change in the position of the robot results in the need to offset the system or touch up the points of all the robot's programs, thus the desire for a firm, unchanging base. We find solid bases anchored to concrete floors, secured to building walls, mounted on overhead structures, or even secured inside of machine systems, all depending on the requirements of the job.

No matter how the base is mounted, there are few key points to remember:

- Make sure that whatever holds the robot in place is robust enough to bear the forces and weight of the system. Bolts, nuts, mounting plates, and so on all have limits to the amount of force they can withstand.

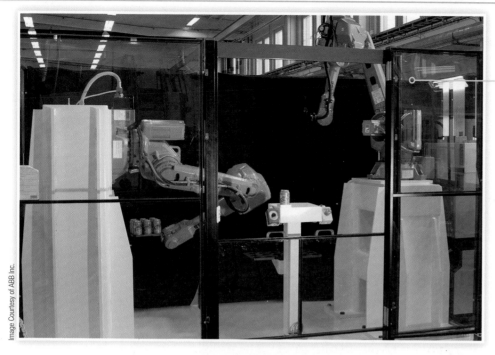

Figure 3-24   This example shows two ABB robots: the one on the right is mounted parallel to the floor, and the one on the left is mounted to the side of the pedestal on a wall-type mount.

- Make sure that what you mount the robot on can handle the weight of the system *as well as* whatever load it will be maneuvering.

- Make sure to check the security/tightness of any mounting hardware periodically, paying special attention to any noted wear.

- Crashes are conditions where the robot endures unexpected forces, so make sure to check the base when you are inspecting the system, especially for wall or overhead mounts.

If you follow these simple guidelines, you should be able to avoid most worst-case scenarios. As each setup will have its own unique circumstances and requirements, you may need to add to these basic rules on a case-by-case basis.

Mobile Bases

As the name implies, mobile bases are systems used to move the manipulator to various location so that it can perform its functions. Some of these are restricted to liner rails that allow the robot to move back and forth over a finite area, while others give the system a great range of mobility and freedom. The type of mobile base used generally depends on what the robot does as well as the environment in which it works. For instance, a wheeled base would not be the best fit for a robot designed to feed parts precisely into machines from an overhead position. Just like many other aspects of robotics, the base should fit the task.

We often refer to a linear base with a finite reach as a gantry base (see Figure 3-25). The name comes from its similarity to the gantry robot as far as the movement is concerned. Many of these systems cover a few feet to about 30 feet or so, but there are systems that cover larger areas and allow multiple robots to work on one tracked base. In these cases, we must pay special attention to the accuracy and strength of the track's mounting as well as how the robots share the length of that track. Misalignment of the track can cause the robot to alarm out or change its reference position. Improper programming or robot management can cause the robots to impact one another with all the damage that could entail.

mobile bases
A mounting system that gives the robot mobility options, often wheeled or tracked in nature

gantry base
A linear base with a finite reach

Figure 3-25 This cell has a standard solid-mount robot and an overhead gantry-mount robot working together to get the job done.

Figure 3-26 Here is the Robonaut 2 out and about on his mobile base, known as the *Centaur*

Another common mobile base is the wheeled or tracked system. These systems give the robot the ability to cover large areas and even traverse difficult terrains. We use these systems for material transportation, gathering information, hazardous material disposal, and general human space navigation. The famous Robonaut system uses a wheeled base to navigate the space station (see Figure 3-26), while Baxter uses his wheeled base to get around in industry. The fact that these bases are robust and able to conquer obstacles makes them a favorite among many hobbyists.

A type of mobile base that is getting a lot of attention these days is the legged system. This base uses two or more legs to move the system (see Figure 3-27). When only two are used, they often work and resemble human legs; however, balance is an issue. It seems that walking on two legs is harder than one might think, especially if the system moves on anything besides a level surface. The systems that have more than two legs tend to emulate various animals, but some are unique designs found nowhere in nature. This is not new technology, as Ralph Mosher was working on a walking system for GE in 1968, but it seems that recently science has really begun to conquer organic motion for walking robots.

There are many specialized mobile bases for robotics as well, dependent on need (see Figure 3-28). We

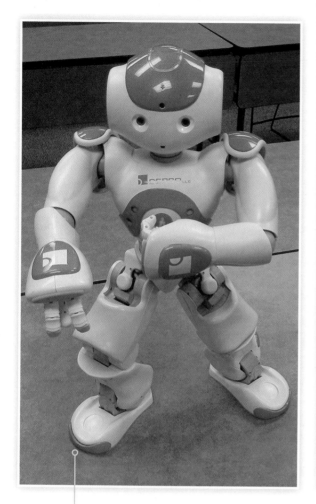

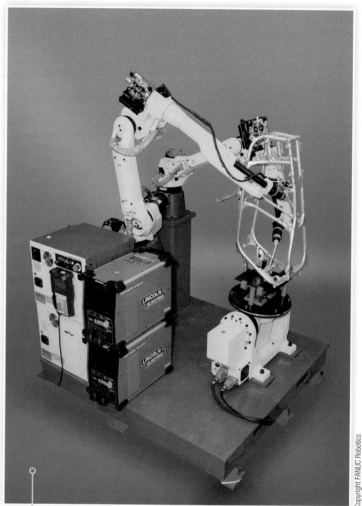

Figure 3-27 Here is the NAO robot, with its feet positioned to support the motions of the upper torso. Even with all the development and advanced programming of this system, it still falls over from time to time from the momentum of its motions or because of environmental factors such as carpet or uneven surfaces.

Figure 3-28 Here are all the components of a solid-mount robotic system combined. The controller, teach pendant, and welding power supplies are on the left side (front). The manipulators are behind that, and if you look closely, you will notice that there are two arms in this picture. The yellow unit on the right is what we call an external axis, and its job is to move the part around. This whole system is designed to ship as a single unit for floor mounting at the customer's site.

have aerial systems for the skies, boating systems for the waters, pressure-resistant systems for the murky depths of the sea, and even systems designed for other planets to expand our knowledge of the universe. As we continue to find innovative ways to navigate various environments, you can expect robots to follow on specialized mobile bases, doing the dull, dirty, difficult, or dangerous work that we humans would rather avoid.

review

By now, you should be familiar with the basic systems of the modern robot and have an understanding of the role of each system. Later on in this book, we will take a closer look at the various sensor systems used by robots to give the controller the information that it needs to interact with the world. As the world of robotics continues to evolve and grow, so too will the devices that fit into each of the categories we have examined. During our exploration of the robot in this chapter, we covered the following topics:

- **Power supply.** This section was about the common forces used to run robots and covered information about how each of these power sources worked.

- **Controller/logic function.** Here you learned about the brain of the robot and its importance.

- **Teach pendant/interface.** This section was about how we communicate with the robot and direct actions or make changes to operation.

- **Manipulator, Degrees of Freedom (DOF) and axis numbering.** Here you learned how to number axes, how we move the robot around, and what Degrees of Freedom are.

- **Base types.** This section was about what we mount the robot to and why as well as some other things to keep in mind.

key terms

Alternating current (AC)	Dead man's switch	Major axes	Program
Amperes/amps	Degree of Freedom (DOF)	Manipulator	Relay
Amp-hours (Ah)	Electricity	Minor axes	Relay logic
Axis	Electromotive force (EMF)	Mobile base	Resistance
Base	Electron flow theory	Muffler	Root Mean Square (RMS)
Controller	External axis	Neutral wire	Single-phase AC
Conventional current	Gantry base	Normally closed (NC)	Solid-mount base
flow theory	Ground	Normally open (NO)	Teach pendant
Cycle	Hertz (Hz)	Pneumatic power	Three-phase AC
Direct current (DC)	Hydraulic power	Polarity	Voltage

review questions

1. What is the difference between AC and DC electricity?

2. One amp is equal to _____ electrons flowing past a measured point in one second.

3. What could happen if you reversed the polarity of DC components?

4. What is the difference between conventional current flow theory and electron flow theory?

5. How would you connect a group of batteries to increase both the voltage and amp-hours?

6. Describe what happens to AC power through one complete cycle.

7. What is the difference between single-phase and three-phase AC?

8. What did Blaise Pascal discover in 1653?

9. What was the event that revitalized the use of hydraulic power?

10. List at least three things to remember when dealing with hydraulic leaks.

11. What did Robert Boyle discover?

12. What do we have to be careful of when venting pneumatic power, and how do we avoid these dangers?

13. Describe the operation of a pneumatic robot controlled by a drum.

14. What is the function of the robot controller?

15. How do relay logic systems work?

16. What are some of the things that we can do with the teach pendant?

17. Why does the dead man switch kill the robot's manual actions when released or pressed too hard?

18. What is the benefit of having more degrees of freedom in a robot?

19. How do we number the axes of a robot?

20. What are the two main groupings of axes, and what is the function of each?

21. Why are solid-mount bases a preferred type in industry?

22. What are some of the key points to remember when solid mounting robots?

23. Do we count external axes in our DOF count? Why?

Classification of Robots

What You Will Learn

- How to classify robots by their power source
- How to classify robots by their work envelope and the kind of reach they have
- How to classify robots by their drive system
- How belt systems work, and the math that goes with them
- How chain systems differ from belt systems
- The different types of gears used in gear drives
- The math that goes with gear systems
- How the ISO classifies robots

overview

We have only reached Chapter 4, and already you have seen and learned about many robotic systems. You may now be thinking, "How can we sort all these systems?" The answer is as diverse as the field of robotics itself. When it comes to robotic systems, we have many options for grouping, sorting, or categorizing robots. In the early years, many of the companies that sold robots provided a classification system, as the consumer had little knowledge of what options were available. However, as the field of robotics has matured, user and manufacturer alike have come to a consensus on the common ways to classify robots. In the last 10 to 15 years, the International Standards Organization (ISO) has begun to congeal the common sorting methods into one standard set of classifications to simplify the world of robotics. The main problem with classifying robots is the fact that the field is evolving so quickly that it is hard to keep up. As we look at the various ways to classify robots, we will cover the following:

- How are robots classified?
- Power source
- Geometry of the work envelope
- Drive systems: Classification and operation
- ISO classification

How Are Robots Classified?

If you read over the list of topics we cover in this chapter, you already have some clues as to how we classify robots. As mentioned in the overview, we have many options for sorting robotic systems. We can classify them by their power source, how we program them, what they do, internal systems, how they move, or any other criteria that we deem important. The main reasons that we classify robots are to group like with like in an effort to make decisions easier and to convey information about the system. For instance, if I say that the task requires the power of a hydraulic robot, then I would look at robots powered by hydraulic systems or sort the robots by power source. Perhaps the engineer in charge of a project has years of experience with FANUC robots, so he would filter robots by manufacturer first. The needs of the end user and his or her understanding of robots have created the diversity in robot classification.

As you may have noticed, robot classification is a grouping of robots' like qualities. We can group by many factors, but the goal is usually to sort robots for comparison. Another way to think about robot classification is to compare it to a trip to the store. We generally do not find one aisle with only General Mills products and another of 3M products; instead, we find separate sections for bread, milk and milk products, paper goods, pet food, and so on. Stores tend to group items by what they are, what they do, or their type—and not by who made them. It is easier to look at all the store's bread choices when they are grouped in one place versus having to look at each company's options to see if it makes bread. This is the same thought behind classifying robots. Instead of spending hours and hours looking at all the robotic options available, we can focus on a robotic system to find what we need or want.

The sections in this chapter will cover some of the common ways to sort robots and give you information about each of those categories. As you explore these classifications, I encourage you to consider other ways to sort robots. Keep in mind that the purpose behind robot classification is to sort like with like and make it easier to compare systems by first determining the user's needs and then deciding what category of robot best fits that need. As the world of robotics continues to grow and change, so too must the classifications for robots. Your idea today could well become the seed for a new way of classifying robots tomorrow.

Power Source

Sorting robots by their power source is easy and obvious. As the name implies, the group consists of robots that use the same type of power to perform their functions. This method of sorting works great for robots and End-of-Arm tooling alike, as everything that performs work needs some kind of power to make that happen. We will look at the major divisions in this section as well as some of the subgroupings therein.

Electric

DC brushes
Items made of carbon to transfer electricity from the power wires going into the motor to the rotating portions of the motor

Stepper motor
A motor that moves a set portion of the rotation with each application of power

Electric robots use electricity to run motors of various designs that move the robot and do whatever the systems is designed for (see Figure 4-1). When it comes to electricity, we have the options of alternating current (AC) or direct current (DC). (If you have forgotten the difference, feel free to refer back to Chapter 3.) DC provides a greater amount of torque, but if the motors use brushes, they will require more maintenance (see Figure 4-2). DC brushes are made of carbon and transfer the electricity from the power wires going into the motor to the rotating portions of the motor. Because they are in contact with the rotating portion of the motor, they do wear down over time and generate sparks that could become an ignition hazard under the right conditions, such as a flammable atmosphere. There are DC brushless motors, which negate these problems, but they are higher in price. DC has become the choice for many mobile robots due to the variety of battery packs and renewable sources they can pull this power from.

Figure 4-1 A FANUC robot stacking bags of chocolate on a pallet. The black objects with the red cap are the servomotors, and the large gray square under the robot with the red cables and the slits is the AC power supply for the robot.

Copyright FANUC Robotics

Figure 4-2 A couple of smaller all-DC robots that you might be able to build in your classroom.

servomotor
A continuous rotation type of motor with built-in feedback devices called encoders, often designed for use with variable frequency drives

encoders
Devices tied to the shaft of the motor that provide information about the motor's movement, such as direction, speed, and location in the rotation, by breaking the complete rotation of the motor into a specific number of measurable units

AC is the common choice for industrial systems as it is readily available, is low maintenance, and drives both stepper and servomotors. A **stepper motor** moves a set portion of the rotation with each application of power; the more steps per rotation, the finer the position control. Stepper motors were the common motor for early robots and are still favored for many applications where there are other means of verifying position. A **servomotor** is a continuous-rotation-type motor with built-in feedback devices called **encoders**, which provide feedback about the motor's rotational position. High-end encoders can provide information about

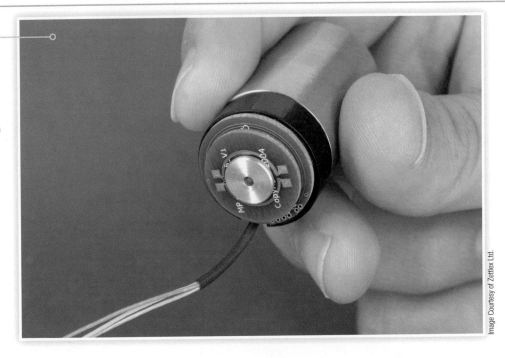

Figure 4-3   The cover has been removed to expose the encoder on a small precision motor. These types of encoders allow for high levels of precision in hobby and small robotic systems.

Image Courtesy of Zettlex Ltd.

the speed and direction of rotation as well as the total number of degrees that the motor has moved (see Figure 4-3). The robot controller uses the feedback from encoders to know when to stop the motor for proper positioning and to verify that the system is working correctly.

Hydraulic

A hydraulically powered robot uses a noncompressible fluid given force through velocity to drive its moving parts. The name of this category can be misleading, as there will be some electrical power required, primarily for the hydraulic pump, valve control, system controls, and any additional sensors. The main reason we call it a hydraulic robot is because hydraulic power is what moves the robot and allows it to generate great amounts of force. Improvements in the AC servomotor have eroded the number of hydraulic robots in use, but these powerhouses of the robotic world are still found in industry today. The great power of the hydraulic robot does come at the following cost:

- Hydraulic leaks
- Cost of oil
- Fire hazard
- Increased maintenance
- Increased noise

It is a simple fact that a hydraulic system will eventually leak. When that happens, there will be a mess to clean up, at best; at the worse end of the scales lurks the potential of fire and equipment damage. When it comes to maintenance cost, the dollars add up: mechanics replace parts worn out from normal operation; test the oil to make sure that nothing is going wrong; and change the oil yearly or every six months on average, depending on the number of running hours and type of oil used. Last, but not least, the hydraulic pump and any cooling system create added noise for the work environment.

safety note

A pinhole leak is the most dangerous type of leak in hydraulic systems, as it can turn the oil into a flammable mist and generate forces great enough to cut through metal.

Pneumatic

A pneumatic system works basically the same way as the hydraulic, with one key difference: a pneumatic machine uses compressible gas instead of a noncompressible liquid. One of the main problems with a pneumatic system is positioning. Because gas is compressible, the only way to hold position is at the extremes of travel or by keeping the system under constant force. You can stall out a pneumatic motor and not damage the system, but as soon as you stop applying pneumatic force, the motor can drift. Pneumatic robots are fine for areas that have physical stops that the robot can work with, but they are not good at holding position without additional equipment.

Pneumatic systems are noisy too. However, they are generally cheap to run, as many industries already have a supply of compressed air to tap into. Many of the early industrial robots used pneumatics for the prime mover, but this power source has also given way to the AC servomotor. One area where pneumatics has really taken root is in tooling (see Figure 4-4). Given that most grippers have two positions, open or closed, pneumatics has proven to be a perfect fit for this need. Because of this, many electrically powered robots have tooling that requires air to operate.

Pneumatic systems do require some extra maintenance, but not as much as their hydraulic counterparts. The main concerns are protecting air lines from damage, the compressibility of gas, and the noise that the system generates. Since most of the lines for a pneumatic system are thin-walled rubber or plastic, it does not take a lot of force to damage them. Any holes or leaks create a drop in pressure, can send chips flying, and if the one of the lines becomes severed, it will whip around, creating a dangerous situation. Since we do not return compressed air back to the storage tank, the used air has to go somewhere. That means that at some point in the system, air returns to the atmosphere, usually under pressure. Escaping pressurized air creates a lot of noise, even with a muffler in place to diffuse it, which is a concern with pneumatic power.

Nuclear

When we talk about nuclear robots, we are talking about systems that carry their own nuclear reactor to create the electricity they need. Several of NASA's robotic endeavors, including the Mars rover *Curiosity*, have used nuclear reactors as the main power source. This is a technology that we have worked with for years. We have successfully used nuclear power in submarines and other such vessels that need power for long-term use without refuel, making it the perfect fit for endeavors in space. This technology may be the go-to source as we send more and more robots into space for research, construction, or mining.

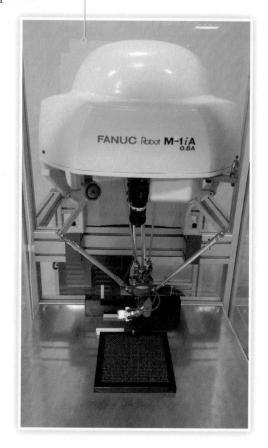

Figure 4-4   Two blue air lines run to the gripper on this robot to provide the gripping action.

Green

While direct sources of green power may not be the choice of industry, there are many other robotic fields that may invest in and use this power source. Green power is derived from a source that is easily renewable and often has little or no environmental impact. In Chapter 1, we briefly mentioned a robot designed to run off broken-down organic material, the robotic version of an animal eating grass. While still in the development stages, researchers are looking at similar systems for deep exploration of the sea. Solar power is another evolving green power source and may be the prime mover for the "lawnbot" of tomorrow. BEAM roboticists prefer the sun as a power source since it gives long, albeit sometimes jerky-motioned, life to their creations. While green power is a budding field and mostly found in hobby and research robotics, it could easily one day become a big part of the robotic power supply family.

Any new power source that becomes a norm in robotics will add to the number of power source classifications. Each classification holds the potential for subdivisions based on the amount of power that the system uses or the generated force of the system. Sorting robots by power source is a very broad method and, thus, usually provides just a first step in narrowing down the choice of a robot for a particular task. This brings us to the next common way to sort robots: by their work envelope.

> **Green power**
> Power derived from a source that is easily renewable that often has little or no negative environmental impact

Geometry of the Work Envelope

Another popular way to classify robots is by the area they can reach, or their "work envelope." This has been one of industry's favorite ways to classify robots for a while now, as it gives the user a good idea of how the robot will move and how it can interact with the world around it. This is where an understanding of cubes, cylinders, spheres, and other geometric shapes comes in handy.

Cartesian Robot

Cartesian robots have a cubic or rectangular work envelope. Many gantry-type robots fall into this category and tend to move in a linear (or straight line) way. These robots often have two or three major axes to move in: X front to back, Y side to side, and Z up or down (see Figure 4-5). When there are only two major

Figure 4-5a   Gantry-type robots are great for working with large parts, as this robot welding a large tank demonstrates.

Image Courtesy of Miller Welding Automation

axes, X is often the one omitted. These robots are popular for loading and unloading parts as well as moving materials over large distances with the system mounted over the equipment they serve, which saves floor space.

Cylindrical Robot

Cylindrical robots, as the name implies, have a cylindrical work envelope (see Figure 4-6). In many ways, they are the rotary cousin of the Cartesian robot we just discussed. These robots commonly have a rotary axis on the base to spin the robot, two linear axes to move the tooling into the general work area, and then two or three minor axes for tooling orientation. The Y-axis of the Cartesian robot has been replaced with the rotary axis of the cylindrical robot. These systems are good for reaching deep into machines, save on floor space, and tend to have the rigid structure needed for large payloads. The only downside is the loss of Y-axis travel, but the addition of a mobile base can take care of that.

Spherical Robot

Imagine the robot in the middle of a ball instead of a cylinder. Spherical, or polar, geometry gives the user a wide range of options for robot positioning but not the complete range of a sphere because of the physical restraints of the robot's construction and the surface on which it is mounted (see Figure 4-7). We get this geometry by taking the cylindrical robot geometry, and then replacing the linear Z motion of axis 2 with a rotational axis. This robot has the sweep and reach of a cylindrical robot with the addition of angular positioning for the major motions of the arm. The primary difference between cylindrical and spherical robots is that the spherical units have a long reach with a smaller size. This geometry was more of an evolutionary step and is thus not often seen in use today.

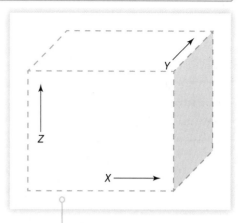

Figure 4-5b   The Cartesian system for robot movement creates a square work envelope.

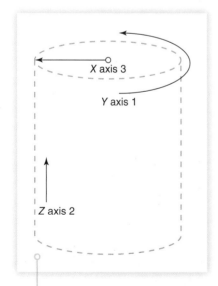

Figure 4-6   This is what a cylindrical work envelope looks like as well as how the X, Y, and Z axes relate to the motions and axis.

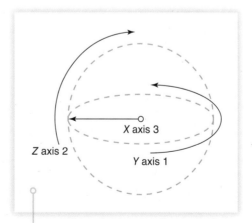

Figure 4-7   Here are the X, Y, and Z axes for a spherical robot, with the related axis. Notice how the bottom is cut off near where the robot mounts.

Articulated Robot

Articulated robots have a spherical-type envelope that is constrained by the construction of the robot. The articulated robot leaves linear motion behind for rotational motion at the various axes. This robot is also known as jointed arm, revolute, and even anthropomorphic, because in many cases, its motions look very organic and lifelike. This robot features the most common geometry due to the flexibility of its design and ability to replicate a wide range of human motions. The work envelope for this system looks like the spherical robot's area with some odd bites taken out of it. The places it cannot reach are due to factors such as the base cannot rotate 360 degrees and places where parts of the robot prevent full rotation of an axis (see Figure 4-8). These systems require the most complex controllers in the industrial world, putting them at the top of the scale in system cost. Sometimes they have an extra axis between the major and minor axes, numbered 4 in the robot numbering system, which adds either rotation or extension as well as another DOF.

SCARA Robot

Selective Compliance Articulated Robot Arm (SCARA) is unique in that it combines Cartesian linear motion with the rotation of an articulated system, creating a new

anthropomorphic
Resembling the way people or animals move; in this context, giving a robot a very organic and lifelike appearance

Selective Compliance Articulated Robot Arm (SCARA)
Robots that blend linear Cartesian motion with articulated rotation to create a new motion type

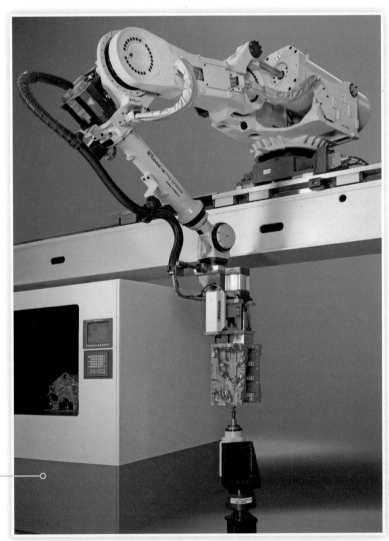

figure 4-8a   This FANUC articulated robot arm is mounted on a gantry base, combining the benefits of the articulated robot work envelope with the distance versatility of a gantry-style system.

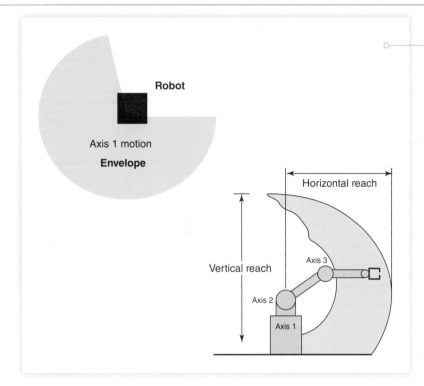

figure 4-8b   The unique envelope created by using only rotational axes.

motion type (see Figure 4-9). SCARA has a cylindrical geometry with axes 1 and 2 moving in a rotational manner and axis 3 moving in a linear vertical way to manipulate the tooling into position while applying force. The orientation of axes 1 and 2 provides horizontal rotation versus the vertical rotation of the other systems that we have discussed, in a similar fashion to axis 1 of the articulated geometry. Another difference is that the wrist (or minor axes) usually only has one, rotational axis. SCARA robots are popular in the electronics field, where their motion and strengths seem to be a good fit for the tasks required.

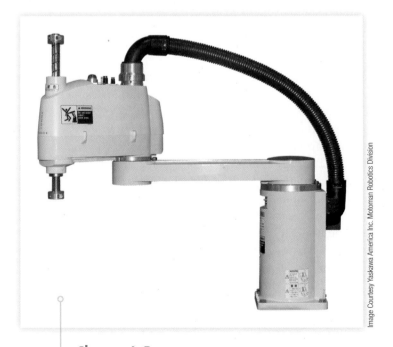

Image Courtesy Yaskawa America Inc. Motoman Robotics Division

figure 4-9a   An example of a SCARA robot with a cylindrical work envelope.

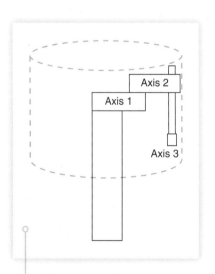

Figure 4-9b   While cylindrical in nature, the depth of the cylinder of the SCARA work envelope is limited by the depth of travel of axis 3 and the reach of the tooling.

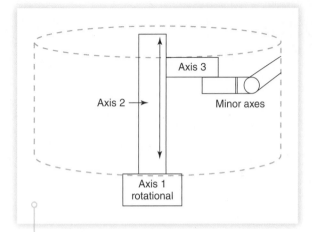

Figure 4-10 While the horizontally base-jointed arm has the same basic geometry of the SCARA, the design gives the robot a much larger work area as well as options in how the tooling is orientated or twisted.

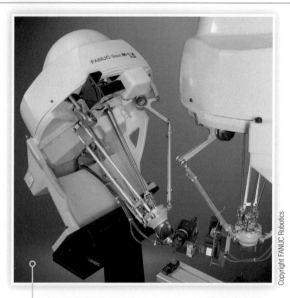

Figure 4-11 Two delta-style robots work together, complete with vision systems attached to the tooling. Notice the configuration of the arms used to create the unique motions of this style of robot.

Horizontally Base-Jointed Arm Robot

This is an adaptation of the SCARA system, with axis 2 as the linear axis instead of 3. Instead of the tooling rising up and down, as with the SCARA, this system moves the whole arm up and down (see Figure 4-10). These robots also tend to have a normal minor axes complement of two or three versus the single rotational of the traditional SCARA types. This configuration provides the power of the SCARA robot in the vertical direction with flexibility in tooling orientation that rivals any of the other systems we have looked at.

Delta Robot

Delta robots have become popular in industry and 3D printing over the past few years due to their speed and unique design, and with that unique design there is a unique geometry. Delta robots mount over the workspace, like Cartesian systems, but that is where the similarities end. If you look at Figure 4-11, you can see that the system is made up of three vertical arms coming to a pyramid-type point at the tooling below. The three major axes still drive the tooling around, and the three minor axes orientate the tooling, but the work envelope does not resemble those we have discussed. The result of this arrangement, due to the sweeping motion of the three major axes, is a cone similar to an acorn or the nose cone of a rocket. The majority of the work envelope is closer to the base of the robot; the envelope narrows as the tooling is moved farther away from the overhead unit (see Figure 4-12). These systems sacrifice a large portion of their work envelope for speed and the benefit of mounting over the work area. Delta systems have become a popular choice for part sorting when coupled with machine vision. The mechanical speed of the system gives the controller more time to process images and determine any necessary offsets without sacrificing cycle time or the amount of time it takes a robot to complete its programmed actions (see Figure 4-13).

Figure 4-12 This graphic shows the unique shape of the delta robot configuration's work envelope.

Image courtesy of ABB Inc.

Figure 4-13 This is one of ABB's delta-style robots. If you look closely, you can see the large airline used for tooling applications.

As we continue to refine the current systems and create new ones, we can expect the list of geometries or work envelopes to change as well. The delta systems are a great example of this, given that most manufacturers did not even consider these robots 20 years ago and today they are a major player in the field. It will be interesting to see how the next 20 years affect the geometry of the robot and the options available.

Drive Systems: Classification and Operation

drive system

A combination of gears, sprockets, chains, belts, shafts, and other power transmission equipment that transmits power from a generation source, such as electric motors, to where useful work is performed

Another way of sorting robots is by the method used to connect the motors to the moving parts, or the drive system. This is another broad classification (similar to grouping by power source) that will result in many different robots being grouped together. Because of this, drive system sorting is often more of a criteria in specifying a robotic system than a sole sorting method. As we look at the types of drive systems used by robots, we will also look at how they work and the forces involved.

Direct Drive

direct-drive systems

Systems that have the motor shaft connected directly to the robot for motion

Direct-drive systems have the rotating shaft of the motor connected directly to the part of the robot they move. This system was first patented in 1984 by Takeo Kanade and Haruhiko Asada. Since this system greatly increased the speed and accuracy of the robot, it has become a favorite among industrial as well as non-industrial robots. Because this design couples the robot directly to the output shaft of the motor, there is a 1:1 ratio of movement. In other words, a full rotation of the motor shaft creates a full rotation of the robot joint. One consideration with this method is that there is no mechanical amplification of force, thus the torque (the rotational force generated by the motor) is the limiting factor in the payload of the robot. Luckily, we have developed electric motors capable of generating large amounts of torque to help with this limitation, making the direct drive suitable for loads of up to 250 pounds. One thing to remember is to deduct the weight of the robot from the payload of the motor. To do this, we first figure out the payload from the motor carrying the bulk of the load and then check the subsequent motors of the system to ensure that no part of the system is overloaded.

torque

A measure of the work potential of rotational force

Example 1

For this example, we have a robot with the following servomotors:

Axis 1 – 250 ft. lbs. of torque

Axis 2 – 175 ft. lbs. of torque

Axis 3 – 100 ft. lbs. of torque

Axes 4 and 5 – 75 ft. lbs. of torque

Axis 6 – 25 ft. lbs. of torque

For the sake of our example, we will ignore the complexity of figuring accelerated forces into the mix and work instead with just the pure weight of the system.

Part/Portion of the Robot	Weight
Tooling	10 lbs.
From axis 5 to 6	15 lbs.
From axis 4 to 5	20 lbs.
From axis 3 to 4	50 lbs.
From axis 2 to 3	70 lbs.
From axis 1 to 2	70 lbs.

(All the weights in this example include the weight of the robot parts as well as any motors and other equipment installed on the portions of the robot.)

First, we have to see if there are any places where the torque is insufficient for the weight.

For axis 1, we add all the moving weights together to see what the total is:

$$10 \text{ lbs.} + 15 \text{ lbs.} + 20 \text{ lbs.} + 50 \text{ lbs.} + 70 \text{ lbs.} + 70 \text{ lbs.} = 235 \text{ lbs.}$$

Looking at the torque for the axis 1 motor, we are good to go with 15 lbs. (250 lbs. − 235 lbs. = 15 lbs.) to spare. Next is axis 2. At this point, we could add all the weights up again, leaving out the weight for the portion between axes 1 and 2, or simply take our answer from above and subtract that weight from it.

$$235 \text{ lbs.} - 70 \text{ lbs.} = 165 \text{ lbs.}$$

Again, we are good to go with 10 lbs. to spare. (175 lbs. − 165 lbs. = 10 lbs.) Now we move to axis 3.

$$165 \text{ lbs.} - 70 \text{ lbs.} = 95 \text{ lbs.}$$

We are under the limit again, but only by 5 lbs. (100 lbs. − 95 lbs. = 5 lbs.) this time. Next is axis 4.

$$95 \text{ lbs.} - 50 \text{ lbs.} = 45 \text{ lbs.}$$

Here we have 30 lbs. to spare and no issues. (75 lbs. − 45 lbs. = 30 lbs.) Next is axis 5.

$$45 \text{ lbs.} - 20 \text{ lbs.} = 25 \text{ lbs.}$$

Here we have 50 lbs. to spare and plenty of power. (75 lbs. − 25 lbs. = 50 lbs.) Last, but not least, is axis 6.

$$25 \text{ lbs.} - 15 \text{ lbs.} = 10 \text{ lbs.}$$

We are left with 15 lbs. of force, so that would make our payload after this set of tooling 15 lbs. right? (25 lbs. − 10 lbs. = 15 lbs.)

Wrong. Look back to what we had to spare with axis 3. We only had 5 lbs. left over after the weight of the robot with the 10 lbs. tooling in place. In other words, if we pick up too much weight, axis 3 of the robot will likely stop due to an over-current alarm or something similar, and there is a chance that we will damage the motor.

Again, we did not add in any force multiplication due to inertia or the deeper math that that involves, but this gives you an idea of how a robot can alarm out on an axis that appears to be fine. For those of you who would like to delve deeper into this kind of math, a course in physics would be a good place to start.

Reduction Drive

Reduction drive systems alter the output of the motor shaft via mechanical means. As a rule, these systems slow the speed of the rotational shaft in order to increase the torque or force of the system, but they can also change the direction of rotation or turn rotational motion into linear motion. These systems often require more maintenance, as they have additional moving parts, and the more complex the system, the greater the chance something will need repair. At the very least, they require more preventative maintenance than their direct-drive counterparts. Given the variety of ways in which we can reduce the motor output, there are several subgroups of this drive type for further sorting of robotic systems.

reduction drive
Systems that take motor output and via mechanical means reduce or alter it

Belt Drive

In this reduction-drive system, both the motor and the portion of the system we wish to move with that motor have a pulley attached and are connected by a belt. The common belts used are the V-belt, which is shaped like a V; the flat belt, which is a flat band of material; and the synchronous belt, which has teeth at set intervals along its length (see Figure 4-14). The flat and V-belts rely on friction to prevent slippage, with the teeth of the synchronous belt maintaining positional integrity. slippage is when some of the rotation of the pulley attached to the motor, known as the drive pulley, is not transmitted to the pulley attached to the system, known as the driven pulley. Flat belts have the highest chance of slippage, followed by V-belts and synchronous belts. V-belts tend to slip more often as they wear out and the tension on the belt loosens or if the belt is not set to the proper tension to begin with. The synchronous belt is the least likely to slip, but often receives damage and must be replaced anytime there is a crash or unplanned stopping of motion. (See Figure 4-15 for a V-belt system example and Figure 4-16 for a synchronous belt.)

There are some interesting calculations that accompany pulley systems, which allow us to determine the forces involved, belt speeds, and ratio of the pulleys. These will come in handy for those of you who go on to build your own robots or help to design various mechanical systems. Remember, robots are not the only systems out there to use pulleys and belts, as seen in Figure 4-15. These calculations can also allow you to explore the effects of changing pulley diameters or the speed of motors in systems already in use.

Figure 4-14   A close up of a V-belt with its slanted sides and a synchronous belt with its nonslip teeth. A flat belt would look like a flat band made out of the same material or the synchronous belt without the teeth.

V-belt
Rubber belts with steel or fiber reinforcement that are shaped like a V with the point cut off that are used to transmit force

flat belt
A belt made up of a flat band of rubber reinforced with steel or fibers that transfers power from the drive pulley to the driven pulley

Torque in Belt-Driven Systems

The base formula for torque is $T = F \times d$, where T is torque, F is force, and d is distance. We can adapt the torque formula for rotation as follows:

$T = F \times R$, where

T = torque

F = force (note: the force at the drive pulley will be the same as the force at the driven pulley)

R = radius of the pulley (radius is half the diameter of a circle; diameter is the distance from one side of the circle to the other, with the line passing through the center of the circle).

EXAMPLE 1 For our first example, what would the torque need to be if the system had to move 10 pounds and the drive pulley had a diameter of 6 inches?

Chapter 4 Classification of Robots 103

synchronous belt
A belt that has teeth at set intervals that is used to transfer power without slipping

slippage
Loss of rotation between the drive and driven elements, typically used to reference the loss in belt driven systems

drive pulley
The pulley is attached to the motor or force in a system

driven pulley
The pulley attached to the output or load of a system

figure 4-15   The air compressor from my workshop. While this is not a robotic system, it has the same parts and function as those found in robots. The smaller pulley on the left is the drive pulley, and the larger pulley on the right is the driven pulley.

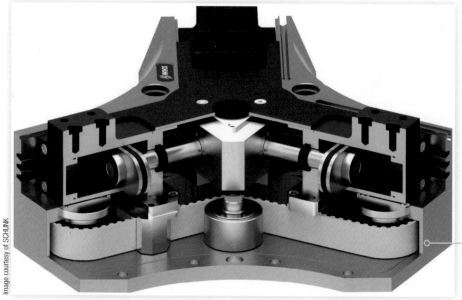

Image courtesy of SCHUNK

figure 4-16   Can you identify the belt used to drive this gripper system for a robot? Notice the teeth on the white belt at the bottom of the cutaway image.

First, we have to find the radius by taking 6/2, which equals 3 inches.

Next, we plug in the values and solve:

$T = F \times R$

$T = 10$ lbs. $\times 3$ in.

$T = 30$ in. $\cdot$ lbs. of force is required for this system to function properly.

EXAMPLE 2 If we know the torque that the system needs and the amount of force or weight involved, we can use the same formula to figure out the size of the pulley required. For this example, we will use a weight of 15 pounds and a required torque of 60 inch pounds.

$T = F \times R$ or $R = T/F$

$R = 60$ [in. $\cdot$ lbs.] / 15 lbs. (The lbs. of the 15 cancels out the lbs. of the 60)

$R = 4$ in.

EXAMPLE 3 If we know the force involved and the diameter of the drive and driven pulleys, we can compare the torque generated by each. For this example, we will use a force of 10 pounds, a drive pulley diameter of 4 inches, and a driven pulley diameter of 8 inches. Remember, the force is equal at both pulleys due to the fact that the belt links the two and thus transmits the force equally, minus friction and other detracting forces. For the sake of clarity, we will exclude friction and detracting forces from these examples.

$T = F \times R$

For the drive belt, $T = 10$ lbs. $\times 2$ in. or $T = 20$ in. $\cdot$ lbs.

For the driven belt, $T = 10$ lbs. $\times 4$ in. or $T = 40$ in. $\cdot$ lbs.

As you can see, by doubling the size of the driven pulley, we get twice the torque out of the system; however, that comes at a cost of the driven pulley running at half the drive pulley's speed.

revolutions per minute [RPM]

Commonly used to measure the speed of rotating systems

Pulley Ratio and Speed in Belt-Driven Systems

Next, we will look at how to find the ratio of the drive pulley to the driven pulley and use this to determine the speed of the driven pulley. Since the drive pulley attaches to the motor, the revolutions per minute [RPM] of the drive pulley is determined by the motor (see Figure 4-17).

Figure 4-17 The drive pulley in this picture is attached to the motor, and the driven pulley is on its own dedicated shaft. In the field, the motor and driven shaft would be mounted and set up to do work.

$R_s = D_2/D_1$, where
R_s = speed ratio
D_1 = diameter of drive pulley
D_2 = diameter of driven pulley

Also, Rs = RPM_1/RPM_2, where

R_s = speed ratio
RPM_1 = revolutions per minute of the drive pulley
RPM_2 = revolutions per minute of the driven pulley

EXAMPLE 4 For this example, we will use a drive pulley with a diameter of 4 inches and a driven pulley of 8 inches with a motor speed of 1,250 RPM on the drive pulley. First, we plug in the numbers to our equation and then reduce to the lowest terms.

$R_s = D_2/D_1$
R_s = 8/4 or 2/1 (this gives us a ratio of 2:1 or two rotations of the drive
   gear to get one rotation of the driven gear)
$R_s = RPM_1/RPM_2$

If we plug in what we know, we get:

2/1 = 1,250 RPM/RPM_2

If we simplify this, we get:

2 = 1,250 RPM/RPM_2

Multiply both sides by RPM_2 and you get:

2 RPM_2 = 1,250 RPM

Divide by 2 and you get:

RPM_2 = 625 RPM
RPM_2 = 625 revolutions per minute of the driven gear, or one-half of
   the drive gear speed. This demonstrates the math behind the
   doubling of torque from example 3, costing half the speed at
   the output side of the system.

Velocity in Belt-Driven Systems

Velocity is a measure of how fast something (in this case, the belt) is moving. This calculation will come in handy when we look at how to figure out how much power is used by the system. For this, we will use the following equation:

velocity
A measure of how fast
something is moving

$$BDV \left[\frac{ft}{min} \right] = D_1 \times \pi \times RPM \times \frac{1}{12} \left[\frac{ft}{in} \right]$$

BDV = belt-drive velocity in ft./min.
D_1 = drive pulley diameter
π = pi, the constant 3.14
RPM = revolutions per minute of the drive gear
1/12 = conversion factor for ft./min. If you want the belt-drive velocity
   in inches per minute or the pulley diameter is already in feet,
   leave this out of the equation.

EXAMPLE 5 For this example, we will figure out the velocity of a belt of a system with a 4-inch diameter drive pulley that is running at 1,250 RPM.

$$\text{BDV}\left[\frac{\text{ft}}{\text{min}}\right] = D_1 \times \pi \times \text{RPM} \times \frac{1}{12}\left[\frac{\text{ft}}{\text{in}}\right]$$

$$\text{BDV}\left[\frac{\text{ft}}{\text{min}}\right] = 4 \text{ in.} \times 3.14 \times 1,250 \times \frac{1}{12}\left[\frac{\text{ft}}{\text{in}}\right]$$

BVD = 1,308.3 ft. /min. or 15,700 in. / min. without the conversion to feet.

Power of Belt-Driven Systems

Power is a measurement of work, and we can use the formulas for power to find information we are missing. There are several formulas for power as it relates to the belt-driven system, such as $P = F \times v$, which is power (P) equals force (F) times linear velocity (v) and $P = 2\pi \times T \times \text{RPM}$, where power (P) equals two times pi (π) times torque (T) times revolutions per minute (RPM). However, the formula we are most interested in is:

$$\text{HP} = \frac{T[\text{lb} \cdot \text{ft}] \times \text{RPM}}{5252}, \text{ where}$$

$$\text{HP} = \text{horsepower}$$
$$\text{RPM} = \text{revolutions per minute}$$
$$T = \text{torque in ft. lb}$$
$$5252 = \text{conversion constant for lb} \cdot \text{ft of torque.}$$
$$63025 = \text{conversion constant for lb} \cdot \text{in of torque (substitute this for the}$$
$$5252 \text{ if using lb} \cdot \text{in. of torque instead of lb} \cdot \text{ft)}$$

EXAMPLE 6 In this example, we will examine what would happen if we were using a 1.5 hp motor with two speed options, 1,250 rpm or 1,750 rpm. In each case, we will use the formula to determine the torque of the system.

$$\text{Speed } 1 = 1,250 \text{ rpm}$$

First, we will manipulate our formula to find torque instead of HP.

$$\text{HP} = \frac{T[\text{lb} \cdot \text{ft}] \times \text{RPM}}{5,252}$$

becomes

$$T[\text{lb} \cdot \text{ft}] = \frac{5,252 \times \text{HP}}{\text{RPM}}$$

$$T[\text{lb} \cdot \text{ft}] = \frac{5,252 \times 1.5}{1,250}$$

$$T = 6.3 \text{ lb} \cdot \text{ft}$$

$$\text{Speed } 2 = 1,750$$

$$T[\text{lb} \cdot \text{ft}] = \frac{5,252 \times 1.5}{1,750}$$

$$T = 4.5 \text{ lb} \cdot \text{ft}$$

As you can see, the increase in speed causes a decrease in torque.

EXAMPLE 7 We can also use our power equation to determine the horsepower needed for the motor of a system. For this example, we will take the torque we figured in example 1, 30 in. · lbs., and use the RPM from example 5, 1,250 rpm, to figure the horsepower required by the system.

$$HP = \frac{T[lb \cdot in] \times RPM}{63025}$$

Note: We use 63,025 instead of 5,252 because the torque is in. · lbs., not ft. · lbs.

$$HP = \frac{30 \times 1,250}{63,025}$$

HP = 0.60 or 3/5 hp motor. As 3/5 is not a standard horsepower for motors, we would likely use a ¾ or 0.75 hp motor instead. Note: When facing a choice of which motor to use for a nonstandard requirement, always go with the larger motor, as this will put less strain on the motor and the motor will last longer.

Also, if we use the horsepower to watts conversion of 1 HP = 745.7 watts, we can convert the 0.60 hp. required by the system to watts and have an idea of the power consumption involved.

0.60 × 755.7 = 447.42 watts of power consumed.

This calculation can be vital for systems that run off batteries in order to determine how long the system can run before recharging. For systems tied to the electrical grid, this information is crucial in figuring out the amp draw of the system, fusing, and operating costs.

Chain Drive

For the most part, chain-driven systems work in the same way as belt-driven systems, with a few exceptions. These systems use **sprockets**, which have teeth designed to fit into the links of the chain instead of pulleys, and a chain, usually made of metal, which connects the drive sprocket to the driven sprocket. Sprockets look a lot like gears: they have teeth that are widely spaced apart and longer than standard gear teeth (see Figure 4-18). Like the synchronous belt,

> **sprocket**
> A circular metal device similar to a gear, with teeth spaced to fit inside the links of a chain and used to transmit force

Figure 4-18 A drive chain, an idler on the far left, a sprocket that could be drive or driven in the middle, and a gearbox reducer ready to either drive a chain or be driven by one on the far right.

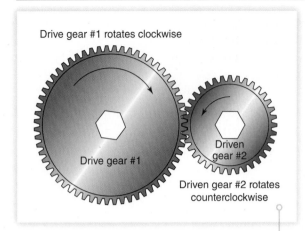

Drive gear #1 rotates clockwise

Drive gear #1

Driven gear #2

Driven gear #2 rotates counterclockwise

Figure 4-19   Whenever you have an even number of gears, the last gear will rotate in the opposite direction of the first gear.

chains do not slip, but they do wear out. Over time, a chain's links will become longer due to the stress on the chain and wear down from the contact with the sprockets. This will result in a change in the length of the chain, which affects the relationship between the drive and driven sprocket; when it becomes severe enough, the chain may jump a tooth on the drive or driven sprocket. In effect, this is the same as slippage, but it generally happens slowly, over time, instead of all at once. Another major difference between a belt-driven system and a chain system is maintenance. We *do not* oil belts, as this would increase the chance of slippage. However, most chains do require lubrication to work properly. Chain drives combine the rigidity and force transmission of a gear system with the flexibility and forgiveness of a belt system.

Gear Drive

mesh

The mating of gear teeth to transmit power through physical contact

gear train

Two or more gears that are connected together and used to transmit power

transmission

A group of gears, often including compound gears and multiple gear ratios, that are used to transmit power

idler gear

A gear on a dedicated shaft used to change the direction of the output gear and/or to link two gears together when there is a large distance between the gears

While it is difficult to determine when we first started to use gears, we have in fact been using gears to transfer force for thousands of years now. Gears come in many shapes, sizes, and varieties, but they all have cogs, or teeth, which are projections that match up or mesh with similar projections on other gears to transmit force. We call the gear tied to the power supply the "drive gear" and the one tied to the output the "driven gear," similar to the method used to label pulleys in a belt system. We call two or more gears connected together a gear train or transmission, as they connect the output(s) to the prime driving force.

Because of how gears connect and the way in which they transmit force, the driven gear rotates in the opposite direction to the drive gear (see Figure 4-19). If this is a problem, we can use a gear known as the "idler." Idler gears are extra gears added to a system to change the direction of rotation on a dedicated shaft, not an output shaft (see Figure 4-20). If there is an even number of total gears in a drive system, the last gear will rotate opposite the input gear. If there are an odd number of total gears in a system, the last gear will turn in the same direction as the input gear. The same system works for gears in the middle of a gear train as well, as long as you count the number of gears from the drive gear to the gear in question, including the gear for which you are trying to determine the direction of rotation in the count. For instance, if there is a transmission with seven gears in the system, and you want to know how the fourth gear in the system rotates, you will count from the first gear to the fourth gear. The answer is four, which is an even number. That means that gear four is rotating opposite the drive gear.

Figure 4-20   To get the output or last gear in a gear train to rotate in the same direction as the drive or input gear, there must be an odd number of gears in the system.

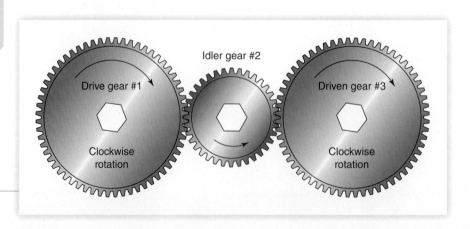

Idler gear #2

Drive gear #1

Clockwise rotation

Driven gear #3

Clockwise rotation

Figure 4-21 This set of compound gears has three gears on one shaft, creating a set of options for power transference.

compound gears
Two or more gears on the same shaft, often made from one solid piece of material

spur gear
A gear made from a round or cylindrical object with teeth cut into the edge

Figure 4-22 In the lower left corner is a set of spur gears used to transmit force to the robot tooling. If you look closely, you can see the compound gear in the middle creating two separate gear trains in this system.

We often combine multiple gears to create complex gear trains that have compound gears and multiple output points with their own torque and speed. Compound gears are two or more gears on the same shaft, often made from one solid piece of material (see Figure 4-21). One of the gears in a compound gear will be a driven gear, while the other will be a drive gear; systems of more than two gears in a compound arrangement will have multiples of one or both. This is because one or more of the gears in the compound arrangement serves as the power source for a completely new gear train, thus making it the drive gear for that transmission system. When you are trying to figure out the rotation of gear systems using compound gears, there are a couple of things you need to remember. One, multiple gears on the same shaft must rotate in the same direction. Two, when you are counting gears to determine their rotation, always start at the drive gear. I recommend figuring out the rotation of the driven gear of the compound set first, as the drive gear will have to rotate in the same direction (since it is on the same shaft). Treat each drive gear as the start of a new gear train in order to determine its rotation.

Gears come in many varieties, but the most common, and the one that you are likely familiar with, is the spur gear. Machinists and industry create spur gears by taking a round or cylindrical object and cutting teeth into the edge. The teeth are not square, but tapered and rounded at the end to reduce friction and other stresses that occur as the teeth mesh with other gears. They are parallel to the shaft running through the gear. Spur gears only mesh in parallel with one another (see Figure 4-22).

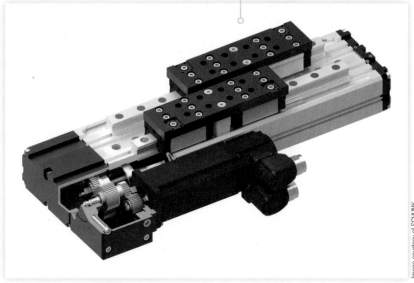

Figure 4-23 This set of bevel gears has a light coat of rust due to improper maintenance. If you fail to oil or grease the gears for your robot, your robot's gears could end up looking like this.

helical gear
A gear similar to a spur gear that has teeth set at an angle along the edge instead of parallel

helix
A smooth space curve

skew gear
Another name for helical gear

bevel gear
A gear that has its teeth cut along a tapered edge that would make a pointed cone if it were not flattened on the end

miter gear
Bevel gears with equal numbers of teeth and the shafts at a 90-degree angle

Helical gears are similar to spur gears, but their teeth are not parallel to the shaft of the gear; instead, they are set at an angle along the edge making them part of a helix or smooth space curve (see Figure 4-23). Because of this unique shape, two gears can mesh with their shafts in parallel or at a 90 degree-angle from each other, which is where their other name, skew gears, comes from. Because of their shape, they tend to engage more slowly than spur gears, making them smoother and quieter in comparison.

Bevel gears have their teeth cut along a tapered edge that would make a pointed cone if not flattened on the end. We often find bevel gears in the wrists of robots to rotate the force 45 or 90 degrees, but they are capable of any angle between 0 and 180 degrees with proper construction. When bevel gears have an equal number of teeth and the shafts at a 90 degree-angle, they are referred to as miter gears. By changing from a square-tooth design, we get variants of the bevel

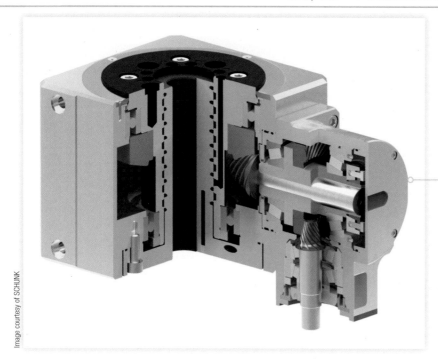

Image courtesy of SCHUNK

Figure 4-24 When looking at the gearing on the right side of the picture, the gear closest to the middle is a spiral bevel gear. The one further out is a Zerol bevel gear with its trademark flat face. Half of the mating gear for this set is cut away for the purposes of illustration.

spiral bevel gear
A type of bevel gear with the tooth curved to reduce noise and smooth out its operation

zerol bevel gear
A gear with the same curved tooth of the spiral bevel gear, but without angled sides

hypoid bevel gear
Similar to the spiral bevel gear; however, if you draw a line from the shaft set at an angle, it will not meet the shaft of the other gear

worm gear
A cylinder that has at least one continuous tooth cut around it

gear with new properties. To reduce noise and smooth out operation, we curve the tooth creating the spiral bevel gear. Zerol bevel gears use the same curved tooth without the angled sides. These gears have a flat face instead of the tapered, coned shape of the other bevel gears (see Figure 4-24). Hypoid bevel gears are similar to the spiral bevel gear, but if you draw a line from the shaft set at an angle, it will not meet the shaft of the other or mating gear. These gears are almost always set at 90-degree angles, can generate gear ratios of 60:1, and, if properly designed, can be quieter than spiral bevel gears.

Worm gears are made up of a cylinder that has one tooth cut around it, similar to what we see on screws and bolts, but much larger and we call this part the worm. We finish the set with a spur or helical gear of the desired design (see Figure 4-25).

Figure 4-25 This set of worm gears is similar to what you will find inside many power transmission gearboxes, especially ones that rotate the output 90 degrees.

figure 4-26 This rack and pinion system generates a long stroke for the robot tooling. Notice the gears in the cutaway on the left hand side.

Image courtesy of SCHUNK

rack and pinion
Systems that consist of a spur gear and a rod or bar that has teeth cut along the length

harmonic drive
Specialized gear systems that use an elliptical wave generator to mesh a flex spline with a circular spline that has gear teeth fixed along the interior

backlash
The distance from the back of the drive gear tooth to the front of the driven gear tooth

These systems turn the force 90 degrees like the bevel systems mentioned above but are simpler in construction and can generate torques up to 500:1. This means that for every foot pound of force in, you get 500 foot pounds of force out! The worm can always be the drive gear, or the one tied to the motor; however, in some configurations, the helical or spur gear cannot. In these cases, the non-worm gear simply cannot overcome the friction of the system and the gear set locks in place, which can be beneficial for holding the position of a system. When used as a break or lock, there must be enough force on the spur or helical gear to hold the system in place, but not so much that the worm gear cannot get the system going once more.

Rack and pinion systems consist of a spur gear and a rod or bar that has teeth cut along the length (see Figure 4-26). This system converts rotational force into linear force and is favored for moving systems over long distances, as in gantry systems or mobile robots. You will find this type of gear system in the tooling of various robots where there needs to be a large amount of travel, such as grippers with a wide range of motion.

A harmonic drive is a specialized gear system that uses an elliptical wave generator to mesh a flex spline with a circular spline that has gear teeth fixed along the interior (see Figure 4-27). The circular spline is typically the driven portion of the system, and the flex spline only contacts the circular spine at two points that are 180 degrees apart. The wave generator is inside the flex spine, but is usually separated from the flex spine by ball bearings. The flex spline has fewer teeth than the circular spine. This system can generate torques up to 320:1 and has no backlash! Backlash is the distance from the back of the drive gear tooth to the front of the driven gear tooth that represents loss of motion in the system. Most gear systems need a certain amount of backlash to function properly. This is why the harmonic drive is unique.

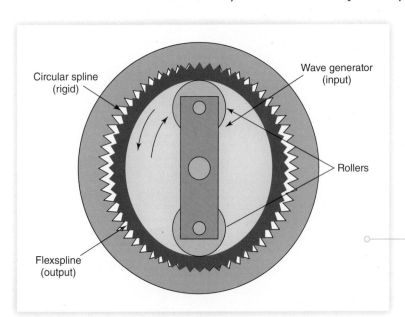

Circular spline (rigid)

Wave generator (input)

Rollers

Flexspline (output)

figure 4-27 This graphic shows the various parts of a harmonic drive system and how they work together. This is the gearing used for axis 6 of many robots.

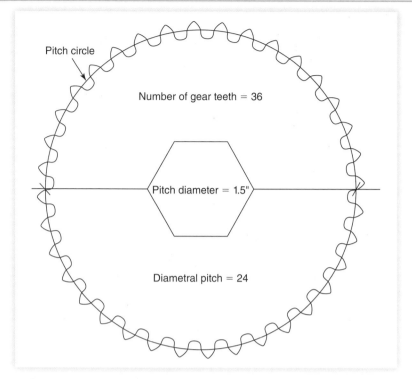

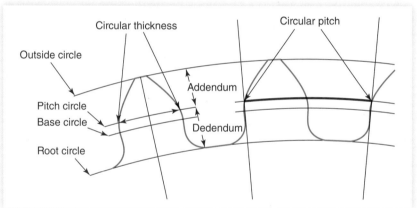

figure 4-28   Look closely at these two images, as they show where we derive the information needed for our various gear calculations.

Now that you have an idea of how gear drives work and the various gear types that you may encounter, let us look at some of the math involved with these systems (see Figure 4-28). There are several calculations that can be of benefit, especially when you trying to figure out torque, velocity, or other important measurements in a gear train. We will first look at the pitch diameter, which is the diameter of the imaginary circle used to design the gear. This circle cuts through the middle of the gear teeth, where the smooth side starts to taper at the top, and is designed to contact the pitch circle of another gear when they mesh. We express the pitch diameter with the following equation:

$$D = \frac{N}{P}, \text{ where}$$

D = pitch diameter
N = number of teeth on the gear
P = diametral pitch or gear size

pitch diameter
The diameter of an imaginary circle used to design a gear

diametral pitch
The ratio of the number of teeth per pitch diameter that describes the size of the gear teeth

(**Diametral pitch** is a ratio of the number of teeth per pitch diameter and describes the size of the gear teeth with a smaller ratio, denoting larger teeth with more space in between then.)

EXAMPLE 8 For this example, we will figure the pitch diameter for a gear with 30 teeth and a diametral pitch of 20.

$$D = \frac{N}{P} \quad D = \frac{30}{20} \quad D = 1.5 \text{ in.}$$

Once we know the pitch diameter, we can use this in a calculation to determine how far apart to place two gears for proper meshing. This is critical, as placing the gears too close will cause binding (the gears do not turn properly) or make it impossible to mount the gears, while placing them too far apart, can lead to gear damage, loss of power, and excessive wear. When you are meshing two gears, the diametral pitch and pressure angle must match for proper operation. The **pressure angle** refers to how the forces interact between two gears and at what angles and determines how a gear tooth is rounded or shaped. When it comes to the pressure angle, the two choices are 14.5 degrees, which was the standard for many years and is still readily available, and 20 degrees, which is able to transmit greater loads and is thus the new favorite. If you mix and match the diametral pitch or pressure angle of gear sets, you are asking for inefficiency and problems at best and utter failure at worst.

pressure angle
How the forces interact between two gears, at what angles, and how a gear tooth is rounded or shaped

To find the center-to-center distance of two gears, we use the following formula:

$$CCD = \frac{D_1 + D_2}{2}, \text{ where}$$

CCD = center-to-center distance
D_1 = pitch diameter of gear 1
D_2 = pitch diameter of gear 2

EXAMPLE 9 For this example, we will determine the center-to-center distance for two gears, using the information from example 8 for gear 1; for gear 2, we will use a gear with 60 teeth and a diametral pitch of 20. First, we must find the pitch diameter for gear 2.

$$D = \frac{N}{P} \quad D = \frac{60}{20} \quad D = 3 \text{ in.}$$

Next, we plug the pitch diameters into the center-to-center distance equation.

$$CCD = \frac{D_1 + D_2}{2} \quad CCD = \frac{1.5 + 3}{2} \quad CCD = 2.25 \text{ in.}$$

gear ratio
The ratio used to determine what happens with the driven gear in reference to the drive gear in terms of torque and speed

Therefore, by our calculations, we know that the center of the shaft for gear 1 should be 2.25 in. from the center of the shaft for gear 2 for proper operation. This is very important information for those who design their own gear trains or modify existing systems.

Just as we use pulleys of different sizes in belt-driven systems to increase torque or speed, we can do the same with gears by increasing or decreasing the number of teeth on a gear. We use the **gear ratio** to determine what happens with the driven gear in reference to the drive gear and thus the torque and speed changes as well. We use the following equation to find the ratio:

$$GR = \frac{N_2}{N_1}, \text{ where}$$

GR = gear ratio
N_1 = total number of drive gear teeth
N_2 = total number of driven gear teeth

EXAMPLE 10 We will now find the gear ratio for the two gears we used in example 9. Remember, gear 1 has 30 teeth and gear 2 has 60 teeth.

$$GR = \frac{N_2}{N_1} \quad GR = \frac{60}{30} \quad GR = 2{:}1$$

You can see that it takes two full rotations of the drive gear to equal one rotation of the driven gear. This means that the driven gear turns at half the speed of the drive gear, but it has twice the torque.

EXAMPLE 11 This time, we will look to see what would happen if we had a drive gear with 48 teeth and a driven gear with 24 teeth.

$$GR = \frac{N_2}{N_1} \quad GR = \frac{24}{48} \quad GR = 0.5{:}1$$

We see that for every half rotation of the drive gear, the driven gear makes a full rotation. In this case, the torque of the driven gear is half that of the drive gear, but the speed of the driven gear is double that of the drive gear. (Remember, a reduction in speed equals more torque, and faster speed equals a reduction in torque.)

Like the belt-driven system, the velocity or distance moved over a set period of time will be the same for all gears connected together. If this were not true, the faster speed of one gear would rip the teeth off of the other gear. This is exactly what happens when one part of a gear train stops suddenly, such as in the case of a robot crash. (Some systems may have enough flexibility to reduce the damage this causes while others can stop quickly enough to minimize damage.) We use velocity to figure out how fast a system moves via the following equation:

V = pitch circle circumference $\times$ RPM, where
V = velocity
Pitch circle circumference = $D \cdot \pi$ and D = pitch diameter
RPM = revolutions per minute of the gear

EXAMPLE 12 For this example, we will find the velocity of a 40-tooth gear with a diametral pitch of 20 rotating at 50 rpm.

$$D = \frac{N}{P}, \text{ where}$$

D = pitch diameter
N = number of teeth on the gear
P = diametral pitch or gear size

$$D = \frac{N}{P} \qquad D = \frac{40}{20} \qquad D = 2 \text{ in.}$$

Pitch circle circumference $= D \cdot \pi$ pitch circle circumference $= 2$ in. $\cdot 3.14$
Pitch circle circumference $= 6.28$ in.
$V =$ Pitch circle circumference $\times$ RPM $V = 6.28$ in. $\times 50$ RPM
$V = 314$ in. per min.

Ball Screw

ball screw
A large shaft with a continuous tooth carved along the outer edge with a nut or block that moves up and down the length of the shaft

Another common type of drive that you may encounter in robotics is the **ball screw**. Ball screws consist of a large shaft with a continuous tooth carved along the outer edge and a nut or block that moves up and down the length of the shaft. The prime mover connects to the shaft either directly via a coupler or through a belt, chain, or gear-drive system, with all the options that creates. The block or nut that moves along the ball screw usually rides along the tooth via ball bearings and attaches to whatever is being moved. In essence, these systems are a combination of the best of both worlds from a worm gear and rack and pinion system. Ball screws are highly efficient and precise to one-thousandth of an inch per foot or finer, making them a favorite for precision motion. Ball screws do suffer backlash, like gear systems, but only when they change directions. A small portion of the direction change takes up the backlash, and from that point on, backlash is no longer a concern. In cases where the block has to carry heavy loads, the system may have a guide rod as well to insure that the block travels straight and does not bind up.

You have a wide range of options under reduction drive for sorting robots, while direct drive is pretty much an all or nothing category. Remember, the two main drive types for robots are reduction or direct drive, with reduction drive having multiple subcategories. As you read at the begging of this section, drive type is a better filtering agent than a full category for finding the robot that fits your needs. This type of categorization does work well for robot tooling, so that may be something to keep in mind.

ISO Classification

International Standards Organization (ISO)
An organization that develops, updates, and maintains sets of standards for use by industries of the world

The **International Standards Organization (ISO)** is an organization that develops, updates, and maintains sets of standards for use by the industries of the world. An ISO certification guarantees that a company is making its products according to a defined set of specifications for quality, safety, and reliability, giving customers peace of mind. A company has to go through a lot of preparation to receive ISO certification and then periodic inspections to insure it is still meeting the requirements of any certifications that it holds. This is not a free service; however, the money spent on ISO certification becomes a great marketing tool to attract new clients and keep current ones.

ISO groups robots into three broad classifications: industrial, service, and medical. The industrial robotics classification is the one that ISO has worked with the longest; the other two are newer. Let's look at the classifications in order of longevity.

ISO defines an industrial robot as an "actuated mechanism programmable in two or more axes with a degree of autonomy, moving within its environment, to perform intended tasks" (Harper 2012). Initially, ISO only dealt with industrial robots, so the current definition was updated to deal with the new types of robots on their radar. Since 2004, ISO has classified industrial robots by their mechanical structure only. The categories are as follows:

- Linear robots
- Articulated robots

- Parallel robots
- Cylindrical robots
- Others

I hope you noticed similarities between the ISO robot classification list and this chapter's section on robot classification according to their work envelope geometry. ISO includes Cartesian robots and gantry robots in the linear group; delta and parallel robots are the same robot; and the way ISO defines its articulated robot category would include the spherical robot described in this chapter. The other ISO categories include new styles of industrial robots that do not fit into any of the other categories; this gives them a classification until a new category can be created.

The ISO defines a service robot as a "robot that performs useful tasks for humans or equipment excluding industrial automation applications" (Virk 2003). As you can see by the definition, the ISO has specifically excluded the industrial robot from this field. This major classification is made up of one category with three subcategories:

- Personal care robot

  - Mobile servant robot
  - Physical assistant robot
  - Person carrier robot

Personal care robots, as defined by ISO, are those that can come into contact with people to help with or perform actions that improve the quality of their life. These are not just your run-of-the-mill hobby robot, but rather high-end systems that can assist with the day-to-day tasks of life, especially for those with physical impairments. The subcategories give us a better idea of the type of systems that the ISO had in mind for this field. The mobile servant robots are designed to move about a space carrying objects and interacting with the environment (a robotic butler, if you will). Physical assistant robots boost the capabilities of people. Exoskeletons fall into this category, as they help the weak or disabled to walk; however, they can also help a normal person carry more weight than possible otherwise. Personal carrier robots are devices that take people from point A to point B. Robotic wheelchairs and self-driving vehicles fall into this subcategory.

ISO defines medical robots as "a robot or a robotic device intended to be used as medical electrical equipment" (Virk 2013). This standard is still under development as of the writing of this book, but we can expect something along the lines of service robot classifications and categories. I am sure systems such as the *da Vinci* operating robot will shape these upcoming categories as well as other complex equipment currently in use in the field. It will be interesting to see how ISO determines the difference between medical robots that help patients recover and the service robots that ISO has already recognized.

As the field or robotics continues to evolve, so too will ISO definitions. Just as the ISO definitions evolved with the industrial robot and have expanded to encompass the new systems in use, we can expect these definitions to grow with the expansion of robotic use. I doubt we will see a day when the ISO is classifying hobby robots or entertainment system, but I believe any of the rest to be fair game. Because of this, you may want to check into the ISO robotic information from time to time to see how its definitions and classifications evolve.

review

Clearly, there are other ways to classify robots, and there is a good chance you will run across some of these methods as you explore the world of robotics. With the way the robotics field is changing and evolving, it is highly likely that new robotic systems will not fit into the current classification methods, causing the development of new categories or ways of grouping robots. Over the course of this chapter, we covered the following topics:

- **How are robots classified?** This section discussed why we classify robots and some possible broad categories.

- **Power source.** We examined this group of robots and looked at some of the strengths and weaknesses of each category within.

- **Geometry of work envelope.** This section showed how we can group robots according to their work envelope and discussed axes of movement.

- **Drive systems: Classification and operation.** This section was one part classification, one part drive systems. We also explored some of the math involved with drive systems.

- **ISO classification.** We explored how ISO groups robots; we discovered that industrial robots are classified by how they work mechanically, giving us nearly the same groupings as we present in the geometry section of the chapter.

formulas

$T = F \times d$, where
T = torque
F = force
d = distance

$T = F \times R$ for a belt system, where
T = torque
F = force
R = radius of the pulley

$R_s = D_2/D_1$, where
R_s = speed ratio
D_1 = diameter of drive pulley
D_2 = diameter of driven pulley

$Rs = RPM_1/RPM_2$, where
R_s = speed ratio
RPM_1 = revolutions per minute of the drive pulley
RPM_2 = revolutions per minute of the driven pulley

$$BDV \left[\frac{ft}{min} \right] = D_1 \times \pi \times RPM \times \frac{1}{12} \left[\frac{ft}{in} \right]$$

BDV = belt-drive velocity in ft./min.
D_1 = drive pulley diameter
π = pi, the constant 3.14
RPM = revolutions per minute of the drive gear
$1/12$ = conversion factor for ft./min.

$$HP = \frac{T[lb \cdot ft] \times RPM}{5,252}$$

HP = horsepower
T = torque
RPM = revolutions per minute
π = 3.14
$5,252$ = conversion constant for lb · ft of torque
$63,025$ = conversion constant for lb · in of torque
$1\ HP = 745.7$ watts

$$D = \frac{N}{P}$$

D = pitch diameter
N = number of teeth on the gear
P = diametral pitch, or gear size

$$CCD = \frac{D_1 + D_2}{2}$$

CCD = center-to-center distance
D_1 = pitch diameter of gear 1
D_2 = pitch diameter of gear 2

$$GR = \frac{N_2}{N_1}$$

GR = gear ratio
N_1 = total number of drive gear teeth
N_2 = total number of drive gear teeth
V = pitch circle circumference $\times$ RPM, where
V = velocity
Pitch circle circumference = $D \cdot \pi$
RPM = revolutions per minute of the gear

key terms

Anthropomorphic	Flat belt	Miter gear	Slippage
Ball screw	Gear ratio	Pitch diameter	Spiral bevel gear
Bevel gear	Gear train	Pressure angle	Sprocket
Compound gears	Green power	Rack and pinion	Spur gear
Backlash	Harmonic drive	Reduction drive	Stepper motor
DC brushes	Helical gear	Revolutions per minute	Synchronous belt
Diametral pitch	Helix	(RPM)	Torque
Direct-drive systems	Hypoid bevel gear	Selective Compliance	Transmission
Drive pulley	Idler gear	Articulated Robot	V-belt
Drive system	International Standards	Arm (SCARA)	Velocity
Driven pulley	Organization (ISO)	Servomotor	Worm gear
Encoders	Mesh	Skew gear	Zerol bevel gear

review questions

1. Why do we normally group robots?
2. What is the difference between a DC brushless motor and a motor containing brushes?
3. Do hydraulic robots require any electricity?
4. What are some of the costs of using a hydraulic robot?
5. What is the primary difference between hydraulics and pneumatics?
6. What is the major problem with using pneumatics for robots, and how do we correct it?
7. What is a nuclear-powered robot?
8. What are the benefits of a Cartesian work envelope?
9. What are the pros and cons of a cylindrical robot?
10. What is the work envelope of an articulated robot?
11. Which geometry is the most common in robots, and why?
12. What is the difference between a SCARA robot and a horizontally base-jointed arm?
13. What is the tradeoff for the work envelope of the delta robot?
14. Describe the operation of a direct-drive robot.
15. What is the difference between a V-belt, a flat belt, and a synchronous belt?
16. In example 3, what was the benefit and cost of using a driven pulley that was twice the diameter of the drive pulley?
17. What is the speed of the drive pulley, regardless of the pulley ratio?
18. What is the difference between a chain-drive and a belt-drive system?
19. Define the motion of the last gear in a transmission that has an even number of gears as well as the motion of the last gear in a transmission that has an odd number of gears.
20. How is it possible for compound gear systems to have both drive and driven gears on the same shaft?
21. How do we determine the rotation of gears when compound gears are involved?
22. What is the benefit of a helical gear over a spur gear?
23. What are the benefits of harmonic drives?
24. What happens when we place gears too close together or too far apart?
25. How does a ball screw work?
26. What is the ISO definition of an industrial robot?
27. What are the categories used by the ISO for industrial robot classification?
28. What does the ISO consider as a personal care robot?
29. How does the ISO define a medical robot?

End-of-Arm Tooling

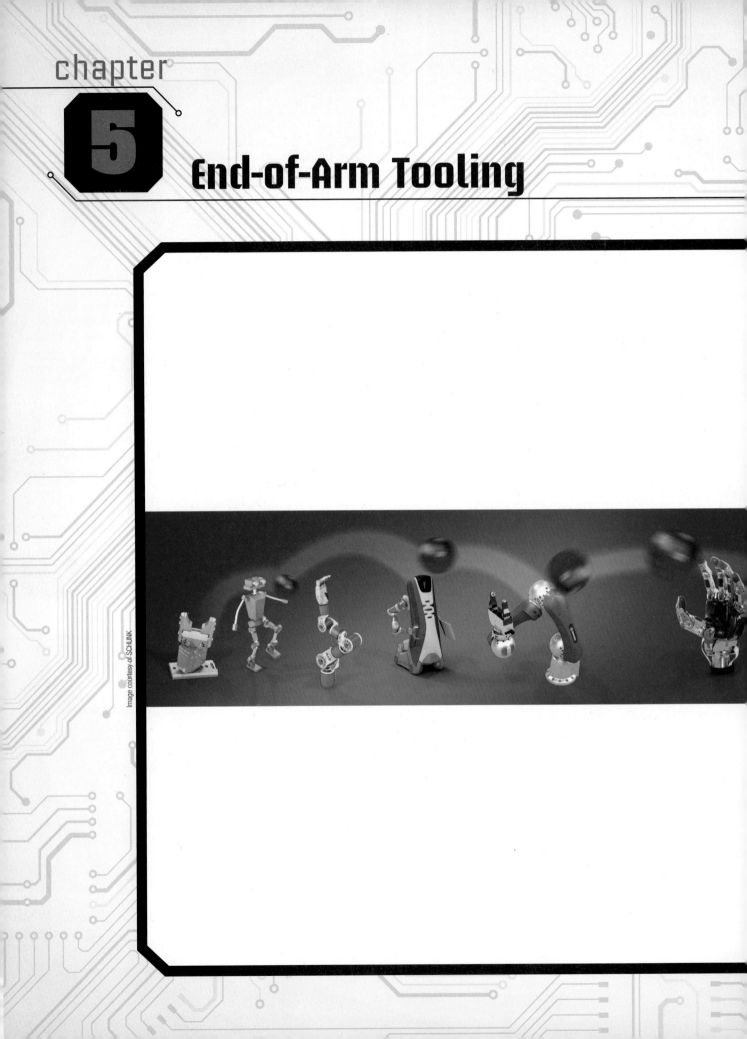

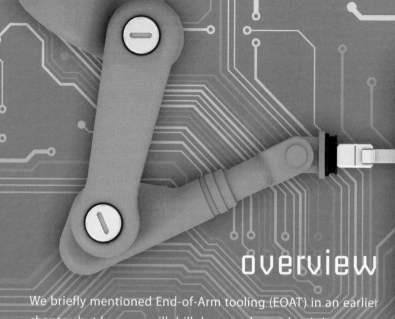

Overview

What You Will Learn

- What robot tooling is and ways in which we use it
- The common types of tooling encountered in robotics
- The common varieties of grippers
- The math behind figuring out the force a gripper needs to generate
- How robots use multiple tooling
- The ways in which tooling can flex to align with parts out of position

We briefly mentioned End-of-Arm tooling (EOAT) in an earlier chapter, but here we will drill down and see what is important about EOAT while exploring some of the different tasks this equipment performs. We will look at common tooling types, specialized tooling, and some of the specialty equipment that assists with tooling changes and other tasks. The goal of the chapter is to give you a general working knowledge of various types of tooling available to the robotic world and to give you an idea of what you will be working with. To that end, we will cover the following:

- What is EOAT?
- Types of tooling available
  - Grippers
  - Gripper force
  - Payload
- Other grippers
- Other types of EOAT
- Multiple tooling
- Positioning of EOAT

What Is EOAT?

End-of-Arm tooling (EOAT) consists of the devices, tools, equipment, grippers, and other tooling at the end of a manipulator that the robot uses to interact with and affect the world around it. The EOAT is the "doing" portion of the robot that works in conjunction with the positioning aspects of the system. EOATs attach to the end of the robot's wrist or minor axes and in many cases are a separate purchase from the robot. It is common to purchase a robot and then spend several hundred to several thousand dollars more, depending on the device, to get the tooling necessary to perform the desired task(s). There are companies, such as SCHUNK and SAS, which specialize in providing the tooling for robot systems. Some of these companies produce robots as well, but often tooling companies focus primarily on creating tooling for other companies' robots. Robotic tooling is a booming field, with new types and configurations constantly under development to meet the ever-changing needs of industry.

While industry has a wide range of tooling developed for its needs, it is not the only player in the market. Researchers, hobbyists, the military, and the entertainment industry also require their robotic systems to interact with the world; thus,

> **EOAT (End-of-Arm Tooling)**
> Tools, devices, equipment, etc. at the end of the robotic manipulator that are used to accomplish tasks

grippers
Tooling that applies some force in order to secure parts or objects for maneuvering

parallel grippers
Tooling with fingers that move in straight lines toward the center or outside of the part

angular grippers
Grippers with fingers that hinge or pivot on a point to move the tips of the fingers

they have a need for tooling. While much of the tooling for industry is derived from equipment already in use, the sky is the limit once you get outside the walls of production facilities. Many research robots have EOAT that resembles hands or other organic manipulators. Hollywood is putting cameras with pan, tilt, and gyroscopic stabilization on its robots. The military and police use robots to gather data about criminals, seek out explosive materials, and, in some cases, neutralize whatever threats may be present. In the world of hobbyists, tooling is only limited by human desire, imagination, and ingenuity. In fact, innovative tooling from these diverse fields of robotics often finds its way into the industry world once the technology has the bugs worked out.

When it comes to EOAT, it seems like there is always something stronger, lighter, and better coming out to solve a problem or replace current tooling in use. As you start to conduct your own research into robotics, I believe that you will discover just how diverse the world of tooling is. Entire books have been written on tooling, creating a snapshot in time of what is out there and how it works. In the next section, we will talk about some of the common types of tooling. By no means will we cover them all.

Types of Tooling Available

As previously mentioned, EOAT is a very diverse field that continues to expand as we find new applications and tasks for the robot. In Chapter 1, we discussed the history of robots and proved that robotic systems predated their industrial use. However, it was when industry bought into robotics and the cash started to flow that the age of the robot truly began to take shape. Because of this, a large variety of tooling for robots is industry related, more so than tooling designed strictly for other fields. We will start our exploration with the type of tooling that does predate the industrial robot, but has become a favorite in industry.

Grippers

When early roboticists were working on their creations, they gave them tooling that often had the shape and/or function of the human hand (see Figure 5-1). We refer to this tooling as a gripper: it is a type of EOAT that applies some force in order to secure parts or objects for maneuvering. This is a simple definition for a complex group of tooling that serves as a common workhorse for all fields of robotics! If you have any experience with robotics, have seen robots in movies, or have built any robotic arms at home, you have likely seen or worked with a gripper. C-3PO, ASIMO, and the Robonaut all have grippers for hands. The B-9 Environmental Control Robot from *Lost in Space* had a claw-type gripper, while Robby the Robot from *Forbidden Planet* had three-fingered grippers. As you look back over the timeline from Chapter 1, look at the pictures in this book, search the Internet for images, or pay attention to robots anywhere, you will find a wide array of grippers in use.

Figure 5-1 The Robonaut manipulates a cell phone with his hand-style grippers.

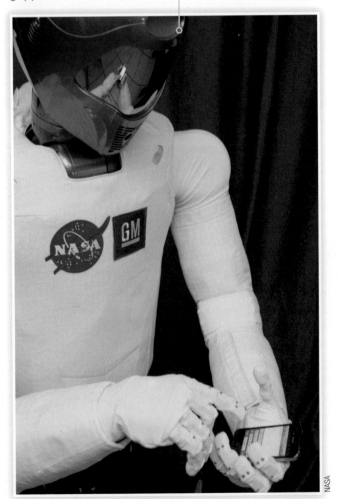

NASA

Image courtesy of SCHUNK

figure 5-2 This FANUC robot is using a set of SCHUNK parallel grippers to get the job done.

figure 5-3 This set of angular grippers has an extra joint on each finger to make them more like human fingers.

Image courtesy of SCHUNK

Most grippers currently use one of two motion types to actuate or open and close the tooling. Parallel grippers have fingers that move in straight lines toward the center or outside of the part to close and grip, or move in a straight line toward the outside of the tooling to open and release the part (see Figure 5-2). Angular grippers have fingers that hinge or pivot on a point to move the tips outward to release parts or inward to grip parts (see Figure 5-3). The gripper's fingers, sometimes called jaws, move to hold parts, while the tooling base attaches to the robot and holds the mechanisms to move the fingers.

Angular grippers work well for parts or items that are of consistent size; parallel grippers have the ability to grip a wider range of parts, depending on how far the fingers can travel. We power these motions by the standard power sources mentioned earlier, electricity, hydraulics, or pneumatics. The task at hand, the readily available power supply, and the environment all play a role in the choice of power source for gripping action. Hydraulic and pneumatic grippers are great for wet, dusty, or explosive environments; hydraulic power is generally reserved for heavy parts. Electrically driven grippers have grown in popularity with improvements in torque and reductions in size. The two movements of the gripper, opening and closing, can be powered by the sources as mentioned above, but to cut down on weight and cost, grippers can have one of the motions powered and the other controlled

fingers
The metal projections of the gripper that move to hold parts; often machined for the specific application that they are used for

jaws
Another name for tooling fingers

tooling base
The part of the tooling that attaches to the robot and holds the mechanisms for movement

odd-shaped parts
Parts with unique shapes and proportions that are not symmetrical in nature

by passive means such as spring pressure, mechanical tension, or even gravity. With passive methods, as soon as you remove the power source, the spring tension or some other force will cause the gripper to return to its unpowered state. When working with any type of gripper, take special care to ensure that no damage occurs to parts or people during shutdowns or unexpected power loss. The last thing we want is for a robot to start throwing 100-pound parts into populated areas!

Many of the grippers sold for industrial applications have generic fingers that are a solid piece of metal. This allows the industry or company buying the gripper to machine the fingers as needed to match the shape of their parts. Tooling suppliers will machine the fingers to customer specifications; however, this increases the cost to the gripper system, which is the reason why many companies prefer generic fingers. Companies that have internal machine shop operations sometimes order the gripper base alone, with no fingers, and then create their own. Whether modifying or creating fingers, end users must ensure that they have enough material left in the fingers to support the part and withstand the forces involved with moving that part as well as to provide enough friction to prevent the part from slipping out. We will dig deeper into these factors shortly.

The number of fingers that a gripper has gives you information about how it positions a part when grasped. Two-finger grippers give side-to-side centering, but nothing else. With these systems, some other means is necessary to provide centering along the length of long or odd-shaped parts. Three- and four-fingered grippers are great for circular or other standard geometry parts, as they will center the part in two directions consistently (see Figure 5-4). This explains their common use in industry. Certain four-fingered grippers and grippers having five or more fingers are great for working with **odd-shaped parts** or parts with unique shapes and proportions (see Figure 5-6). These grippers are

Image courtesy of SCHUNK

Image courtesy of SCHUNK

figure 5-4   This three-finger parallel gripper centers the part, thus making it easier to move and place precisely.

figure 5-5   Here is another example of human hand–style grippers.

Image courtesy of SCHUNK

Figure 5-6 Here is a four-finger system centering a tire rim. This gripper can handle two parts at a time and has a four-finger system to center the tire rims that it works with. Notice how the fingers are specialized to the work they perform and not just an angled piece of metal.

often highly specialized and more expensive than the common three-fingered or four-fingered varieties. Grippers that resemble human hands are less about how they center the part and more about manipulating parts as a person would (see Figure 5-5). These grippers tend to be versatile, but often come with a high price tag, low payload capacity, and complexity of control and operation. We usually use hand-style grippers for intricate tasks, such as part manipulation for assembly or the use of tools and devices designed for people, as opposed to raw-strength tasks or generic part lifting.

At this point, you may be asking yourself, "How do I know which gripper to use?" The answer depends on what you want it to do and the system you are using. In the hobby and entertainment world, the gripper is often used just to pick up or move things out of the way, so almost any simple gripper will do. The industrial world has more rigid requirements. Because of this, industry uses these filtering rules:

1. The tooling must be capable of holding, centering, and manipulating the range of parts that the robot works with.

2. There must be a way of sensing when a gripper has closed on the part or if it is empty. This can be internal or external to the gripper, depending on the system. Newer grippers can determine the amount of force exerted on the part as well, which is very beneficial for delicate materials (see Figure 5-7).

3. The weight of the gripper should be kept as light as possible, as this weight is deducted from the robot's total payload.

4. Proper safety features must be built in for the environment it works in and the parts it works with. (For instance, the gripper must be able to hold parts even when power is lost or have fingers made from nonsparking materials for flammable environments.)

The least expensive gripper that meets these four criteria is usually the one chosen. A fact that is often overlooked is that EOAT is a growing and evolving field, thus the solution today may not be the best option in as little as three to six months. As you look over these guidelines for industry, feel free to adapt and modify them as needed for any other applications with EOAT needs. If you end up working in industry with robots down the road, keep in mind that from time to time, you may want to reevaluate your tooling to ensure that you have the best

Figure 5-7   Too much force here would turn this light bulb into a pile of broken glass; this demonstrates the benefits of force control monitoring.

option for your application instead of settling for the best of the less-than-great tooling options chosen years ago.

Gripper Force

No matter how specialized or advanced a gripper, if it cannot create enough force to hold the parts during normal operation, it is going to be a source of frustration, downtime, and cost. In this section, we will look at some of the math involved in selecting robot tooling. For those of you entering the field of engineering, consider this the tip of the mathematical iceberg: those who design tooling for robots work with power source torque, force vectors, lever multiplication of force, and a multitude of other considerations during the process. Yes, there truly are real-world applications for the math you have spent years learning (or in some cases, avoiding).

While working to determine the force required from the gripper, we need to look at several factors: part size and shape, the direction that we are moving the part, friction, size of the gripper, and any safety factor we want to build in. The part size and shape are fairly straightforward, but you must remember that anytime the center of gravity is outside of the gripping area, the part will act like a lever and multiply any forces created by movement. The center of gravity of a part is where we consider the mass to be centered, and if we support the part at this point, we consider all the forces to be in balance or equilibrium. As we move the part, if the center of gravity is outside or out of line with the grippers, it will cause excessive force issues that we need to factor in. Friction is the force resisting the relative motion of two materials sliding against each other, and it acts as a resistance to slippage of the part. The greater the resistance or friction, the more force it takes for the part to slide in the gripper. However, keep in mind that once the part starts to move, the friction actually decreases, thus making any slipping of the part very dangerous. The larger the gripper, the larger the applied force area, and the greater the surface is that interacts with the part on a friction basis. The larger the size of the gripper contact area with the part, the greater the frictional forces to prevent slippage. Safety factor is the margin of error we build into the process. In other words, we must account for variance in part weight, length, errors in math, and other unaccounted for forces that we have not accounted for. The larger the safety factor, the more room there is for error in the calculations; however, a large safety factor often equates to a higher cost for the tooling. In the case of fragile parts, too much force can be just as bad, if not worse, than not enough force. Let us look at a couple of examples to see how this all fits together.

center of gravity
Where we consider the mass to be centered and all forces are in equilibrium

friction
The force resisting the relative motion of two materials sliding against each other

safety factor
The margin of error we build into a process or system to ensure that accidents or dangerous situations do not occur

Example 1

For this example, we will use the application pictured in Figure 5-8 with the following specifications:

- The part weighs 1 pound.
- The grippers jaws are parallel.
- The part is gripped 2.5 inches from the center of gravity.
- The gripping surface is 0.75 inches long.
- The part is 0.375 inches wide where being gripped.
- The part is being lifted with a maximum acceleration of 2.5 Gs, including normal gravitational force.

Figure 5-8   Here is a two-finger gripper working with a part that is fairly long and has a high probability of a misaligned center of gravity.

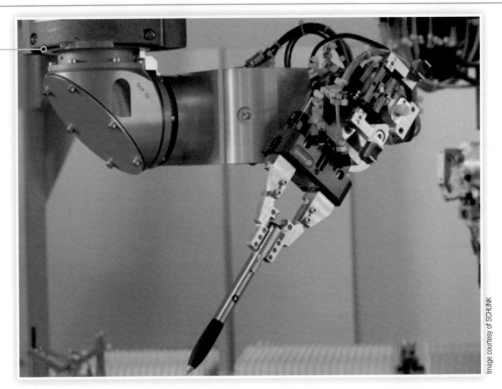

Image courtesy of SCHUNK

- The coefficient of friction between the gripper and the pen is 0.80.

- The engineer in charge wants a safety factor of 2.

Because the part is being lifted/moved sideways, we must take the torque of this action into account, based on the differences in the center of gravity. The overall torque (T) on the jaws equals weight (W) times distance (d). The weight equals mass (m) times gravity (G), or in this case, 1 pound. The distance (d) equals 2.5 inches. The width of the fingers' grip (b) equals 0.75 inches. The width of the part (p) in the jaws equals 0.375 inches.

$$T = m \times G \times d$$
T = torque
m = mass
G = gravity, which is 9.8 m/s^2 or 32.2 ft/s^2
d = distance
$$W = m \times G$$
W = weight
m = mass
G = gravity, which is 9.8 m/s^2 or 32.2 ft/s^2

This means we can substitute the weight of the part for the m $\times$ G portion of the formula as needed.

Because the gripper has a center of force, there are two forces involved in the torque, the force of the part below the center of force and that exerted by the part above the center of force. This changes the torque equation in this way:

$$T = F_2(b/2) + F_1(b/2)$$
$$T = (b/2)(F_1 + F_2)$$

With algebraic substitution for T or torque, we can get to this:

$F_1 + F_2 = 2(m \times G \times d)/b$

F_1 = force below the gripper

F_2 = force above the gripper

m = mass

G = gravity, which is 9.8 m/s² or 32.2 ft/s²

d = distance

b = width of the gripping contact

When we add in p, or the width of the part in the fingers, we get the following (because of force vectoring):

$$F_1 + F_2 = \frac{2(m \times G \times d)}{\sqrt{(b^2 + p^2)}}$$

p = part width in the gripper

By Newton's first law, or the law of inertia, $F_1 = F_2$; and since the gripper provides both of those forces, $F = F_1 + F_2$

Thus,

$$F = \frac{2(m \times G \times d)}{\sqrt{(b^2 + p^2)}}$$

$$F = \frac{2(1 \times 2.5)}{\sqrt{(0.75^2 + 0.375^2)}}$$

$$F = 5.96285 \text{ pounds of force}$$

Next, we have to factor in the acceleration. We do so by multiplying the force times 2.5 Gs of acceleration for 14.907125 pounds of force. We then factor in friction by dividing the new force answer of 14.907125 by 0.80, the coefficient of friction, giving us 18.63391 pounds of force. Last, but not least, we need to factor in the safety factor we specified earlier, so we take the force with friction factored in and multiply by 2, giving us a total need for 37.26781 pounds of force for our gripper. If we just look at the pen in our example, most would never dream that the specification for the gripper for this system would need to be nearly 40 pounds of force exerted by *each* of the jaws!

Example 2

For the second example, we will use pen setup from example 1, but only move in the vertical direction, with no side swing. We will use the same basic set of data:

- The part weighs 1 pound.

- The grippers jaws are parallel.

- The part is gripped 2.5 inches from the center of gravity.

- The gripping surface is 0.75 inches long.

- The part is 0.375 inches wide where being gripped.

- The part is being lifted with a maximum acceleration of 2.5 Gs, including normal gravitational force.
- The coefficient of friction between the gripper and the pen is 0.80.
- The engineer in charge wants a safety factor of 2.

Because this will be a vertical move, we do not need to worry about the part's center of gravity: it will be in line with the motion. We also do not need to know how much of the gripper is in contact with the part, nor the width of the part, so we can ignore the part center of gravity distance of 2.5 inches, the gripping surface of 0.75 inches, and the part thickness of 0.375 inches.

First, we will determine the force needed for the weight of the part under acceleration, which is:

Fr = W × G
Fr = required force
W = weight of part
G = gravitational force, including acceleration
Fr = 1 pound × 2.5 Gs
Fr = 2.5 pounds of force

normal force

The force that an object pushes back with when acted on by a force and that is necessary to prevent damage to the object

Normal force is the force that an object pushes back with when it is acted on by a force. If the object cannot generate enough normal force, it will receive damage in some manner as the force acting on it overcomes the structure of the item. Friction, or the resistance of two items sliding past or against one another, also figures into the normal force. To relate friction and normal force we use the following equation:

F = μN or N = F/μ
F = friction force
μ = coefficient of friction
N = normal force

Previously, we figured that the part exerts 2.5 pounds of force during the move. Because two jaws hold it, we divide this by 2 to get the requirement of each jaw:

2.5/2 = 1.25 pounds for each jaw.

We use this to figure out the normal force required by each jaw:

N = F/μ = 1.2/0.80 = 1.5 pounds of force by each jaw on the pen.

Of course, we cannot forget our old friend, the safety factor. When we multiply the 1.5 by the safety factor of 2, we find that each jaw needs 3 pounds of force for the vertical move.

In the first example, we figured that we needed 37-plus pounds or over 12 times the amount of force needed for a vertical move! This really illustrates the importance of knowing your part, the movements involved in the process, and the forces involved with nonlinear movements. Imagine what would happen if we had a gripper that could only generate 3 or 4 pounds of force (which is fine for vertical moves like example 2) grip the pen during a swinging motion (like example 1). From our calculations above, we can almost guarantee that at some point in the swinging motion the pen will slip from the grippers and become a projectile! This is why it is important to do the math *before* buying or specifying a gripper for the robot.

Payload

While we are on the subject of math, this is a good time to talk about payload and the effect that tooling has on it. Payload is a specification of a robotic system that informs the user how much the robot can safely move. Payload is often specified in kilograms, as metric is the standard measurement system of all but a few countries (the United States being one of the holdouts). If a robot has a payload capacity of 25 kg, then the maximum weight it can safely move, in addition to the weight of the robot, is 25 kg. Anything we add to the robot, such as EOAT, reduces the amount of force that we have to move parts. Therefore, the equation for payload looks like this:

$$Ap = P - Wt$$

Ap = available payload
P = specified payload capacity for robot
Wt = weight of the EOAT and any added peripheral systems or equipment attached to the robot besides the core system

Example 3

For this example, we will use the following information:

- Payload for the robot is 25 kg.

- Weight of the EOAT is 5 kg.

- An additional sensor for the End-of-Arm tooling is required and it weighs 1.3 kg.

  First, we need to add up the total weight added to the robot:

  $$5 + 1.3 = 6.3 \text{ kg}$$

  Next, we plug the information into our base formula:

  $$Ap = P - Wt = 25 - 6.3 = 18.7 \text{ kg of available payload}$$

Of course, this does not take into account any calculations for acceleration; however, as a rule, the manufacturer of the robot will have made those calculations in advance. If you remember the previous examples, you know that how we pick up the part and how we move it does affect the system. Therefore, while you may be able to pick up and slowly move a part that is heavier than the stated payload, any swinging or high-speed motions have a high probability of causing problems with the system. At best, you will get an alarm that one of the axes is overloaded; though system damage or the part coming free is not out of the question. In addition, working a robot close to or at its rated payload will wear out the system faster than working it at 75 percent or less of the rated payload. This might be a way to justify the purchase of a larger robot, as it could save the company in the long run.

Other Grippers

In the industrial world of the robot, there are a couple types of tooling that are lumped in with the gripper family that you may not expect, as they bear no resemblance in shape or movement to the human hand or fingered grippers. Because industry works with materials that are not always easy to grasp with a closing motion, they have come up with grippers that use electromagnets to attract ferrous metal parts balloon types that inflate or deflate to grip as well as suction devices that use a <mark>vacuum</mark> (any pressure that is less than the force exerted by the air around an object) to secure parts. Remember, if the primary design is to pick up and move items, we generally refer to it as a gripper.

We use magnetic grippers in industries that work with metals. They come in a wide range of sizes and shapes, though the basic operation is the same. When a coil of copper wire has current running through it, it generates a magnetic field. If we coil this wire around or inside a metallic frame and pass current through the wire, we create an electromagnet that attracts <mark>ferrous metals</mark>, or metals that contain iron. Once the electricity stops flowing, the magnetic force dissipates and the gripper releases the ferrous metals it attracted. We use magnetic grippers in steel mills, junk yards, sheet metal manufacturing facilities, and other places where iron-based metals need to be moved from one point to another.

The suction cup or <mark>vacuum gripper</mark> works by creating a pressure that is less than atmospheric or an area of low pressure. Because of this lower pressure, the pressure of surrounding atmosphere will exert an upward force on the object and hold it against the suction cup. The process is similar to what happens with the wing of an aircraft: the low pressure over the wing creates lift under the wing and keeps the plane in the air. Suction grippers are great for heavy and light objects alike and are common in the glass, food, and beverage industries. Because there is no closing mechanism, there is no crushing motion that might damage the items moved (see Figure 5-9). Vacuum grippers are useful for moving large, flat items like sheets of metal or glass that would be difficult to pick up by other means. Oftentimes, when used in this way, the tooling will use multiple vacuum grippers (see Figure 5-10) to create a lifting grid. Many of these types of grids have individual control of the vacuum grippers, allowing the programmer to turn on only the grippers needed, saving on both operating cost and maintenance.

Balloon-type grippers work by inflation or deflation to grip the part. These grippers once worked by placing the deflated balloon or flexible bladder inside of some portion of the part and then inflating it with air. The balloon would take the shape of the part and exert force, allowing the system to pick up the part for manipulation. A recent advancement in this technology has been to fill the balloon or flexible bladder with coffee grounds or a similar substance. When used in this way, the air-filled balloon allows the material inside free movement as the system presses the balloon against the part. While holding the balloon against the part, the system vacuums the air out, locking the grounds together in a tight matrix. The effect is a gripper that holds parts securely in much the same

Vacuum
Any pressure that is less than the surrounding atmospheric pressure at that location

ferrous metals
Metals that contain iron

vacuum gripper
A device that works by creating a pressure that is less than the atmospheric pressure of 14.7 psi and is often used to pick up large and flat or delicate items

Figure 5-9 This vacuum gripper can be used singularly or as part of a grid for moving larger/heavier objects.

Image courtesy of SCHUNK

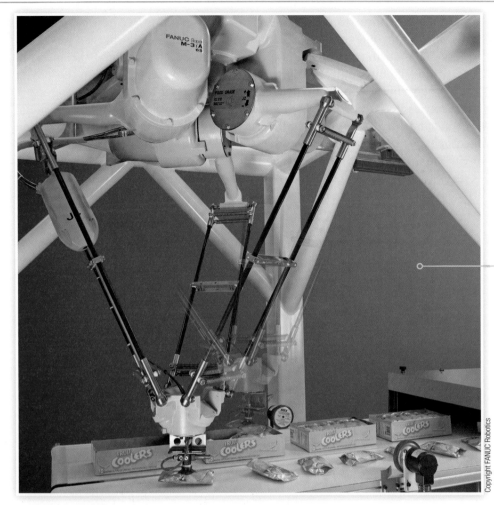

figure 5-10   A vacuum gripper at work picking up packages and putting them in boxes.

way that a human hand does in terms of rigidity and flexibility. The newer filled balloon system has proven effective for both hard and soft objects, earning it a special place in the world of grippers.

Pin and mandrel grippers are similar to the balloon- or bladder-type gripper. These grippers work by inflating a bladder or bellows with air to create pressure against the part. The pin gripper fits around a projection on the part and then inflates the bladder to create pressure against the part. The mandrel gripper works in the same fashion, only it fits on the inside of the part and presses outward against the walls of the part to create the gripping force. With the advancements in gripper technology, these grippers seem to be on the wane; they are limited in the type of parts they can work with and have a limited range of movement for part variation.

Other Types of EOAT

While grippers are one of the more common forms of tooling used with robots, they are by no means the only tooling you will encounter. If there is a specialized tool that we use for production, there is a high probability that someone makes a version of it for the robot. (As mentioned earlier, we will not cover all existing tooling but will review a few of the common nongripper types.)

welding guns
Tube-like tools used to direct the charged wire of welding operations as well as the shield gas used to prevent weld oxidation

MIG welders
A machine that uses wire feed through the system and high-current electricity to join two pieces of metal

laser welders
Systems that use intense beams of light to create the high temperatures needed to join metal

Welding guns are a popular tooling for robots in industry. These are the tube-like tools that people or robots use to direct the electrically charged wire used to fuse metal during welding operations (see Figures 5-11 and 5-12). This tooling requires its own power supply, in addition to or separate from the power requirements of the robot, and uses either MIG or laser technology. MIG welders use electrically charged metal wire, fed through the welding gun, to fuse metal together and provide shielding for the wire from oxygen by an inert gas like carbon dioxide (CO_2). The current causes high levels of heat at the point of contact that melts the wire and both pieces of metal to create a new, solid metal connection. A laser welder uses intense beams of light to create the high temperatures needed to melt and fuse two pieces of metal together, without the need for the traditional

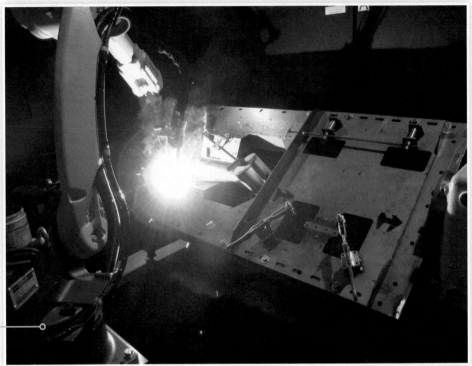

Figure 5-11   This picture shows the Panasonic welding robot in action. Notice how bright the welding process is. You should always wear proper darkened eye protection when watching or working around welders.

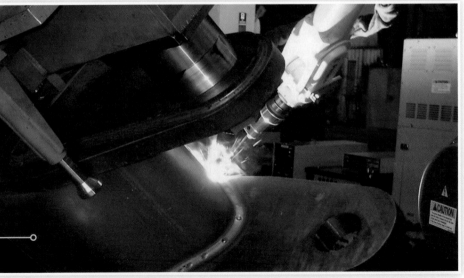

Figure 5-12   This image shows both the welding process and the fused metal left in its wake between the curved and flat pieces of metal. Notice how uniform and even the weld is.

Image Courtesy of Miller Welding Automation

Image Courtesy of Miller Welding Automation

figure 5-13   Spray-painting, as seen here, is one of the first areas in which robots gained popularity in industry and remains a common application to this day.

welding gun. While they do not add metal to the joint, laser welders need specialized power supplies and other equipment to generate the intense beam of light and may consume a gas like carbon dioxide in the process. The benefit to these systems is the fact that they are consistently repeatable, fast, reduce the required raw materials, and can reach places difficult for the human hand.

Sprayers are another tooling common in industry (see Figures 5-13 and 5-14). Originally, these systems applied liquid paint in the auto industry, but the technology has broadened considerably. Today, spraying applications apply both liquid and powder-based paints, adhesives, lacquers, and any other sprayable substance desired for use in industry. Because of the robot's precision, these systems save thousands of dollars each year in wasted raw material while providing the high-quality bonds and coatings that customers demand. The application of these materials is a task that can be both dull and dangerous in many cases, making it a perfect fit for robotic systems.

Add a motor to the EOAT and you now have a drill, threading tool, or milling setup ready to create, modify, or finish parts. In essence, this type of tooling turns the robot into a very flexible industrial machine. I have seen facilities use this type of setup to create control panels for boats, allowing for quick changes of programming and a smaller footprint than most industrial CNC machines. With the wide range of bits and accessories for this type of tooling, there is very little the robot cannot do! This is just another example of the flexibility that robots bring to the industrial world.

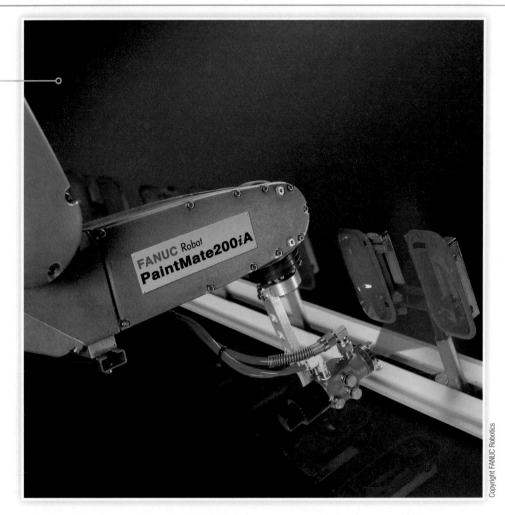

Figure 5-14 There is very little wasted paint in this spraying application.

Inspection is another field where robots are commonly found. The tooling for inspection may include specialized sensors, cameras, infrared systems, or any other sensing device (see Figure 5-15). The idea behind this type of tooling is to use the robot for either quality control or to help ensure the safety of humans. A company I once worked for used a small robotic arm with a specialized gas sensor to check the welds of the product by detecting leaks of the pressurized nitrogen gas inside. If the robot detected the gas, then it would kick the part off to the side for an operator to inspect more closely. This is but one example of the thousands of uses for inspection tooling.

As you explore the world of robotics, you will find variations and adaptations of the tooling we have discussed in this section of the chapter as well as completely new ways for the robot to interact with its environment. When this happens, I encourage you to take a few moments to figure out how it works, what its limitations are, and how it could help you in your chosen field of robotics. The more you understand about tooling, the better your chances of picking the right system for whatever job you want your robot to perform.

Multiple Tooling

What do we do when we need to use more than one type of tooling on a robot? The simple answer is that we change it out or use multiple tools at once. There are several options. To determine which one is best, you first must determine how many different tools you need and how often you need to change the tooling. Let us look at the pros and cons of these options.

Figure 5-15 Here is a robotic system using special lighting and a camera to check the quality and features of a part.

A low-tech method to use multiple tooling involves having the operator or appropriate party physically remove the current tooling and put on the new tooling. The prerequisite here is that the new tooling must align with the mounting plate of the robot and work with the robot's systems. If this is a first-time use of the tooling, some modifications to the tooling, the robot, or both may be required. This method is suitable for systems that are rarely changed or for adapting the robot to a new task that it will perform for a long time to come. It is also common to use this method to replace tooling damaged by normal wear or damage due to unprogrammed contact. The downside to this method is the robot will likely be out of production for a fair amount of time; further, it requires someone to perform all the work of changing the tooling plus making sure everything is set up properly. This may take hours, or in severe cases, days of work.

Another low-tech option is to mount multiple tools to the robot simultaneously. This prevents the need to change tooling out and provides for increased functionality; however, it comes at a high price. First, this reduces the payload of the robot by the weight of two or more tools instead of just one, which decreases the force left to move parts and other materials. Second, this method often requires specialized tooling bases or tooling systems to allow for the multiple functionality. Anytime the tooling is specialized, the price increases accordingly. In situations where the robot does not need to move parts and performs the same operations repeatedly, this option can be worth the cost. This is why you will find these setups in industry.

The next step up is to use quick-change adaptors (see Figure 5-16). The alignment pins provide consistent placement for the tooling, and the coupling system provides for positive locking while making it easy to detach the current tooling and swap in new. These setups can transfer various power sources as well as make communication connections for any sensors that the tooling may have. The system works by attaching one plate to the robot, usually at the end of the wrist; the other goes on the base of the tooling. When it is time to change the tooling, a release mechanism allows the two plates to separate. As long as the new tooling has a proper adaptor plate, all you have to do is attach the new tool and enter any offsets needed to get the system running once more. These units are great for systems that require frequent changes, such as daily or even hourly. It is also a crucial part of the automatic systems that we will discuss next. The down side is the cost of the connection plates and the initial setup for attaching the base plate to the robot and the quick-change plate to new tooling.

In an automatic system (see Figure 5-17), the robot requires multiple tooling to perform each cycle of its operation. This type of operation may have the robot picking up parts from a conveyor or bin, loading them into a machine, taking finished parts out, deburring or threading holes, and then placing the parts on another conveyor. If the robot only needs two or three different tooling options, it may have them mounted on the robot continuously; however, remember that this cuts into the maximum payload of the system. More often, the robot will use quick-change adaptors (Figure 5-16) and simply change them out as needed, like the system pictured in Figure 5-17. In the picture, the tooling is stored on a rotary wheel and rotated into position as needed: an empty spot takes the current tool after it is detached, rotates in the new tooling needed, and holds it in position while a positive lock with the robot is completed (see Figure 5-18). Other systems may store the tools in a stationary location and let the robot do all the positional work of dropping off the current tooling and then picking up the new. The only downside to this type of process is the fact that in each cycle, a certain amount of time is lost to changing tools. However, when compared to the flexibility it gives the system, this is usually a small price to pay. Misalignments of these systems can lead to crashes, where part of the robot impacts something solid or other alarm conditions, so it is wise to add some type of sensor system to ensure proper tooling position before detachment and attachment.

As you can see, there are multiple options to handle the need for multiple tools. Often, more than one of the methods mentioned will fit the tooling change need, making it a case of which one has the most benefit for the least cost. When it comes rapid changing of the tools during the production cycle, some form of automatic tool change is a necessity since shutting the machine down every few minutes for a person to change the tool is just not economical. For the rest of the conditions out there, weigh the pros, cons, and cost of each method to determine which one best fits your specific need.

Figure 5-16 A robot base plate (top) and tool adaptor plate (bottom) make up a quick-change adaptor set.

Image courtesy of SCHUNK

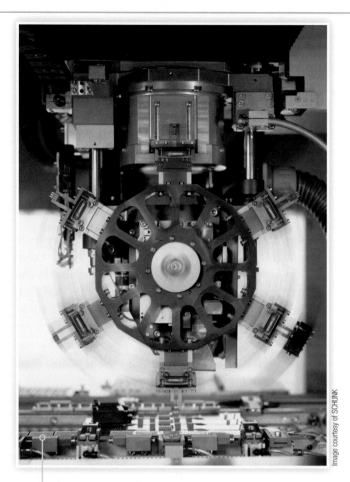

Figure 5-17   This tool-changer system removes and replaces tooling for the robot as needed.

Image courtesy of SCHUNK

Figure 5-18   With this quick-change tooling, a coupler attached to the robot connects around the specially machined base of the tooling. This type of tooling would be perfect for a system where the robot does all the moving and the tooling remains static.

Image courtesy of SCHUNK

Positioning of EOAT

The best tooling in the world becomes useless weight if you cannot get it into position to perform its tasks. Sometimes we run into trouble due to the fact we cannot move the robot into the position needed, while other times it has to do with inconsistency in the parts. Whatever the case, we need to make sure that the robot can do its job without the operator making adjustments every few minutes or we might as well get rid of the robot altogether. This section will focus on some of the ways we get around the problem of positioning tooling.

The common industrial robot has six or seven degrees of freedom (DOF). Have you wondered why? The reason is the three major and three minor axes of movement in a six axes robot allows for a wide range of positioning options and often is all we need for industry. The three minor axes are there to orientate the tooling into the proper position for the task it must perform. The only time we tend to have problems with getting the tooling into position, from the robot perspective, is when we have only one or two minor axes or parts that are not consistent in position or dimension. When the problem is the parts, we have several options, but if the robot simply does not have the DOF to reach the position we only have two options, adapt the process to the robot or change the robot. Let us look at a few options to help the robot out with inconsistent parts.

Remote Center Compliance (RCC) is a simple way for tooling to respond to parts that are not always in the same position. RCC devices allow the tooling to shift a little from the center position without causing the robot to alarm out or exerting excessive force on the tooling. By using springs or materials that can flex (see Figures 5-19 and 5-20), systems achieve passive RCC. Often passive RCCs will have the addition of a shear plate that breaks loose in the case of excessive side force to preserve tooling and prevent damage. RCCs are good for reaming, taping, or other such operations requiring some flexibility. In these instances, the tooling comes in, makes contact with the part, and flexes in the required direction to complete the programmed task.

Many of the modern RCC devices are active RCCs, meaning that they have sensors inside the unit to detect how much side shift is occurring. While these RCCs are more complex, they do alleviate the need for a shear plate inside the setup. If the tooling is being forced too far out of position, the sensor detects this, stops the robot, and sends an error to the teach pendant screen. Some of these systems can also detect the amount of torque that the tooling is using to perform its tasks, giving valuable feedback to the operator. With proper programming, the robot can alert the operator to dull tools, holes that have not threaded properly, or other situations that require operator intervention.

A modern solution to the problem of part alignment is the use of vision systems. With vision systems, the robot uses a camera to take a picture of the part and then compares this picture to sorting criteria predefined by the programmer (see Figure 5-21). Once the system has filtered the visual information, it runs a subroutine that allows it to offset the tooling in relation to the new part position. In the last 10 years, vision technology has truly advanced and is

> **Remote Center Compliance (RCC)**
> A simple way for tooling to respond to parts that are not always in the same position by mechanical means of flexing to adapt

Image courtesy of SCHUNK

Figure 5-19 If you look closely, you can see the springs between the tooling base and where it attaches into the tooling holder on the robot, allowing for some side-to-side motion of the tooling without incident.

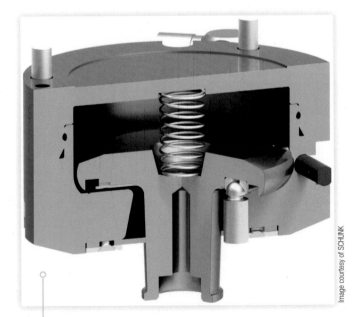

Image courtesy of SCHUNK

Figure 5-20 This cutaway shows a type of RCC unit that allows the tooling room to compress and move while engaging the part. It centers up once the pressure is released via the pin visible on the right side interior.

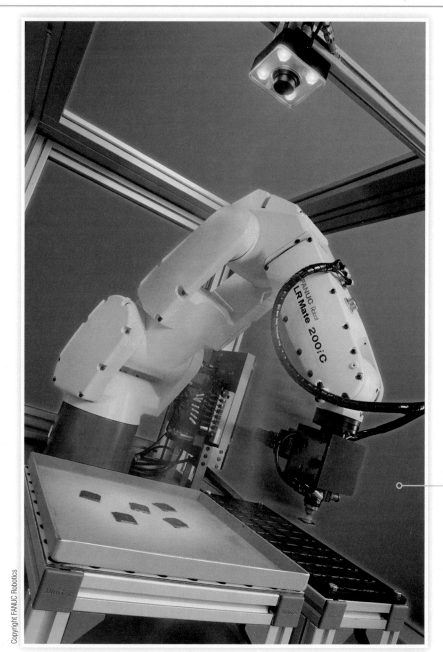

Copyright FANUC Robotics

Figure 5-21   This vision system identifies parts and their location so that the robot can pick them up with a vacuum gripper and place them in the tray next to the robot.

becoming the go-to standard for dealing with part positioning variance. The use of vision systems increases the overall cost and complexity of the robot, but the versatility provided far outweighs the cost. Not only can we use the vision system to adjust the robots position, we can sort items by color, pick specific pieces out a pile of parts, perform quality checks, and carry out many other functions that require visual information.

Just as we did not cover every type of tooling out there, we have only scratched the surface of ways to adjust for changes in part position or misalignments. While vision systems and advanced programming are the common solution of the modern robotic world, there are many mechanical systems out there performing the same functions with consistently reliable results. There is a good chance that you may find yourself working with systems in the field that use some combination of the mechanical, sensor, and vision to meet the specific challenges of the tasks your robots perform.

review

By no means have we covered all the various tooling and options you will encounter in the world of robotics, but you should have a better understanding of the diversity of the field. While there seems to be a new system or process coming out for robotic tooling weekly or monthly, the truth is there are a lot of old, tried, and true systems out there that will be around for many years to come. I encourage you to take a bit of time and research some of the advances in robotic tooling so you can stay current in the field. Here is a quick recap of what we covered in this chapter:

- **What is EOAT?** We discussed what tooling is and why it is important to the world of the robot.

- **Types of tooling available.** This section talked about grippers, welders, spraying guns, suction cups, and other various tooling for robots as well as looking at some of the math involved with tooling.

- **Multiple tooling.** This section of the chapter was about the options for attaching tooling and what situations these options best serve.

- **Positioning of EOAT.** Here we wrapped up our discussion on tooling by discussing how to deal with misalignments of the part and ways we can shift the tooling to fit the task at hand.

formulas

Payload

$Ap = P - Wt$

Ap = available payload

P = specified payload capacity for robots

Wt = weight of the EOAT and any added peripheral systems

Vertical grasp formula

$Fr = W \times G$

Fr = required force

W = weight of part

G = gravitational force, including acceleration

Normal force

$F = \mu N$ or $N = F/\mu$

F = friction Force

μ = coefficient of friction

N = normal force

Torque

$T = W \times d$

T = torque

W = weight

d = distance at which weight is applied

Weight

$W = m \times G$

W = weight

m = mass of item

G = gravity

Required force for horizontal grasp

$F = 2(m \times G \times d)/\sqrt{(b^2 + p^2)}$

F = force required

m = mass

G = gravity ($m \times G = W$)

d = distance from Center of Gravity, CG, to Center of Force, CF

b = width of jaws contact

p = thickness of part in gripper

key terms

Angular grippers	Grippers	Parallel grippers	Vacuum
Center of gravity	Jaws	Payload	Vacuum grippers
End-of-Arm Tooling (EOAT)	Laser welder	Remote Center Compliance	Welding gun
Ferrous metals	MIG welder	(RCC)	
Fingers	Normal force	Safety factor	
Friction	Odd-shaped parts	Tooling base	

review questions

1. Where do we attach the End-of-Arm tooling?

2. What is the primary tooling type that predates the industrial robot?

3. What is the difference between angular grippers and parallel grippers?

4. If you need a gripper for a wide range of part sizes, what would be better, parallel or angular and why?

5. What types of environments are hydraulic or pneumatic grippers well suited to?

6. What precaution should one take regardless of the type of gripper or power source for the gripper?

7. To center a part in two directions at once, what is the recommended number of fingers on a gripper?

8. What is the purpose behind human hand–style grippers?

9. What are the industrial rules for determining which gripper to use?

10. When determining the force required from a gripper, what factors do we look at?

11. From the two examples in the chapter, what can you say about the effect of moving the part where the center of gravity is out of line with the motion?

12. How does tooling affect the payload of a robot?

13. How does an electromagnet work?

14. How does a vacuum gripper lift parts?

15. What are the two drawbacks to mounting multiple tools on the robot at the same time?

16. What are the benefits of quick-change adaptors?

17. Describe the operation of an automatic tool change system.

18. What is the difference between active and passive RCC units?

19. What is taking the place of the RCC system in modern robots, and how does it work?

6

Sensors and Vision

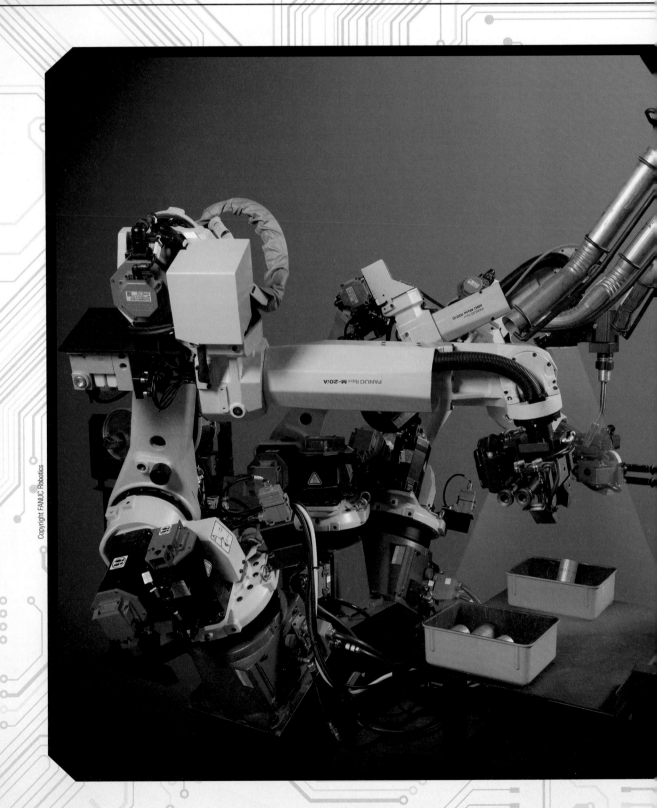

overview

Up to this point, we have spent a lot of time looking at how the robot has evolved, power sources, classification, components, operation, and safety, but we have yet to truly delve into robotic sensors. Without sensors, robots have no information about the world around them and thus no real way to react to changes in their environment. Imagine for a moment that you are in the middle of a conversation with a friend when suddenly you go deaf, blind, and completely numb. You cannot see or hear your friend, you cannot feel where your body is or anything touching you, you have suddenly stopped receiving any information about the world around you. You could send the command to your arm to move, but you would have no idea if it moved as you wanted or hit an object along the way. You could try to speak, but you would have no idea if words or gibberish came out. This is what a robot is like without sensors. We put sensors on robots so they can determine how they are moving, in which direction, at what speed, when they have reached the specific point, and any other data that is pertinent to their operation. In fact, without the advancements in sensors creating the options we have today, it would be near impossible for complex systems to perform the tasks that make them cutting edge. In this chapter, we will examine the basic operation of common sensors in the following categories:

- Limit switches
- Proximity switches
- Tactile and impact
- Position
- Sound
- Vision systems

Limit Switches

The limit switch is a straightforward basic device; because of this, it has been helping the robot to gather information about the world around it for many years now. From the simple hobby robot that uses the micro limit switch for impact detection to the industrial robot that uses the limit switch to ensure that all the safety doors are closed before operation—these devices help the robot to understand more about the world around it. Not all robotic systems use limit switches, especially internally, but they are very useful for providing information about the state or position of objects that concern the robot.

A limit switch is a device activated by contact with an object that changes the state of its contacts when the object exerts a certain amount of force. The limit switch consists of a body, which houses and protects the electrical contacts, and some type of actuator, which moves with physical contact (see Figure 6-1). Actuators can use pressure or rotational motion to change the contacts of the switch. Many of the larger units have their actuators attached as a separate unit on the limit switch body, known as a head unit, allowing for options in actuator orientation and allowing one limit switch base to serve many functions (see Figure 6-2).

limit switch
A device activated by contact with an object that changes the state of its contacts when the object exerts a certain amount of force

Figure 6-1 This Cutler-Hammer limit switch has the arm mounted for contact on the right side of the unit.

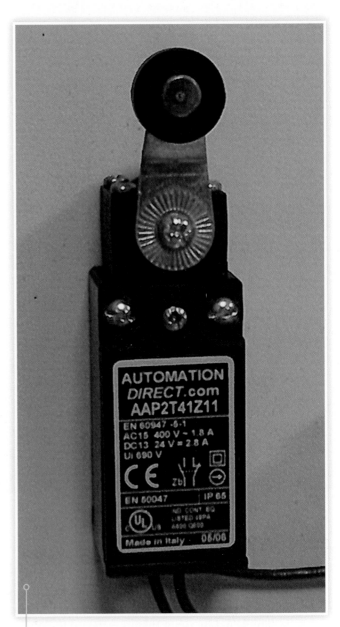

Figure 6-2 Here is an Automation Direct limit switch with a much smaller arm and a wheel on the arm to help reduce the connection impact. This type of prox would work well with a door or other movable guarding.

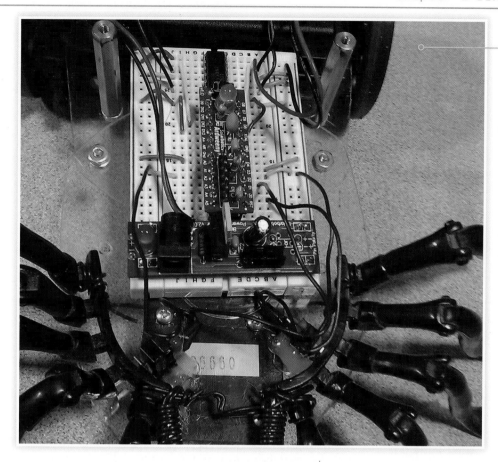

Figure 6-3   With the covers off, you can see the two micro limit switches mounted just behind the legs of my Spiderbot. The silver piece of metal is the actuator arm. When the legs hit an obstacle, they compress and activate the micro limit switch for that side, sending a signal to the Arduino controller.

When changing out the head unit on a limit switch, make sure that the new switch unit matches the limit switch in question as well as the task. Another great feature of this type of limit switch is the ability to rotate the head unit of the switch by removing the appropriate screws and then turning the actuator as need, thus adapting the switch to the various situations in the field without needing a separate switch to handle each mounting situation. The range of motions and force needed to activate the switch can range from slight movements requiring 0.2 newton of force (see Figure 6-3), to large movements requiring about 20 newtons of force. (A newton is a common measurement of force and the metric equivalent of the English pound.)

When selecting a limit switch, you will need to know how much amperage the contacts must handle, how fast the switch needs to respond, and how much force the object contacting the limit switch can generate. Standard limit switches and their smaller cousins, the micro limit switch, can usually handle up to 10 or 20 amps of current, while the smallest of the family, the subminiature micro switch, is good for 1 to 7 amps on average. Each switch takes a certain amount of time to make (activate) or break (deactivate), so make sure that the switch can respond fast enough for the conditions under which you will use it. For instance, if you use it to help the robot keep track of parts on a conveyor line, make sure that there is enough space and time between parts for the limit switch to reset. Some limit switches require a fair amount of force to activate, making them a poor choice for light parts. You must also consider the application. If we are using a micro limit switch to prevent the robot from going past a certain position and prevent damage, the best option is likely a small-movement, low-force limit switch that can react quickly and shut down the robot before any mechanical damage occurs.

proximity switch
A solid-state device that uses light, magnetic fields, or electrostatic fields to detect various items without the need for physical contact

When used properly, limit switches are a valuable information-gathering tool for robots and other applications. They were one of the primary sensing systems in early robotics and are still a staple of the modern robot's sensors. Today, we use them to prevent the robot from traveling too far, monitor the doors on safety cages, confirm raw parts ready for loading, verify that the machine the robot works with is in position, and to answer other yes/no type questions. The micro and sub-miniature micro limit switches are popular in the hobby robotics community for a variety of functions, as they are simple to use, rugged, and reliable. While there may not be as many limit switches today as there once were in industry, they are still a vital part of many guarding and sensing systems that work either directly or indirectly with the robot.

Proximity Switches

A proximity switch is a solid-state device that uses light, magnetic fields, or electrostatic fields to detect various items without the need for physical contact. A solid-state device is a unit made up of a solid piece of material that manipulates the flow of electrons without any moving parts. Because there are no moving parts, solid-state devices are capable of performing millions of operation without wearing out. However, the downside is that most solid-state devices never fully stop the flow of electrons—though often it is only a very small amount that leaks through. They are also vulnerable to electromagnetic pulses (EMP) and other strong magnetic fields. Because the proximity switch (or prox) does not touch the object that it is sensing and has no moving parts due to its solid-state components, these switches can work for years with only the barest of preventative maintenance. Proximity switches send low-amperage signals to controllers and often only have one or two signal outputs, but the outputs come in a variety of voltages. A common range for DC prox switches is 10 to 30 VDC, and a common range for AC is 20 to 264 VAC. Some prox switches can work on either AC or DC voltage; a common range for these is 24 to 240 V. The main reasons that prox switches fail are because of physical damage to the switch or voltage that is greater than the prox's voltage rating. During my time in industry, I have seen prox switches damaged to the point where you could see the internal components that still worked to some extent, which is proof of the tenacity of this sensor. (See Food for Thought 6-1 for a story about one tough Balluff prox.)

Each of the three main types of prox switch has a set of tasks and objects that it works best with. The first group we will explore is the inductive proximity switch, which uses an oscillating magnetic field to detect ferrous metal items (see Figure 6-5). By oscillating, we mean a field that is going on and off. In this case, the field creates a wave that looks just like the sine wave of AC voltage. Unlike AC power, inductive prox switches

solid-state device
A device made up of a solid piece of material that manipulates the flow of electrons without any moving parts

inductive proximity switch
A switch that uses an oscillating magnetic field to detect ferrous metal items and send out signals accordingly

Figure 6-4 This drawing represents the magnetic field generated by an inductive prox that interacts with magnetic metals.

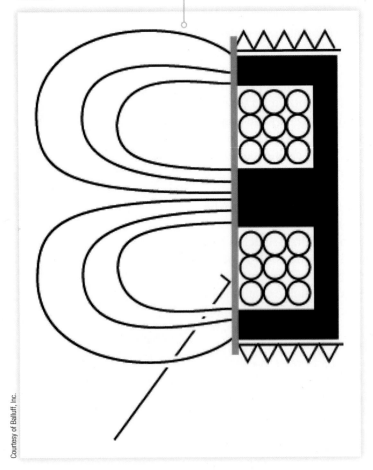

Courtesy of Balluff, Inc.

food for thought

6-1 Balluff Prox Toughness

We often hear companies talk about how great their products are or how well they hold up under adverse conditions, but many of us wonder just how much of that is hype. While I cannot vouch for all the claims out there, I can tell you my Balluff prox story and what I have seen Balluff prox switches survive.

The prox in question sensed the motion of sanding tape in a polishing machine. This system polished the part in two directions, and the purpose of the prox was to ensure that a proper amount of sandpaper advanced with each direction change. The sandpaper was on rolls, which threaded through some rollers, including a plastic roller that had a piece of metal embedded inside. The prox was set up to monitor this plastic roller and register when the metal piece rotated by, thereby indicating to the system that the sandpaper had advanced. If this prox did not change states, the system would fault out and an error/alarm light would come on. To reset the alarm, the operator had to hit a reset button and then start the system again.

On the day in question, the operator was having trouble with the system faulting out intermittently. As improperly routed/installed tape rolls were a common problem with this operation, I generally looked there first when troubleshooting the system. This time, I found that not only had the operator routed the tape incorrectly, he had placed it in such a way that the abrasive side of the sanding tape was running across the Balluff prox that was used to insure tape advancement. The entire front of the prox

had been sanded down to somewhere between a 30- and 45-degree angle, with very little of the flat tip remaining. At best, there was maybe a sixteenth of an inch of the flat end surface left. Now, here is the amazing part: the prox was still functional! The operator was only having an intermittent problem with the machine, which means that the prox was still working most of the time. In fact, I verified this by advancing some of the tape by hand and watching the indicator light change state on the prox. Of course, I knew that this was the problem with the system given the damage to the prox and thus began to remove the prox from the system.

This is where I got my second surprise of the day. I had changed many of these prox switches. It was common to change the switch under powered conditions, as most of the switches that we used had handy screw-on connectors that made replacement a breeze. The proper way is to unscrew the connector, thus removing power, and then remove the prox from the bracketing. I was not thinking, and I started by removing the prox from the bracket first. It was at this point when I discovered that not only was a large chunk of the tip of the prox gone, but the live wires inside were exposed. Yes, I discovered this in the way no maintenance person should… by being shocked. Luckily, it was a low-voltage prox and the shock was minimal. While this is an entertaining portion of my story, it demonstrates how much damage a Balluff prox can take and still function, at least intermittently. Once I realized the error of my ways, I removed the connector (and thus the power) from the prox before removing the prox from the bracket. The rest of the story is much more boring and standard as far as replacing a prox goes.

Safety disclaimer: Balluff prox switches may still function with exposed internal electrical components, so always kill or disconnect the power before you remove them. Also, remember that amperages of just .05 amps can be fatal, so always, always, be cautious with electricity.

only create the positive portion of the waveform and have thousands of cycles per second, in contrast with the 60 Hz of American AC power. The oscillator, the part of the prox that creates this rapid positive sine wave pulse, connects to a coil of copper in the end of the prox, generating a magnetic field from the end of the prox switch. This magnetic field interacts with ferrous metals, those that are magnetic, and allows the prox to work. Prox switches with a low-strength magnetic field can sense items at a maximum distance of 0.06 in. or 1.5 mm, while the stronger inductive prox switches can sense metal items up to 0.4 in. or 10 mm away.

When a metal part enters the magnetic field of the inductive prox (see Figure 6-4), it generates eddy currents in the part, which takes energy away from the magnetic field created by the oscillator in the prox. An eddy current is a flow

eddy current
A flow of electrons created by the magnetic field moving across a ferrous metal item that generate their own magnetic field

Courtesy of Balluff, Inc.

Figure 6-5   Here are some truly small prox switches, which would be perfect for robotic systems.

Courtesy of Balluff, Inc.

Figure 6-6   In this cutaway of a prox switch, you can see both the internal solid-state components surrounded by red resin and the coil in the tip surrounded by black resin. The coil is the copper-colored element in the black resin.

of electrons created by the magnetic field moving across a ferrous metal item. In turn, they generate their own magnetic field. The movement of the magnetic field in this case is primarily caused by the magnetic field created by the sine wave of the oscillator that builds to maximum and then collapses back to zero thousands of times per second. The eddy currents and their induced magnetic fields draw enough power from the oscillator that eventually it can no longer create an oscillating field. A separate unit in the prox known as the trigger unit looks for the oscillation to die down to a certain level and it then activates the output(s) of the prox accordingly (see Figure 6-6).

capacitive proximity switch
A switch that generates an electrostatic field; working on the same principle as capacitors, it uses the item sensed to complete the capacitive circuit as well as sense materials at various distances

The oscillation of magnetic fields is a key concept and the heart of many electrical components. For example, transformers use this principle to step up, step down, or clean up and isolate electricity in systems, while the induction motor, which is one of the main motors found in industry, uses induced magnetic fields to cause the rotor to spin. As you learn more about electricity and motors, you will further understand how induced magnetic fields are crucial to the modern electrical world.

photoelectric proximity switch
A switch that sends out a specific wavelength of light and uses a receiver to detect that specific wavelength of light when returned by a reflector or the detected item

Capacitive proximity switches work using the same principle as capacitors (see Figure 6-7). The switch provides the electrostatic field, air is the insulating material or the dielectric, and the object being sensed provides the conductor that finishes the capacitive circuit. Unlike the inductive prox, the capacitive prox will interact with materials that are both magnetic and nonmagnetic. The capacitive prox works just the opposite of the inductive prox in that the oscillation of the circuit *begins* when the item sensed enters the electrostatic field, and it dies off

Figure 6-7   Notice how the body of this capacitive prox switch is plastic instead of metal. It is common to find inductive prox switches with metal bodies and capacitive prox switches with plastic bodies. However, do not assume that all metal prox switches are inductive or that all plastic prox switches are capacitive.

when no object is present. Another important difference is the fact that the field on the inductive prox does not change, due to the construction of the switch, while the capacitive prox has an adjustment on the switch that allows the user to vary the electrostatic field and change the distance from which it interacts with objects. Because of this, it is possible to set the switch to detect items inside cartons, boxes, or other packaging materials!

A `photoelectric proximity switch` detects levels of light, senses objects via reflected or blocked beams, and detects colors via an emitter that sends out a specific wavelength of light and a receiver that looks for that wavelength to return after it has interacted with the surrounding environment. The simple part detection units work in the same way as the photo eyes we examined in Chapter 2 on safety, so we will focus on the other functions of the photoelectric prox in this section.

With the light level detection photoelectric switch, we can tune out or ignore the background, differentiate between parts on a background, and even detect transparent objects. With the simple photo eye, the beam is either returned or not; however, this one analyzes the light coming back to the receiver and determines if it has been affected or altered in any way. A light level detection prox could differentiate between a rough finish raw part and a smooth finish machined part as it passes by, determining the differences in the light returned to the switch. With proper calibration, this switch could monitor the empty spaces between parts on a conveyor line. There are many options with this type of setup and thus many ways to integrate this switch into various systems.

The color detection photoelectric prox switch analyzes the light coming into the receiver to determine color or color difference in parts (see Figure 6-8).

Figure 6-8   This Alpha Rex LEGO NXT robot has a simple color sensor on the back. When properly calibrated, it can tell the difference between a red or blue plastic ball that comes with the kit.

This sensor can find the edge between parts of different colors and thus impart important positioning information to robotic systems. Roboticists have used this sensor to tell the difference between blue and red balls, follow colored lines, and carry out other such tasks involving color recognition. These switches do have drawbacks that must be taken into consideration. Changes in the ambient light will change the light returning to the switch and the reading, thus requiring recalibration. For example, this is why a robot could tell the difference between the red and blue ball in the lab, but fail to work at the science fair in the gym. Also, remember that these sensors have an emitter that sends out a specific wavelength of light. Depending on the light sent out, it may be impossible for the receiver to detect some colors or the difference between certain colors. This is why there is not a one-size-fits-all color detection prox.

Like the inductive and capacitive prox, the photoelectric prox also has a set operating range that is dependent on the strength of the emitter and the wavelength of the light sent out. Photoelectric prox switches have the greatest sensing range of all the prox switches, with ranges from point blank to well over 100 feet, but they are not limitless. The longest ranges are primarily for part present or absent applications; the more complex applications, such as color detection, generally have limited ranges of a few inches to a foot or two.

Proximity switches have evolved considerably over the years. However, they often answer a yes or no question: Is something there, or not? Because of this, the amount of data we get is limited and we can only use it in limited ways. Photoelectric switches give a greater amount of information—this is why we can use them to follow lines or find patterns—but the capacitive and inductive only tell us if they sense something or not. As we continue our look at sensors, we will delve into systems that are more complex and at the same time provide a larger amount of data that we can use to control the actions of robots.

Tactile and Impact

tactile
The ability to sense pressure and impact

impact
A robot's contact with an object in the intended movement path

Other sensing fields for robotics include tactile, or the ability to sense pressure; and impact, or when the robot contacts an object in the intended movement path. These are both reactionary types of sensing and have to do with force, but the applications and techniques for each differ significantly. Tactile sensing is about determining how much force is being applied, what the shape of the part is, how it is gripped, if the part is hot or cold—basically, the same kinds of sensing tasks that skin performs for the human body. Impact sensing is concerned with detection of collisions, determining if forward movement is impeded or stopped, and, perhaps most importantly, shutting down or modifying the motion of the system to prevent damage to both whatever is hit and the robot. As we explore some of the ways in which we sense these forces, we must keep in mind that these are but a few of the many possibilities. Even as you read this, new sensors and new methods are under development.

Tactile sensors are concerned with determining how much force the system applies to an object and the area that force covers. These systems range from simple buttons that activate when depressed and disengage when released, to complex arrays of sensors that closely mimic the function of human skin in the amount of information they provide. For the simple "Is it gripped?" system, all that is required is a switch that will fit in the desired space, usually a gripper, that is capable of being activated by and withstanding the force of the system while making contact with the desired object. A subminiature micro limit switch or push button is often the switch of choice for this type of operation. Those of you who have experience with robots might ask, "What about sensors that detect if a gripper is open or closed, would those be considered tactile sensors?" The answer is no. Tactile means of or connected to the sense of touch; thus, a tactile sensor must "touch" the part it is sensing. Otherwise, it is a type of sensor classification.

Complex tactile sensors use arrays or organized groups of sensing elements to gather information about contact with objects. Each element of an array is an individual sensor that provides information about how it is interacting with the part. The more elements in an array and the smaller the size of the element, the more information the system will have about the part and its manipulation. Simple systems consist of elements that have a digital type of output, either on or off, 1 or 0. With this type of sensor, the system knows if it is gripping the part or not and how many elements are involved with that grip, denoting the surface over which the force is applied, but it has no data on the amount of force the system is using to gripping the part. Depending on the circumstances, this might be enough information to determine part orientation or proper grip of the part. With proper programming, the robot could monitor for any new elements coming on or current elements going off during movements, indicating that the part is slipping.

High-end complex tactile sensors not only give information about contact with the part, but also how much force the system applies to the part. Some even measure the temperature of the part it contacts. This is added data for the robot to use when manipulating parts, allowing robots to handle delicate items such as an egg one minute, and heavy items, such as a case of eggs, the next. Complex tactile sensors have a variety of options for sensing changes in force, such as changes in voltage, resistance, capacitance, and magnetic flux (to name a few). The common factor in all these elements is that change in the shape of the element causes a change in the output of that element. It is as simple as that. The strain gage element changes resistance as the wire inside is stretched or compressed. The capacitive element has a change in the charge it can store, depending on the distance of two electrodes. Piezoelectric elements create electricity with applied force; the greater the force, the more electricity produced, though it is a very small amount in the millivolt range. No matter the type of element used, the result is that a change in some measurable value determines the amount of force applied at the point of contact. The greater the force, the greater the deviation from the base signal. Base signals are those generated by the element under noncontact conditions.

The manufacturer of the tactile sensor will often provide data on the change in force to change in output ratio or some other means for making use of the data from the array. Robot or tooling that comes with complex tactile systems as a standard feature should require little more than calibration from the user. Calibration is a specified process that insures a precision system performs properly and provides for any adjustments needed. In this case, the user would use a verified accurate pressure gage and place it in the tactile sensor area of the robot applying a force. Then the user would compare the reading from the pressure gage and the robot and determine the amount of difference between the two. If the difference is great enough, the user would adjust or offset the robot system to correct for the error of the system. You should calibrate the system during robot installation and verification, after changing sensors, after any damage to the system, and periodically as recommended by the manufacturer.

calibration
A specified process that ensures that a precision system performs properly and provides for any adjustments needed

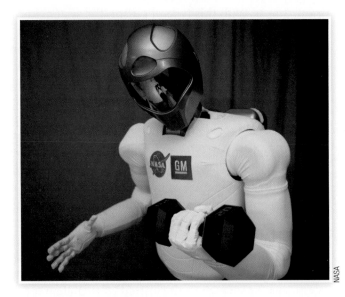

Figure 6-9   These two pictures demonstrate the principle behind tactile, or force sensing. The astronaut shaking the Robonaut's hand would not likely have enjoyed the experience had the Robonaut used the same amount of force as it did lifting the 20-lb. dumbbell.

Impact detection plays a crucial role in minimizing or preventing damage to the robot as well as equipment and people around the robot (see Figure 6-9). Initially, impact detection was all about keeping the robot and other equipment safe, as people were not to be in the path of the robot. In fact, it was not until 2012 that this technology advanced enough for the robot to work without the isolation cage and beside its human counterparts in industry. Sensing an impact with something rigid—say the machine a robot is feeding parts into, for example—is easier than detecting an impact with something lightweight and movable, like a person. Today, we typically either monitor the amperage the motor uses or insert sensing devices designed to detect impacts. The common sensing devices for impact are strain gauges at crucial points, such as the joints of the robot, to detect sudden changes in the forces of motion so that the robot can respond accordingly. Manufacturers could wrap the same tactile skin used to detect parts around the body of the robot to detect impacts as well, though this is rarely the case, as it would cost more for the sensor wrap and the controller would have to handle a larger amount of data.

When the system uses the amperage draw of the motors to monitor for impact, it creates a very accurate system that has great sensitivity but a greater level of complexity, too. There is some high-level math involved (that we will not get into) that relates the programmed path, speed of travel, payload, and type of motors performing the work to determine the amperage draw of the motor(s). The result of this calculation is compared to the actual current draw of the motor(s); if the specified plus or minus error limit is exceeded, the robot responds as programmed. Potential problems with this method include a system shut down due to excessive payload or noise in the system. (We will look at noise shortly.) Since we are monitoring the torque output of the motor via the amount of current used, there is no distinction between high current draw due to a heavy part and an actual impact. This is why some robots alarm out and stop, indicating an impact, when a visual inspection shows no objects in the robot's path. Anything that causes the motor to use excessive amounts of current, such as a too-heavy load, bad bearings, friction, caked dirt in the joints, or something else requiring more force, can cause an impact alarm.

The noise aforementioned refers to the mathematical difficulties in properly calculating the torque required at the start of motor motion for robot movement. Motors require more amperage to start moving than to maintain motion, so we have to figure that into our calculations. Next, our calculations must consider how we are going to move the robot; for instance, the motor will need more amperage to raise the arm up than to lower it down, as gravity is a factor. In addition, we cannot forget to figure in how fast we want the robot to move along with what happens to the motor as the robot accelerates and decelerates. All these factors create some mathematical gray areas, where we must estimate what will happen: we call these estimates "noise." If we are off in our estimates and set the alarm parameters tightly, the system may have false alarms. The biggest danger with false alarms, whether due to noise or mechanical issues, is that someone will just change the alarm parameters, thereby increasing the difference and thus the force required to stop the system instead of fixing the problem or math as needed. This could lead to impacts of great force that could damage man and machine alike.

> **noise**
> Mathematical difficulties in properly calculating the torque required at the start of motor motion for robot movement

No matter how we detect the impact, the robot has to do something to prevent or limit the damage caused. One of the tactics is to E-stop the robot and lock it in place. The thought here is that the robot stops all motion as fast as possible and then freezes in place to let the operator decide what to do next. However, the robot will still be moving after the detected impact for a short period of time before the full stop occurs. Granted, it is usually less than a second, but a fraction of a second can seem like eternity when a robot arm is impacting your body. A secondary concern is the strain this causes to the internal systems of the robot. Think of it as driving down the road and suddenly throwing your vehicle into park. While this may not kill your vehicle, it sure did not do it any favors. Because of these factors, many robots now disengage the motor from the joints until the system can come to a full stop. To continue with the car example, instead of going from drive to park, you now put your car in neutral, take your foot off the gas, and apply the brakes to stop the car. It reduces the force the robot has to stop, reduces the energy of the impact, and reduces the stress on the robot's internal systems. This type of response requires an advanced force reduction system coupled with specialized programming and sensors, but it has brought the robot out of the cage and alongside its fellow workers.

You can expect tactile and impact sensor systems to continue to advance and evolve with the robot, as these systems allow the robot to respond to a greater range of events. Tactile sensors have already ushered in new ways of picking up and working with parts. This technology is also a focus of research and advancement, so you can just imagine what future improvements might bring. When it comes to impact sensors, we often use robots to do difficult and dangerous tasks, where data about what is happening to the robot could be crucial to protecting the system. It is hard to predict what advancements in these fields will bring, but I think the Robonaut and Baxter robots are a good indicator of what we might expect.

Position

open-loop system
A system that assumes that everything is working correctly because there is only limited information on a few positions or no feedback at all to confirm everything is working as directed.

closed-loop system
A system that sends out the control pulse to initiate movement and then receives a return signal that confirms movement, often including the direction, speed, and distance moved

One of the great benefits of the robot is its ability to repeatedly perform tasks with precision. To do this, the controller has to record and track the position of the robot as points in space with some form of reference point, often known as the origin or zero point. To help clarify this, imagine that your friend tells you to meet at point A. Would you have enough information to meet them? Of course not; you need to know where point A is, what point A is, when to be at point A, and other such pertinent data. So, you question your friend. He says, "Oh, sorry, point A is at 1717 Mockingbird Lane." Now do you have enough data? No, you still need to know what town he means as well as when to be there. To convert our example to the robot, the address is similar to a robot's position in space, the town of the address is similar to a robot's home position, and the timing is similar to the orientation of the tooling. If you do not have all the data, there is a good chance you will not wind up at point A; and it is the same for a robot, without all the data, the chances of the robot going back to the same point twice are pretty slim.

When it comes to determining position, there are two main types of control systems, open loop and closed loop. **Open-loop systems** work on the assumption that the control pulse, whatever it may be, activates the motion system, and the robot performs as expected. This type of control was common in the early pneumatic-only robots, since motion depended on valve sequencing, and stepper motor systems, since the motor moves a precise amount each time the voltage is applied. The open-loop systems often included limit switches or prox switches at designated points to check for robot position, but they were few and far between, giving complex programs a large amount of positional error to build up with no correction. This chance for error made the open-loop control system unpopular with industry, especially when the benefits of a closed-loop system became apparent.

Closed-loop systems send out the control pulse, whatever it may be, to initiate movement and then receive a signal back that confirms movement and in many cases in which direction and how far (see Figure 6-10). To go back to our point A example, in an open-loop control, your friend would say, "Go to point A at 7 pm." On the other hand, in closed-loop control he or she would say, "Give me a call when you head to point A, and I will make sure you don't get lost." When it comes to robotics, the preferred feedback sensor of choice for closed-loop control is the motor encoder. **Motor encoders** are devices that directly monitor the rotation of a motor shaft and turn that information into a meaningful signal. The robot controller uses this signal to determine what the motor is doing as well as when it has reached the desired point. Keep in mind that if a robot has six axes of movement, it will need six feedback devices to be a fully closed-loop system. This means six pieces of positional date, one for each axis, to find a specified point in space.

Figure 6-10 In this setup, a prox tracks the metal protrusions as they pass. This could be used as a basic form of feedback for a closed-loop system that provides changes of state (detection of the protrusion), which the controller could count and manipulate into useful data.

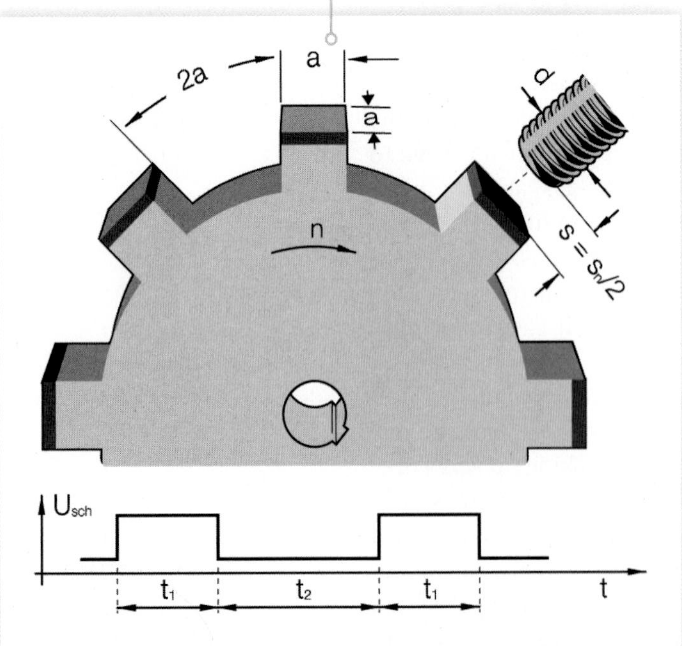

Courtesy of Balluff, Inc.

A simple encoder you may encounter uses a **Hall effect sensor**; it is based on the Hall effect, which was discovered by Edwin Herbert Hall in 1879. Edwin Hall discovered that current flows through a conductor in a magnetic field; that the magnetic field tends to push the negatively charged atoms to one side of the conductor and the positively charged atoms to the opposite side when the magnetic field is perpendicular or at a 90-degree angle to the conductor. Today, we use a semiconductor to allow voltage to pass when a magnetic field lines up with the sensor. By attaching a disk with a specific number of magnets evenly spaced around the edge where they can activate a Hall effect sensor, we can track the rotation of the motor and thus the position of the system. See example 1 to understand how this might work.

Example 1

For this example, we want to move the axis in question 90 degrees, with the motor coupled directly to the robot axis. The motor has a Hall effect encoder with 40 magnets positioned evenly around the edge of a disk mounted on the motor shaft. How many pulses should the Hall effect sensor register if the motor moves the desired 90 degrees?

To determine the number of pulses, we need to determine what portion of a full rotation (or 360 degrees) the axis needs to move. To do this, we will divide 90 degrees by 360 degrees.

$\dfrac{90}{360}$ This works out to 0.25, or ¼ of a full rotation.

Next, we will take this portion of the rotation, 0.25, and multiply it times the number of pulses or magnets on the sensing disk.

$40 \cdot 0.25 = 10$.   Therefore, by our calculations, there should be 10 pulses sent back to the controller if the motor moves the proper amount to change the axis position by 90 degrees.

The downside to this kind of system is that it gives us pulses, but no information on which direction it moved. Yes, we can verify that the system moved the proper distance, but what if it went in the wrong direction? This is where our good friend the optical encoder comes into play.

Incremental optical encoders consist of a disk that has either holes for light to pass through or special reflectors to return light, an emitter, a receiver, and some solid-state devices for signal interpretation and transmission (see Figure 6-11).

motor encoder
A device that directly monitors the rotation of a motor shaft and turns that information into a meaningful signal

hall effect sensor
A sensor that uses a magnetic field to cause voltage flow in a semiconductor used to track rotation

incremental optical encoder
A disk that has either holes for light to pass through or special reflectors to return light, an emitter, a receiver, and some solid-state devices for signal interpretation and transmission

Figure 6-11 Here are several examples of encoders that can be attached to motors.

The principle of operation is simple: the transmitter sends out light, and the light either passes through the holes or reflects back to the receiver, triggering the electronics of the encoder to send a signal back to the controller. Because we are using light, it is possible to make the divisions on the rotating disk extremely small, allowing for a division of 1,024 or greater. In other words, we can take our circle of 360 degrees and divide it by 1,024, giving us movements as precise as 0.35 degrees, or just over 1/3 of a degree. The more divisions on the encoder, the finer the positioning options. "What about the direction of rotation?" you ask. By adding a second row of reflectors or light windows, offset from the pulse count, and the appropriate emitter and receiver, we can now determine the direction of rotation by comparing the signals from the two rings. The offset creates a unique type of signal for each direction of rotation (see Figure 6-12). While we are looking at encoder options, another option is the addition of a zero pulse. This is a specialized area of the encoder disk used to establish a zero or home position. We can use this pulse to count the number of full rotations made as well as to ensure proper positioning of the system should the encoder or motor need adjustment, replacement, or maintenance.

An absolute optical encoder adds enough emitters and receivers, usually four or more in total, to give each position of the encoder its own unique binary address (see Figure 6-13). A binary address is a unique set of 1s and 0s that the controller

absolute optical encoder

An encoder with enough emitters and receivers to give each position of the encoder its own unique binary address

binary address

A unique set of 1s and 0s that the controller can understand and use

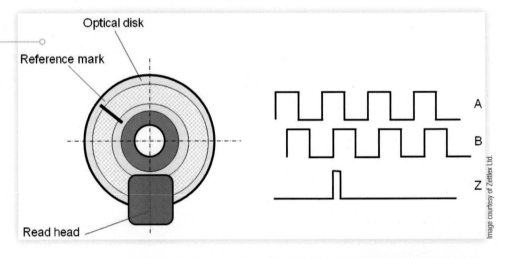

Figure 6-12 Here is an incremental encoder setup with the zero pulse as well as the type of signal that it generates to the controller.

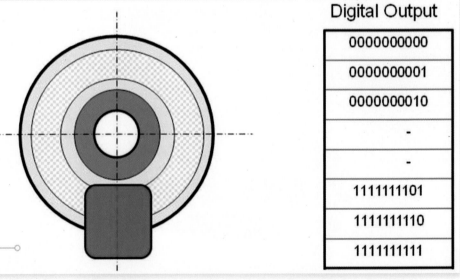

Figure 6-13 This is an example of an absolute encoder setup along with the binary data it generates.

can understand and use. These encoders have more internal parts, cost more, and require more of the controller's processing power, but they give the system accurate data about the specific position of the motor as well as the direction it is moving. When using these encoders, the robot knows where the motor is at any given point instead of just how many pulses it moved from the last point and possibly what direction, depending on the incremental encoder type. This added level of information makes it much easier for the controller to detect when axes are out of position or if the motor is not moving correctly, thus lending a greater degree of control and monitoring.

Since encoders are noncontact sensing methods with a large number of solid-state parts, they are robust and able to last years. The biggest enemy of encoders is oil, metal chips, shavings, or any other contaminant that gets inside the unit and either clouds the disk that interacts with the light or damages it. I have removed many encoders and sent them off to the manufacturer for cleaning or rebuilding due to oil in the unit or damage to the disk during machine repairs.

On the subject of ways to track position, let us take a look at Global Positioning System (GPS). This system determines geographical position based on the time it takes to receive signals from three or four separate satellites in orbit around the Earth. There is a high probability that you have used a GPS system if you have taken a long-distance trip and used an electronic device for navigation. Just as you may have used GPS to make sure that you were going in the right direction and to get turn-by-turn directions, mobile robotic vehicles can use GPS to navigate various terrains or roadways to accomplish their programmed tasks. Drones use GPS to make sure that they are on target and to give their users, who may be miles or countries away, information about where they are physically located in the world. As we continue to explore and perfect the self-driving car, you can be sure that GPS will play a part. With a proper setup, such as a localized system that uses equipment in the facility instead of satellites, warehouses and other large structures with robotic material handlers can utilize GPS for their positioning data. Keep in mind, GPS is usually accurate to within a few feet and only gives information about where you are in the world, so mobile robotic systems will need a suite of local sensors to keep the robot from running over someone or crashing into objects blocking their path as well as fine-tuning movements to accomplish their programmed tasks.

There are more ways to track motor position, but these are the ones that you are most likely to see in the field. Depending on the age of the robot, you may run into resolvers and synchros, which are a type of encoder that works on a principle similar to variable transformers. These were popular for many years, which explains why you may find them on older robots and possibly some newer robots, depending on the environment they work in. Regardless of the positioning sensor that you find in the field, make sure to take some time and understand how the system works and the type of signals it generates. This can save you hours of frustration down the road when you troubleshoot or program the system.

> **Global Positioning System (GPS)**
> A system that determines geographical position based on the time it takes to receive signal from three or four separate satellites in orbit around Earth

Sound

The ability to detect sound is one of the five major senses; we consider it a disability when a person is deaf or can only hear a limited range of sounds. With this in mind, it should come as no surprise that we have figured out ways to give robots information about sound. As with many other areas, the robot's mechanical nature allows for options in sound detection and manipulation that we mere

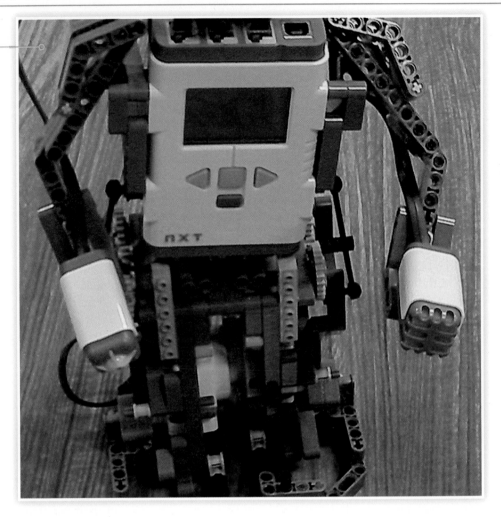

Figure 6-14 The Alpha Rex uses a sound sensor for the hand on the right and a push button or touch sensor on the left (as referenced from our view).

ultrasonic sensor
A sensor system that emits sounds above the normal hearing range of humans and then determines the distance of items by the amount of time it takes for the sound to return to the receiver

mortals can only envy. Obviously, not all robots need to deal with sound in any manner to complete their functions; however, adding sound sensors opens up new options for operation and environmental response, increasing the flexibility of robotic systems.

For sound detection, a robotic system needs a microphone, the proper hardware, and the necessary programming so that the robot can turn the signal from the microphone into something useful (see Figure 6-14). It truly is that simple. With these elements in place, we can record patterns of sound and use them as triggers for robot actions. We can have the robot listen for sounds over a set intensity or decibels, such as what a crash might generate, and use that as an added safety feature to the robot. We can also program the robot to recognize a specific set of sounds or voice commands so that the system can start or stop programs based on verbal commands. Once the robot can hear, we have many options to work with.

What if we need to know where a sound came from? With the installation of a single microphone, it is nearly impossible to determine where a sound came from unless the sound is continuous and there is a chance to move the microphone to different points while taking sound samples. To detect where a sound came from, add multiple microphones to the robot so that the controller can compare signal strengths and determine a probable location for the sound's origin (see Figure 6-15). Often for robots that need to pinpoint or follow sounds, the designers will add multiple microphones positioned strategically around the robot chassis. Designers usually add microphones in even-numbered pairs, with two being the minimum; however, it is not uncommon to have four or six different microphones integrated into the system to improve sound detection accuracy.

So far, we have not looked at anything too radical in the world of sound detection. But what if I told you that a robot could use sound to see? Yep, it's a lot like bats that chase insects at night, or "Daredevil" for you comic book fans. Robots can use ultrasonic sensors to detect or "see" objects via the high-frequency sound that they reflect back. An ultrasonic sensor is similar to the photo eyes or photoelectric prox switches except that they are emitting and receiving sound instead of light. The emitter sends out a high-frequency sound pulse, above the human range of hearing, which strikes objects and then returns to the ultrasonic sensor. The receiver measures the amount of time that it takes the sound wave to return and uses this to calculate the distance. To determine the distance, we take the speed of the sound wave, multiply it by the time it took for the wave to return, and then divide by 2, as the wave had to travel the same distance to the object and back to reach the sensor. The ultrasonic sensor has become a favorite among hobby roboticists when they want their robot to avoid or detect objects (see Figure 6-16).

Another use for ultrasonic sensors is the detection of air leaks. Small air leaks emit a high-frequency sound that we cannot hear, but that the ultrasonic detector can pick up with ease. When used for leak detection, we turn off the ultrasonic emitter on the sensor, as the source will be the item the system is inspecting. To find the leaks, the item being inspected is pressurized and

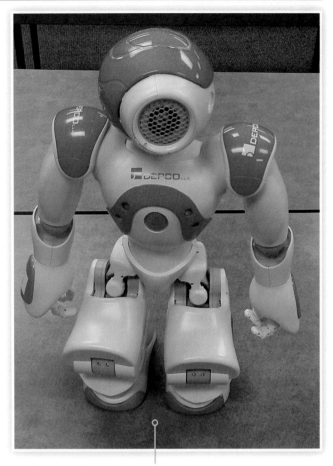

Figure 6-15   Here is the NAO robot. Its head turned sideways, showing off his microphone under the little black dots along the white frame around the speaker. It has the same setup on both sides of its head. The four small dots in the orange center of its chest are ultrasonic sensors for object avoidance.

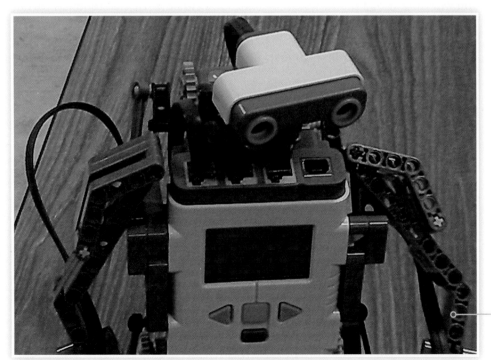

Figure 6-16   Here is the LEGO NXT ultrasonic sensor serving as the head for the Alpha Rex build, enabling the system to detect objects in its path.

Figure 6-17 This picture was taken mid-performance as the NAO robot performs the "Gangnam Style" dance. It even provides the music for the performance via the speakers on either sides of its head.

then the robot moves the ultrasonic sensor around to the key points as programmed.

If there are any pinholes or finer leaks, the ultrasonic sensor will pick up the sound from the leak, enabling the robot to determine its location. This usually requires several spatial points detecting the sound to determine the exact location, even if those points are just before the leak, over it, and just past the leak. This is a crucial inspection for tanks and pressure vessels, where the smallest leak could spell trouble, if not disaster.

While on the topic of sound, I want to talk about the benefits of adding speakers to robotic systems (see Figure 6-17). Even though we can use a speaker to detect sound, it is not a sensor under most conditions; however, speakers give a voice to an otherwise mute robotic system. Speakers are not standard equipment in most industrial systems, but many of the hobby and research robotic systems use them to make robotic systems more interactive and lifelike. There is a great difference between reading a message on a display screen and having the robot tell you what is going on. Movies, TV, and books have often portrayed the advanced robot as a thinking analytical system capable of enhancing human life by delivering key warnings and information through speech. As we break the robotic system out of the protective cage and place it alongside people in various aspects of daily life, I expect that speakers and speech-generation programming will become commonplace in most robotic systems.

Many industrial robots work fine without sound systems, deaf to the noise and sounds of industry, but as we have mentioned before, they are popular with the hobby, research, and entertainment fields of robotics. There is a good chance that you will work with a sound system of some kind either in the classroom, as part of a lab, if you get into building your own robots, or as part of higher-level learning on robotic systems. If this was your first time learning about sound sensors, just remember that you heard it here first.

Vision Systems

vision system
A system that uses cameras and software to process images and provide that information to the robot

Vision systems allow robots to see the world around them, using cameras and software that processes the images taken by the camera (see Figure 6-18). Dr. Robert Shillman truly started the odyssey of robotic vision with the DataMan system in 1982, and this technology has matured since then. Early robotic vision systems consisted of cameras that took images in a very controlled environment and at a set time, which the robot would compare to sample pictures. Color was not an option, changes in lighting had disastrous effects on the chance of recognizing objects, and these systems were prone to failure that required human intervention. The early systems were beneficial to certain industries and certain applications, but many of the early robot buyers felt it was more of a headache and expense than it was worth. To compare the vision systems of today with the early systems is akin to comparing the Ford Model T to a Corvette. We have learned a lot in the years that followed Shillman's DataMan. Thanks to further advancements in the technology, we now have vision systems that are truly a benefit to robots. The modern systems use high-resolution cameras, differentiate between colors, and offset the robot so that it can perform the programmed task even when the item it is interacting with is out of the expected position. We have come a long way from the days of highly controlled conditions and wonky software.

The modern vision system usually consists of a specialized light source, a camera mounted on the robot or at a specific point, and specialized software from either the robot or the vision system manufacturer. The light source is very important in vision applications, as the type of light used to illuminate objects directly influences the type of light coming back to the camera. For instance, if someone uses a red light to illuminate an area, any red objects in that area will be nearly invisible to the camera. Sometimes this is on purpose, so that the system will ignore specific parts; however, if someone simply chose the wrong type of light, it could greatly complicate the process. Some of the modern lighting systems are capable of generating light over the visible spectrum as well as the infrared spectrum, allowing for multiple lighting options. The advances in camera technology have turned the images from grainy and pixelated into high-resolution pictures with sharp detail. Proper lighting with a high-resolution image has created a wealth of useful data for the robot (see Figure 6-19).

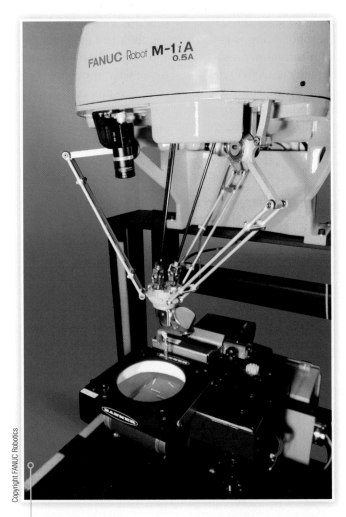

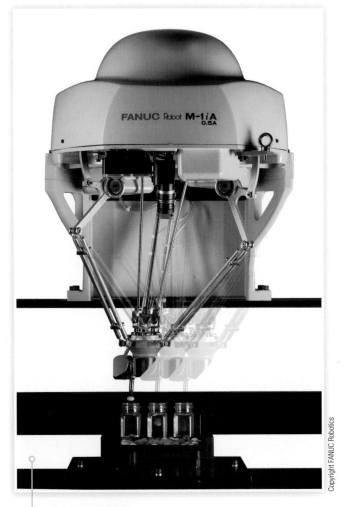

Figure 6-18   A FANUC Delta robot uses its camera and a lit work area to work with parts. The camera is the black device mounted under the FANUC logo.

Figure 6-19   The FANUC Delta robot uses its vision system to determine the different colors of the pills and sorts them into the proper container.

The other key point of a modern vision system is determining where the part is by the image. Once the system knows where the part is, it can make the calculations necessary to offset movements. However, how does it know where the part is? Most robotic systems have one camera. This seems to complicate this question, as depth perception generally requires two points of data. The answer is that we have to calibrate the image from the camera to something that the robot understands. To put it simply, we take an image from the camera and then relate the image to points in space that the robot knows how to reach. The basic process is to place a calibration image in a location that the robot can reach, take a picture, and then use the system software to translate that image into point data. Once completed, the robot has the basic knowledge needed to process image data, but is not yet ready to grab parts or execute other such tasks. The next step is to take an image of what the parts should look like and use this as a reference image. Once this is set up, we can then program the robot to perform the desired functions. During operation, the robot will take a picture of the designated area, compare this to the reference picture, and use the positional data that we gave it at calibration to offset motions accordingly (see Figure 6-20). One thing to keep in mind is that when you mount the camera to something mobile, like the robot, make sure that you take the picture from the same position each time. Changing this position may require you to recalibrate the system.

Sometimes we use vision systems for taking pictures or video of what is around the robot without the robot doing anything specific with this information. This is a common practice for unmanned vehicles, bomb disposal systems, and any other system where a human controller would like to see what the robot sees. These systems may not have lighting components nor the specialized software we discussed earlier, since the point is to provide visual data to the user, not the robot. In the absence of specialized software, the operator is responsible for how the robot interacts with the visual data provided. This will likely require practice for proficiency.

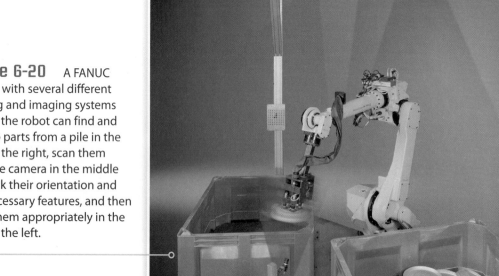

Figure 6-20 A FANUC system with several different lighting and imaging systems so that the robot can find and pick up parts from a pile in the box on the right, scan them with the camera in the middle to check their orientation and any necessary features, and then stack them appropriately in the box on the left.

It is astounding how far robotic vision has advanced since the early 1980s. When vision systems first started, they required a large amount of special equipment and human help. Today, we can add vision to most robots for about the cost of standard tooling simply by adding a camera and some software. You can even get vision systems that provide their own lighting, creating a wealth of options for the user's application. If vision systems continue to advance as they have, I think we will see a day where it is rare for a robot to be without a vision system. It will be standard equipment, similar to the way encoders and prox switches are now.

review

We have examined some of the common sensors in the field today. As you continue along your journey of robotic knowledge, do not hesitate to research the new sensors you encounter and learn more about how they work. As the field of robotics continues to mature and grow, you can expect new uses for current sensors as well as new sensing systems. Without sensors, robots are just lumbering creations that have no idea how they affect the world around them. Sensors transform robots from senseless machines into high-tech systems capable of performing astonishing tasks. During the course of exploring sensors, we covered the following topics:

- **Limit switches.** Here we looked at one of the basic switches used to sense the presence or absence of objects.

- **Proximity switches.** This section detailed the different types of prox switches available as well as the basics of their operation.

- **Tactile and impact.** We looked at two sides of the force-sensing question along with some of the ways of sensing force.

- **Position.** This section was about the sensors that we use to determine where the robot is in space as well as where it is moving to.

- **Sound.** This section covered the standard sound detection sensors as well as ultrasonic sensing.

- **Vision systems.** This section detailed how we let the robot see the world around it and what it can do with that information.

key terms

Absolute optical encoder	Global Positioning System (GPS)	Limit switch	Proximity switch
Binary address		Motor encoder	Solid-state device
Calibration	Hall effect sensor	Noise	Tactile
Capacitive proximity switch	Impact	Open-loop system	Ultrasonic sensor
Closed-loop system	Incremental optical encoder	Photoelectric proximity switch	Vision system
Eddy current	Inductive proximity switch		

review questions

1. What can you do to correct a limit switch in the field that has the actuator in the wrong position?

2. What do you need to know when selecting a limit switch?

3. What are some of the downsides to solid-state devices?

4. What is the sensing range for inductive prox switches from low power to high power?

5. Describe the operation of an inductive prox switch.

6. Describe the operation of a capacitive prox switch.

7. What can we detect with light-level photoelectric prox switches?

8. What types of information are we looking when we use tactile sensors?

9. What types of information do we look for with impact sensors?

10. How could a tactile array that senses 1 or 0 be used to detect slipping or change in grip during the movements of the robot?

11. What kinds of information can a high-end complex tactile sensor give to a robot?

12. Describe how each of the following elements might work in a tactile sensor array: voltage, resistance, and capacitance sensing.

13. How do we commonly monitor impact today?

14. When we use current monitoring for impact, what are some things other than an impact that could trigger an alarm?

15. What is the benefit of disengaging the axes as opposed to the E-stop method of impact halt?

16. What is the difference between an open-loop and a closed-loop control system?

17. What are the parts of the optical encoder, and how does it work?

18. How can we determine the direction of rotation with incremental encoders?

19. How does an absolute encoder work?

20. How do ultrasonic sensors work, and how do we determine the distance from the sensor to the object detected?

21. Who do we credit with starting the field of robotic vision, what did he or she create, and when?

22. What does a modern robotic vision system consist of?

23. What is the process involved in setting up a vision system to offset the program based on images from first getting the robot ready to go?

Peripheral Systems

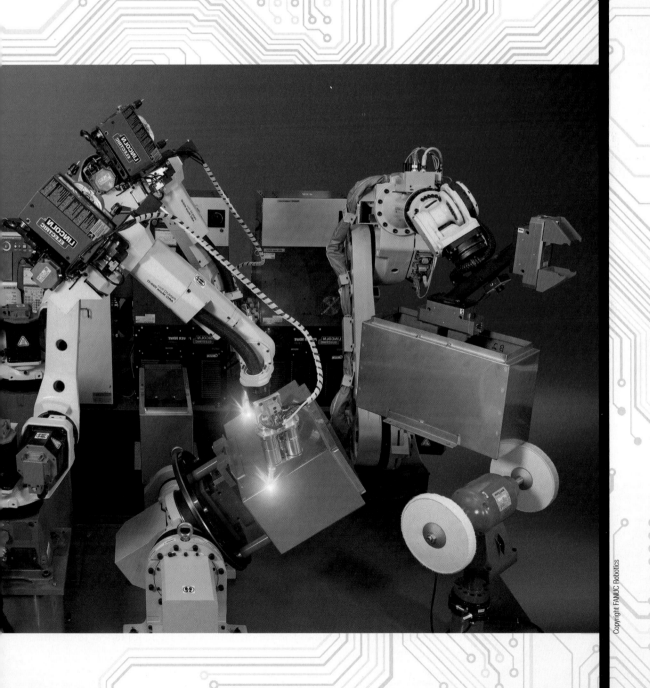

overview

Just as people can do more when they work together (as opposed to working alone), so can a robot expand its usefulness and capability by working with other pieces of equipment or systems. In the industrial setting, the ability to work with existing equipment is one of the primary responsibilities of the robot. The ability to work with other equipment and systems expands the flexibility of a robot and makes it easier to sell the system to potential clients. During our exploration of peripheral systems, we will cover the following topics:

- What is a peripheral system?
- Safety systems
- Positioners
- End-of-Arm tooling peripheral systems
- What is a work cell?
- Nonindustrial peripherals
- System communication

What Is a Peripheral System?

peripheral system
System or equipment that performs tasks related to or involving the robot but that is not part of the robot

A **peripheral system** is a system or equipment that performs tasks related to or involving the robot, but that is not part of the robot. To put it another way, this system works with the robot or performs specific tasks for the robot, but is capable of functioning *without* the robot. Peripheral systems feed data to the robot and often perform tasks based on information from the robot, but you could remove the robot and still have a functioning system. Of course, removing the robot would require rerouting the data from the system and possibly having to find another source of signal to start the system; however, the system could function if it were tied to a machine or PLC instead of the robot. This is the basic logic test to define a peripheral system.

A Programmable Logic Controller [PLC] is a specialized computer with a logical program to direct the action of equipment based on inputs and other information filtered by the program. This is basically an industrial-grade computer that runs one type of programming very well and very quickly to control the operation of industrial machines (see Figures 7-1 and 7-2).

tech note

Programmable Logic Controller [PLC]
A specialized computer with a logical program to direct equipment action based on inputs and other information filtered by the program

Figure 7-1   Here is an example of an Allen Bradley Micrologix 1000 PLC.

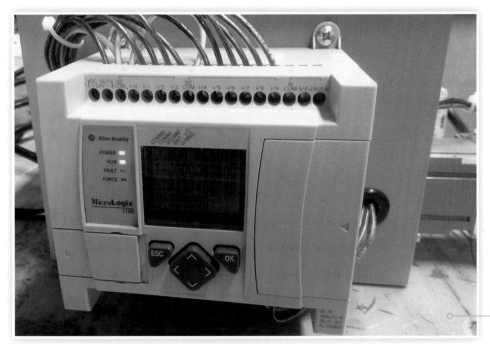

Figure 7-2   Pictured is an Allen Bradley Micrologix 1100 system. You may encounter a wide range of PLC units in industry.

Another way to think about peripheral systems is that they are the equipment that the robot is working with. If you order just the robot, you would get the core system. Anything that is not a part of the core system, such as options or additions, could well fall under the category of peripheral equipment. Keep in mind that the core system consists of the robot, controller, teach pendant, built-in sensors, and power supply. Often, robot manufactures will add in tooling and other equipment to enhance the operation of their robot or to customize the robot's performance for a specific industry or task; however, we consider them peripherals because the robot could work without them. Just because it came on the truck with the robot does not mean it is part of the core system. Keep this in mind as you read through this chapter and the various examples of peripheral systems that we will explore.

Safety Systems

Safety systems are required in industry to ensure the safety of those who work around robots and are thus a common peripheral in the robotic world. Because of the wide range of applications for robots in industry, it falls to the company that is buying a robot to ensure that the proper safety protocols and equipment are in place. These systems often include a barrier between the robot and workers with various sensing devices that transmit data to the robot, warning of the potential for human injury. The power for the sensors may come from the robot's power supply or another source; the data can run straight to the robot or through another control system that communicates with the robot. (We will talk more about communication later in this chapter.) The safety devices that are a part of the robot or built into the robot are considered a part of the core system, and thus not a peripheral.

A good example of this is an articulated robotic arm that is working inside a safety cage to feed several machines. The cage for our example has two entry points, both equipped with sensors to detect when the door opens. There is also a pressure mat inside the cage to detect the presence of people or objects inside the safety zone. All the safety sensors feed signal back to the robot controller through input terminals that were unused by the manufacturer. The system stops the robot if someone opens a door, and the robot moves at a designated safe speed in manual mode only any time the sensor mat detects the presence of something in the safety cage. In this setup, the safety cage, door sensors, and pressure mat make up the peripheral safety system. We could add in light curtains, extra E-stops, or any other safety devices to increase the safety factor for workers around the robot and they would all fall under the category of peripheral systems. However, we would not count the E-stop on the teach pendant or robot controller as part of the safety peripheral system, as they are built-in safety features of the core robotic system.

Figure 7-3   This Panasonic robot comes with its own safety cage setup. This system both ships and unloads as a unit, allowing the customer to get it up and running in a few hours (non-packaged systems can take days or weeks). The cage houses all the operations of the robot, has two sliding doors for access with sensors in place to stop the system if opened, and includes a light curtain in the front to sense when the operator is loading or working with parts. The cage, light curtains, door sensors, and any other bells and whistles for this system beyond the core equipment are considered peripheral equipment.

Image courtesy of Miller Welding Automation

safety note

Remember that it is an OSHA requirement that you take the teach pendant with you any time you are in the danger zone so that you have access to an E-stop. While this is not a requirement currently for nonindustrial robots, it is still a good idea to have some way to stop the system whenever you are close to the action zone.

There is a wide range of items that could fall into the category of safety peripherals. We could use a vision system to watch for objects or people in the path. Light curtains are great at detecting someone entering or working in a danger zone, depending on how we set them up. Pressure mats and overhead laser systems have ensured that people are clear of the danger zone for quite a while now. Remember, if it is not a part of the core robotic system, it is usually classified as a peripheral system. As Figure 7-3 shows, the manufacturer sometimes saves the customer time and effort by putting together a package system that is ready to rock and roll off the truck—saving time, money, and the trouble of proving out a safety system created at the customer's site. Even though it came as a package system, the extras are still peripherals.

Positioners

A common addition to the robot system is a device known as a positioner. This device manipulates the position of the parts that the robot is working with and has nothing to do with the physical position of the robot (see Figure 7-4). We use positioners to rotate parts, swap finished parts with raw parts, secure parts during processing operations, and complete other tasks that assist in the process of producing goods (see Figure 7-5). These positioners often come as part of a packaged setup for the robot but are separate systems, with their own motors, encoders, and other equipment necessary to manipulate parts. The swapping of raw parts for finished parts is probably one of the more common uses of the positioner. When used in this fashion, the system will often have a rotary platform with all the appropriate fixturing and clamps installed on it. The operator will put the raw parts in one set of fixtures, lock them in place, and then press an indicator button that sends a signal to the system controller that the parts are ready. When the system is ready to process parts, say, at start up or after finishing the previous cycle, the positioner

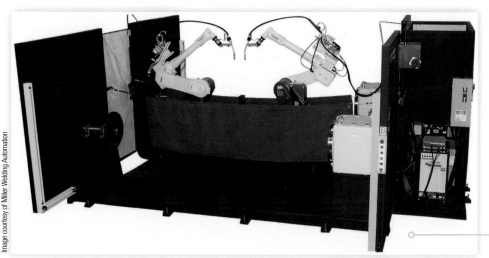

Image courtesy of Miller Welding Automation

Figure 7-4   This Panasonic system has a couple of positioners for part manipulation: they are the white units with the silver metallic ends on the right side of the picture. These precision servomotors have their feedback tied into the robot controller, enabling the system to complete some very intricate moves that would be difficult if not impossible without the external positioners.

figure 7-5 The white external motor on this system is positioned in such a way that it provides the power for swapping raw and finished parts around.

will rotate the raw parts into the work zone and, at the same time, rotate out the other fixture for the operator to remove finished parts and refill with raw parts.

One of my favorite robot positioners is the part positioner on a Panasonic PA 102S welding robot that we had as part of a robotic welding lab. The PA 102S is a robotic welding system designed for a drop-and-go type operation, meaning that you place it in the facility, feed it power and shield gas, secure it to the floor, do a quick calibration weld, and you are off to the races. If requested, Panasonic will set up operational programs so that the system can start producing parts as soon as the calibration weld is completed. The one I worked with was for educational purposes, so we did not have any programs from Panasonic; however, it did not take us long to set up a program and start fusing metal. While the robot often assists with production in industry, the Panasonic PA 102S (see Figure 7-6) is a production robot that welds parts without the need for other equipment (due to the welding tooling and power supply that comes with the system). When welding, the optimal position is on top of the part with about 15 degrees of angle for the torch tip, with the wire feed speed and amperage dependent on the welding wire used and the type of metals fused.

figure 7-6 The PA 102S system.

Now that you understand the basics of the systems, let me outline how the positioner worked. This positioner was a large servomotor capable of generating enough torque to rotate 50-pound parts with ease while maintaining precision and speed. The PA 102S integrated the positional information from this servo with the main system's programming, allowing the robot to monitor the positioner's location at all times and make movement calculations based on its position. The end result was a system where you could teach four points, all on top of the part at the optimal welding position, with the positioner rotating the part, and end up with a program that would weld a circular pipe perfectly as the positioner rotated. For fun, we placed a piece of pipe with a three-inch diameter on the outer edge of a one-foot piece of flat metal stock and then secured the flat metal to the positioner. Once finished, the result was a program that would weld the outside of the piece of pipe, in the optimal position, as the flat stock rotated in a circle on the positioner. Again, we taught four points for this program and set the proper amperage and feed for the material used, but the robot controller and software calculated the rest. This is only one example among the many out there, but I hope it gives you an idea of the versatility and options that a positioner can add as a peripheral system (see Figure 7-7).

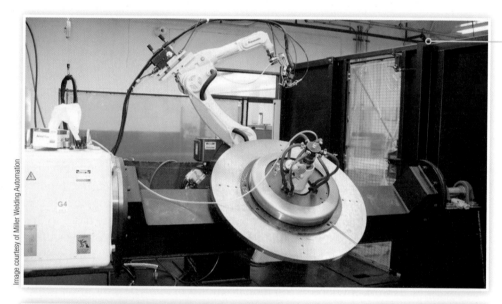

Figure 7-7 A positioner works with the welder to create the optimal welding angle.

figure 7-8 This FANUC robot is mounted on an overhead positioning peripheral, allowing it to work with more than one machine. This type of system may have multiple robots and can take care of many pieces of production equipment.

Copyright FANUC Robotics

NASA

Another type of positioner peripheral that you may encounter is a mobile robot base. We attach the robot to this positioner, and it moves the entire robot from point A to point B, with optional stops in between, to assist the robot in performance of tasks. We use these bases so that robots can keep up with moving parts on chains and conveyors, feed multiple machines in an area, move parts to and from warehouses, and carry out other tasks where the robot needs to cover an area larger than its work envelope. A common positioner of this type is the linear gantry base, so named for the design similarities to the robot of the same name, which moves the robot over a finite linear distance (see Figure 7-8). Wheeled base positioners sometimes fall into the realm of peripheral and sometimes the core system, depending on the construction of the robot. For example, we would consider the wheels of a robotic warehouse attendant part of the core system, where mounting a robotic arm on a generic wheeled base would make the wheeled portion a peripheral. Remember our litmus test: if it can stand separately from the robot it is a peripheral; if not, it is part of the core (see Figure 7-9).

figure 7-9 Since the Robonaut is designed for a variety of bases, this mobile base is considered a peripheral unit for the system and not part of the core system. The robotic handshake image from Chapter 1 shows the Robonaut on a completely different type of base.

Many robotic systems take advantage of the options provided by peripheral positioning systems. Whether we use them to swap raw and finished parts, rotate parts during the process, or move the whole robot, they add flexibility and create options that the robot would not have otherwise. We often refer to these units as external axes because they link to the robot controller, the system stores positional data for them, and we can program them as we would other axes. This functionality allowed the PA 102S system aforementioned to weld in the optimal position while completely welding a circle around the pipe. If not for the positioner, at some point, the system would have welded upside down, which is a difficult position from which to get a good weld. Another benefit is that the robot works just fine without these axes, so we can turn them off or ignore them when they are not needed or are malfunctioning. Many systems in the industrial world use positioners to great advantage, so there is a high probability that you will run across them if you work with robots in industry.

End-of-Arm Tooling Peripheral Systems

Tooling can be a peripheral system on its own and may come with added peripheral systems that relate to tooling itself (see Figure 7-10). Remember, EOAT is an add-on to the robot; though some manufacturers send certain tooling standard with the robot, they are standalone systems. They require a power source and data input, and often create data output, but usually you can swap them between

Figure 7-10 Here you can see an ABB robot with a spot welder attached as the tooling. This is one example of the multitude of choices for tooling peripheral systems.

Image courtesy of ABB Inc.

robots of similar construction with little difficulty or alteration, thus earning them their peripheral status. We focused on many of the types of End-of-Arm tooling in a previous chapter, so in this chapter, we will delve deeper into the support systems that might accompany tooling.

One type of tooling peripheral system that you may encounter is a type of positioner that has nothing to do with the part, but instead helps in changing robot tooling. These systems come in wide variety of configurations and operate differently, but their basic principle is to move the desired tooling into position for exchange with the tooling the robot is currently using (see Figure 7-11 for an example). Before we place new tooling on the robot, the current tooling has to go somewhere. The positioner helps with this as well by presenting an empty slot to the robot. The robot puts the current tooling in that slot and then releases the tooling via a coupling mechanism. Once the robot is clear, the positioner will index the required tooling into position and wait for the robot to connect the quick release to the tooling (see Figure 7-12). Once it makes a positive connection, the robot carefully removes the tooling from the positioner, and it's back to work for the robot. Some systems will perform this operation multiple times in the completion of a single cycle of the program, allowing one robot to take on multiple rolls in the work process.

Image courtesy of SCHUNK

Figure 7-11   Tool changers, like the one pictured, are a common peripheral for robot tooling.

Image courtesy of SCHUNK

Figure 7-12   This tooling is designed to work with quick-release devices. The coupling device clamps around the shaft, making a positive connection in the grooved portion.

Positioners are not the only peripherals helping the EOAT; these are but one type of a wide variety of existing support systems. In the welding world, there are systems to clean the welding gun tip of any slag or stray pieces of melted metal that might accumulate therein. For the drilling world, there are sensors to ensure that the drill bit has not broken or worn down beyond a set point. There are cleaning devices to ensure that spray nozzles do not clog up between uses or to clean out one fluid before using a different fluid. There are force sensors and measurement systems that help determine the quality of parts. If there is a support need for tooling, you can bet there is a peripheral system to fill that role.

The peripheral system that goes with the tooling may perform tasks such as drilling, reaming, polishing, or other finishing actions. Figure 7-13 shows a robot using a peripheral motor with a drill attached to perform tasks. This type of setup not only saves time in the production process, it can also eliminate the need for extra equipment by having the robot perform part of the production process. When we use external motors in this way, the data that the system monitors is usually the amount of torque or amperage that the motor is using as well as the speed of rotation; this is instead of the positional data that we monitor when using motors for part positioning (see Figure 7-14).

The field of tooling peripherals is evolving just as rapidly as the world of robot tooling. As you encounter these devices, make sure that you take some time to understand how the specific system you are working with operates, what commands the robot uses to interact with the system, and how to maintain these systems (see Figure 7-15). Failure to do so could result in improper operation of these systems, crashes, part damage, and other less-than-desirable situations.

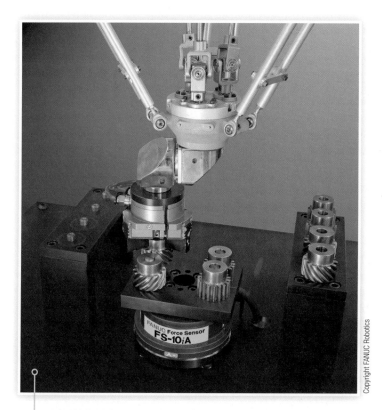

Copyright FANUC Robotics

Figure 7-13 Here is a force sensor set up to check the geometry of gears using a Delta robot.

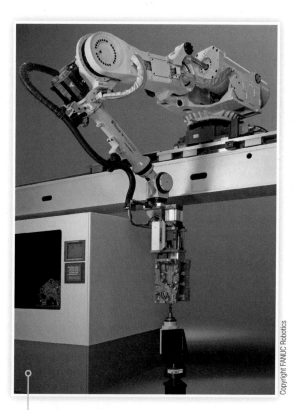

Copyright FANUC Robotics

Figure 7-14 Here you can see a gantry-mounted robot using a standalone motor to perform a drilling function.

figure 7-15 This welding robot has several peripheral systems, include two large pistons that use a set of jaws to hold parts.

Image courtesy of Miller Welding Automation

What Is a Work Cell?

work cell
A logical grouping of machines that perform various operations on parts in a logical order during the production process

lean manufacturing
An initiative that is all about cutting production costs by minimizing wasted time and materials

In industry, one of the current buzzwords to describe a group of machines is cell, or work cell, which is a logical grouping of machines that perform various operations on parts in a logical order as a part of the production process. These machines may work independently of one another, but often share information. The machines are located in close proximity to enhance the production process by creating a smooth, efficient workflow. Work cells are part of the lean manufacturing initiative, which focuses on cutting production costs so that U.S. industry can compete with countries that make products more cheaply than we do.

What does the work cell have to do with peripheral systems? Work cells are groupings of standalone systems, machines, and auxiliary equipment all working together and often they utilize robots for material handling and other operations (see Figure 7-16). Work cells are a great example of how all these various systems come together and interact in order to get the job done. The cells may be as simple as a welding robot, positioner, welding gun tip cleaner, and an operator (see Figure 7-17). Or they may consist of a welding robot, a machine that removes materials from raw cast parts, a conveyor that connects this machine to the welder, and another conveyor that takes the welded parts to a robot that picks them up and uses a pedestal grinder to buff and shape parts after a camera does a visual inspection. Often though, work cells are complex setups consisting of multiple machines all working to process parts, like Figure 7-18, or the robotic department at a car manufacturer that welds the vehicle frame in less than an hour. This cell has hundreds of machines and peripherals all working in unison on the car body from the time it enters the department until it leaves for the next section.

The work cell also demonstrates the difference between a standalone machine and a peripheral system. If we go back to the previous example (the system with a welding robot, a milling machine, two conveyors, a part-handling robot with camera, and a pedestal grinder), we can see the difference between the standalone machines and peripheral equipment. The welding robot, part-handling robot, milling machine, two conveyors, and pedestal grinder are all standalone pieces of equipment. Typically, we do not count standalone machines as peripherals unless they are under the control of the robot. The camera, welding gun tip cleaner, and positioning system for the welding robot are peripherals. If the robots have safety systems or cages, those count as peripheral systems. Any sensors added to the system beyond what came with the core equipment packages would also fall into the

Figure 7-16   Here is a robotic welding cell working on cars as they go by. These robots are not only a part of the work cell, they actually make up the work cell. Each robot has a specific task to perform as the car body comes by that builds on the work of the previous robot.

Image courtesy of ABB Inc.

Figure 7-17   This Panasonic welding cell is ready to go to a customer for use. If you look closely, you can see that the light curtain on this system is mounted at the base of the operator area instead of the entrance. This configuration monitors the entire operator work area instead of just the entrance and thus ensures operator safety at all times.

Image courtesy of Miller Welding Automation

peripheral category as well. Another way to think of peripheral devices and systems is that they augment or enhance the operation of a core piece of equipment.

If you spend any time working in industry, it is a safe bet that you will come into contact with work cells. This method of manufacturing has proven efficient

Figure 7-18 Here are a couple of robots working with various machines in their area to create a work cell.

Image courtesy of ABB Inc.

and withstood the trial of time. Many companies have spent thousands of dollars creating, setting up, and tweaking their work cells to reduce the number of times that they handle the parts, how far parts travel between operations, and how long it takes to process the parts. All of these reductions represent a reduction in the overall cost of parts, and thus a savings to both manufacturer and customer.

Nonindustrial Peripherals

Industry is not the only place we find systems to help the robot perform tasks. We use the same logic test to define these systems; they augment the operation of the robot but are not a part of the core system. Just like their industrial counterparts, the peripherals in other fields of robotics fit a specific need of the robot to increase flexibility and enhance operational capacity. As with the previous sections, we will look at a few examples from the wide gambit of possible options.

A peripheral that is common in the private sector of robotics is the charging station. A popular option for systems like the Roomba and Lawnbot is to have a charging station where the robot can feed itself whenever its batteries are low. Some robots find these systems by carefully tracking the location, others can follow a signal home when in need of a refuel. These additions to the robot arsenal turn service robots into "set it up and forget it" systems that users love. The selling point is to make life easier for the user by negating the need to remember to recharge the robot manually. The advancements in wireless charging is sure to make an impact on this robotic peripheral as the technology gains widespread acceptance and more companies offer these devices.

When is a camera not just a camera? When the camera is the payload for a popular hobby robotic system known as the Quadcopter. Home users are not the only clientele it attracts. Hollywood has been using the Quadcopter to carry expensive cameras on high-tech gyroscopic mounts for a while now, replacing the costly helicopter flybys and capturing camera angles and sweeps that would otherwise be difficult if not impossible with traditional equipment. Fear not, there are hobby systems out there like the AR Parrot Drone, which comes standard with video capability that you can control with an app on your smartphone. You too can have the panning, sweeping shot for your own home movies. Or, you can just use the camera to make sure that you do not fly into anything should the Quadcopter end up our of your line of sight.

For military and police applications, robots are equipped with a variety of peripheral systems, depending on the task. The police may want an offensive system to engage the bad guys without risk to life and limb. The military may be worried about hidden explosive devices and require specific sensing equipment. Other robots are all about gathering data and sending it back to their operators. Whatever the need, many of these systems are designed so that various equipment can be added or removed to customize the robot in the same way we swap tooling in industry. When set up in this way, a robotic company can manufacture a large number of just a few types of robots instead of having to build a few of a large number of different robots. This reduces production costs, saving the customer money, while at the same time giving customers the options they need to get the robot they want. The reduced cost also means that more systems find their way into use, with all the potential lifesaving benefits that go along with that.

This is by no means all the different areas outside of industry that use peripheral systems. However, it does give you an idea of some of the systems that you may encounter. As we continue to invite the robot into our homes, business, and lives, we can expect the various options for support equipment and systems to grow.

System Communications

No discussion about peripheral systems would be complete without discussing the communications aspect. Without proper communications, most peripheral systems are wasted money at best, and a recipe for certain disaster at worst. The worst-case scenario is where some signals make it through while others are lost, creating intermittent operation and functionality that can become erratic and unpredictable. To give you a comparison, imagine that you and a blind individual from another country are alone in a room. You see smoke from a fire coming under the door. You try to tell the blind person what is going on, but you do not share a common language. The blind person cannot see the smoke but can smell it, so he or she knows that something bad is happening but not where it is coming from or how best to avoid it. The chaos that could ensue is very similar to a robotic system trying to respond to intermittent information or information that it can only partially decipher. With this in mind, let us explore some of the ways we make sure that the signal reaches the robot in a timely, understandable manner.

Before we get too deep into the how, we should take a moment to talk about the two main types of signals we use to transmit data. Digital signals have exactly two states, on or off, which translate as 1 or 0 in the binary code. We base these signals on a set voltage or amperage level when they are based on real-world equipment and not just data in the system. Analog signals are a range of voltage or amperage, both in the milli range, that correlate to a set scale inside the system. We often use digital signals to indicate the presence or absence of things with sensors like prox switches and limit switches, or any other place where a yes or no answer will do. Analog signals come in handy when we need to monitor

digital signal
A signal that has exactly two states, 1 or 0

analog signal
A signal with a range of voltage or amperage that correlates to a set scale used to express changing values in the sensors and output devices

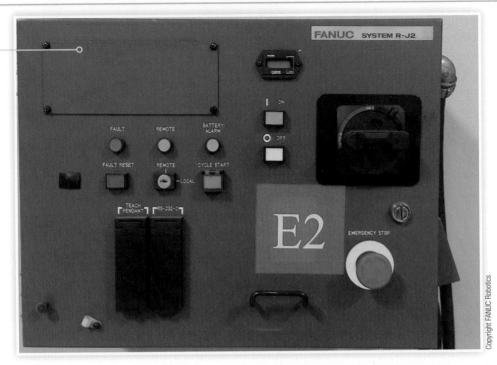

Figure 7-19   If you look closely, you can see two covered communication ports on the R-J2 controller. The one on the left is for the teach pendant, while the one on the right is the RS232 connection.

ranges of data, such as the temperature of a room, the distance between the robot and an object, or how full a tank of liquid is. Hooking an analog signal to digital input and vice versa is a quick way to ensure problems with the system and data interpretation.

There are two main ways to get the signal from sensors, machines, and peripheral systems to the robot. The oldest and most direct method is to bring the signal directly to the robot via wire. In the beginning, this usually involved taking the signal wire from a device and hooking it to an open input of the appropriate type on the robot. Direct communication does eliminate some of the complexity of other methods, but it does require a physical path from device to robot for the wires.

RS232 standard
A communication standard where certain wires are designated to transmit specific signals

The next big advancement in direct communication was the RS232 standard (and others like it), a communication standard in which certain wires are designated to transmit specific signals such as clear to send, request to send, transmit data, receive data, and so on. This set up a specific way for two controllers to share data, allowing for large amounts of data to pass quickly back and forth (see Figure 7-19). Now, instead of having to run 50 wires from the safety system to the robot, all one needed was an RS232 cable from the safety system controller to the robot controller. The downside was the need for more controllers and the specific hardware to make this happen. In addition, RS232 was a standard, not a set law or rule. Some companies liked to switch the function of the pins, creating a proprietary system that required their cables or setup in order to get proper communication. RS232 and similar data communication setups use the nine-pin serial port, which was standard on computers until a few years ago, to communicate between devices and computers (see Figure 7-20). Many modern computers, especially laptops, no longer use the nine-pin serial connector for data transmission, which is why the RS232 standard is falling out of favor for industrial communication.

Because the serial adaptor is no longer a standard option with many computers, USB communication cables have replaced the RS232 communications of

figure 7-20 Here is a close-up shot of the communication side of the controller for a SCORBOT. You can clearly see the connection for the teach pendant, the RS232 connection, the robot communication connection, and the nine-pin connectors for various axes and auxiliaries, such as the optional gantry base.

days gone by (see Figure 7-21). The principle is the same, but the data transmissions are faster and there are fewer problems with pin swapping. An added bonus is that USB is capable of sharing power over the wire, albeit in small amounts. Given that most industries are a mix of old and new systems, there is a good chance that you will see all the various hardwired methods that we have discussed, and a few other direct communication protocols that we did not. No matter what you encounter, make sure you do the proper research to get a good feel for how data should flow and what to look for when communications break down (see Figure 7-22).

As we mentioned previously, systems communicate in two main ways. The second method is via a network. Networking is where two or more systems share information over some form of connection. These systems use ethernet cables for the hardwired setups or some form of wireless technology to transmit information back and forth. Networking is the trend of the day, and industry is really moving toward this style of information sharing. Hardwired systems require a communication cable or wires between the devices that

networking
Two or more systems sharing information over some form of connection

figure 7-21 Look closely to see the USB communication port on the controller next to the three connections used to communicate with and power the servos. This is the port used to update or manipulate the data in the controller, including sending new programs from the computer.

Figure 7-22   Devices like this KEYSPAN adaptor can allow you to adapt RS232 communication to USB communication; however, the problem is that they do not always work with the system you are trying to communicate with or the computer you are using them on. If you go this route, you will likely have to install software on the computer to get the adaptor to work properly.

Figure 7-23   Here is the smartphone peripheral for the Robonaut 2.

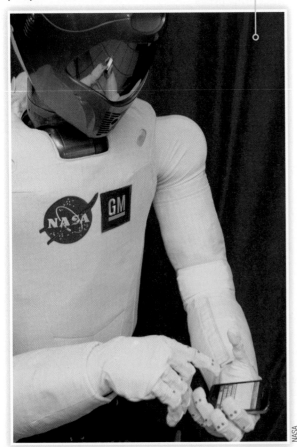

NASA

need to talk, whereas wireless networks only need a device or card that works similar to a Bluetooth device to communicate. Wireless networks allow new equipment to share data with equipment in other areas of the plant without the need to run a lot of wires. Another benefit of networking is the ability to tie computers into the network for real-time monitoring of data. With this kind of setup, an engineer or maintenance technician can turn to a nearby computer and gather information on why a machine has alarmed out before he or she has even set foot on the production floor. When you add Internet connectivity and highly automated systems, you create options such as receiving emails or texts for predefined conditions and the ability to start production runs from the house with nothing more than a smartphone (see Figure 7-23).

Just as information is lost when everyone talks at once in a crowded room, so too do we lose data if all the systems try to transmit and receive data at the same time. Because of this, networks have various ways to make sure that the data gets where it is needed and not lost as a garbled mess along the way. Some systems will check the network for transmissions before transmitting; if information is flowing, they will wait until the lines are clear or a predetermined amount of time has passed and check the line again. Other systems pass a digital token, which is just a bit of code that is the network equivalent of raising your hand to talk. Some systems have a designated leader, the master, and designated followers, the slaves, where the slave systems only transmit data when the master system requests it. These are a few of the ways that we ensure data is transmitted as intelligible information instead of a mess from multiple sources with no way to sort it all out.

For those of you interested in learning more about data communications, I encourage you to research networking and perhaps take a class or two on the subject. With our love of data and all we can do with it, you can rest assured that we will continue to look for new ways for everything in our lives, whether it be at home or the office, to talk effectively back and forth. Wireless technology is the hot field in networking currently and already has industry buy-in, so that might be a great place to start.

review

By now, you should have a decent idea of what a peripheral system is and how it affects the world of robotics. From the safety system in industry to the charging station for cleaning robots, peripheral systems are enhancing the operation of robots and making them more useful to their human operators. The few examples from each section in no way cover all existing possibilities, so you still have lots of room for discovery during your exploration into robotics. During our examination of peripheral systems, we covered the following topics:

- **What is a peripheral system?** We talked about what a peripheral system is and how to tell it apart from the core robot.

- **Safety systems.** This section dealt with the peripheral systems from the safety side of things.

- **Positioners.** This section covered both part positioners as well as ways to move the entire robot.

- **End-of-Arm tooling peripheral systems.** We discussed some of the peripheral systems that help with robot tooling as well as the fact that EOAT is a peripheral system.

- **What is a work cell?** This section presented an overview of what a work cell is as well as how all the systems work together, including the peripherals.

- **Nonindustrial peripherals.** Support systems are not industry-only systems; we discussed some systems used with robots in the home and in other places.

- **System communication.** This section gave a brief overview of how we transmit data back and forth in an intelligent manner.

key terms

Analog signal	Networking	Programmable	RS232 standard
Digital signal	Peripheral system	Logic Controller (PLC)	Work cell
Lean manufacturing			

review questions

1. How do we determine if something is a peripheral system to the robot or a part of the robot?

2. What are the components of the core robotic system?

3. Who is responsible for the safety protocols of an industrial robot?

4. Do we classify the E-stops on the teach pendant and the robot controller as peripheral safety systems? Justify your answer.

5. Describe the basic operation of a positioner system used to swap raw and finished parts.

6. What are some of the tasks associated with mobile robot bases?

7. Describe how a tool-changing peripheral works.

8. When we use external motors to help with the production process instead of positioning parts, what data do we normally monitor at the robot controller?

9. What are some of the benefits of the work cell?

10. What are the criteria for a peripheral system for nonindustrial robots?

11. What is the benefit of building a base model that works with several peripheral systems?

12. When it comes to communication, what do we consider a worst-case scenario?

13. What is the difference between digital signals and analog signals?

14. What was the downside to RS232?

15. What is the difference between a hardwired network and a wireless network?

16. What is the benefit of adding Internet connectivity and highly automated systems to a network?

17. What are some of the ways in which networks ensure that all the machines are not talking at once?

Robot Operation

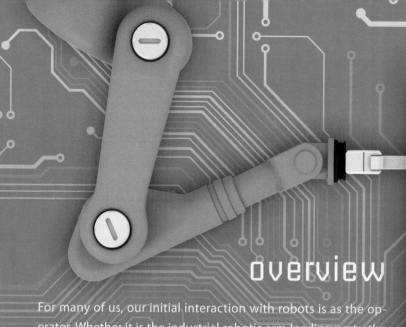

overview

For many of us, our initial interaction with robots is as the operator. Whether it is the industrial robotic arm loading parts, the Roomba robot cleaning floors, or a Robosapien just for fun, many of us will spend less time building, repairing, and programming the robot and more time running the system. That is the focus of this chapter. Many times, we think of robots as a turn-on-and-go type of system, but there are other concerns we should address along the way. During our exploration of robot operation, we will examine the following topics:

- Before powering up the robot
- Powering up the system
- Moving the robot manually
- What do I want the robot to do?
- How do I know what the robot will do?
- So, I crashed the robot… now what?!

What You Will Learn

- Some of the basic checks to perform before powering up the robot
- What to check once the robot has power, including safety checks
- What frames are, and how the robot moves in various frames
- The right-hand rule
- How to determine what program to run
- Precautions to keep in mind before starting the robot for the day
- How to check programs manually when you are unsure of what they do or how they work
- Two common ways to run the robot manually
- The five steps of crash recovery, and what each step entails

Before Powering Up the Robot

In the modern world, many devices need some basic checks before use. The field of robotics is no different. In industry, most operators have a daily checklist of inspections to perform at various times throughout the day; this list often includes tasks that require the robot to be powered down, for safety reasons. The types of checks performed depend on the system and application in question, but some general checks are recommended by most manufacturers.

Checking the robot for damage before applying power is a common task on the checklist. This includes looking for dents, missing paint, broken tooling, cut wires, damaged cables/hoses, or anything else that appears to be out of place or damaged (see Figure 8-1). During this inspection, we want to look at the peripheral systems as well. Is the safety cage intact? Are all the sensors in good physical condition? Are all the pieces of the fixture tight and functional? The point is to make sure that everything is in good physical condition before we apply power and invite a potentially hazardous condition, such as erratic operation or energizing exposed electrical wires. This is also a good time to clean the system if there is a buildup of grime or dust. When cleaning the robot, avoid using

Figure 8-1 This close-up of an ABB robot shows some of the connections and lines that are commonly inspected before startup for the day.

Image courtesy of ABB Inc.

compressed air, as this may drive foreign materials into the joints and bearings of the robot.

Another common check is to ensure that all the connections are tight and secure. Robotic systems have several different parts: the main robot, controller, teach pendant, peripheral systems, and work cell communications. These common components require signal and/or power cables. Vibrations from the normal operation of robotic and nearby industrial systems can cause connections to loosen, and any loose connections between these systems could lead to erratic system behavior as well as damage the equipment. One of the more serious situations is when the robot tries to maintain a designated position with the signal from the positioning device only reaching the system part of the time or corrupted along the way. Under these conditions, robots can move a large distance at full or near full speed, creating an unexpected movement that is dangerous to people and equipment alike. This is why we never remove communication cables from a robot while it is powered up and why we check connections with the system powered down (see Figure 8-2). Of course, the other danger is that the connection may have exposed contacts, creating an electrocution hazard. Depending on the type of connection involved, you may have to physically turn a twist lock or perform a visual check to make sure the locking device is still firmly in place.

safety note

⚠️ You should always plug in devices as if the power were on, making sure to keep your fingers away from the metal connection prongs. Getting in the habit of letting your fingers brush the prongs or metal connections can lead to an electric shock down the road, with all the danger that entails.

Figure 8-2 Here is an example of the cabling that is part of the SCORBOT pre-power checks.

Certain operations will require their own unique pre-power checks, so make sure that you understand how the systems works and what it does. A good example of this is the welding industry, where the robot's tooling is a welding gun that requires its own specific checks. One important check is to make sure that the ground between the welding power supply and the fixturing is secure (see Figure 8-3). If this becomes loose, the welds will not get proper penetration, which will cause problems with part quality and can lead to buildup of slag. Slag is molten metal splatter that has hardened in the welding gun tip; this is another pre-power check and fix issue. If the robot cuts, grinds, or welds metal, you may need to enter the work envelope to clean up waste from the robot's operation to prevent problems. It is much safer to do this with the robot powered down and locked out. Again, the specifics will depend on the task that the robot performs and the equipment involved in the process.

slag
Molten metal splatter that has hardened in the welding gun tip

Figure 8-3 Here is a welding robot in operation; the process generates sparks made of red-hot metal.

Image courtesy of Miller Welding Automation

Many nonindustrial robots require the same inspections before power is applied. You want to make sure that the robot is in good working order, with no visible damage and that all the safety systems are present. For battery-powered systems, I would recommend making sure that the battery is at full charge for maximum performance. Ensure that any cables for the system are securely in place. Just as we saw in industry, certain types of robots will require special attention. For a lawncare robot, check the blades or plastic line, addressing deficiencies as needed. For a floor-care robot that works in zones or designated areas, make sure that the zone markers are in good condition and properly powered. To prevent mayhem, we want to check all that we can to make sure that the robot is in good working order *before* we apply power to the system.

The kind of pre-power check that a system requires will depend on the robot that you work with. Many systems have unique items on the checklist that we have not covered. As robots evolve and become more complex, we are sure to find more checks on the daily list. If you happen to encounter something that needs to be checked, but is not a part of the routine, bring it to someone's attention. The daily checklist is a human creation and as such, there is always room for improvement.

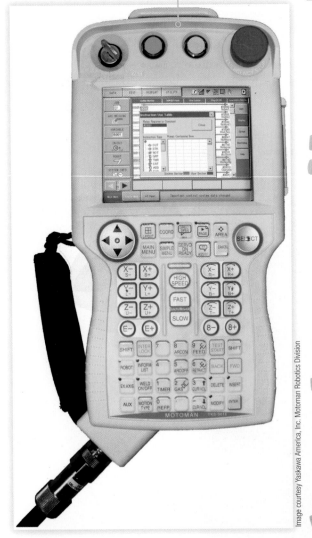

Figure 8-4 As a rule, most industrial systems provide messages on the teach pendant stating when the system is booting up as well as ready to go. Some systems go to a standard menu screen once the system is ready to go, so make sure that you understand the specifics of the robot you are working with.

Image courtesy Yaskawa America, Inc. Motoman Robotics Division

Powering Up the System

Now that we have completed all the pre-power checks, it is time to turn on the robot and begin our next series of checks. (Yes, there are more checks to perform before we just hit the big green button and walk away.) But before we begin the power checks, we have to give the system enough time to boot up and gather data. Just like a computer, a robot must go through a boot sequence, where all the computational systems get up and running, all the available data is checked, and the robot completes any necessary logical sorting so it knows what state it is in, where it is at, and what is going on in the world around it. Just as one computer takes less than a minute to boot up and another computer takes several minutes, each robot will require a specific amount of time to get ready to go (see Figure 8-4). During this time, do not try to start any programs or perform any checks of the safety equipment; that may cause problems with the boot sequence and lead to erratic robot performance or alarm conditions.

Once the system has completed the boot sequence, the first thing we check is the safety equipment. You certainly do not want to find that an E-stop no longer works when it is all that stands between you and serious injury! You should check each E-stop separately and reset the system in between to ensure its proper operation. Note that if there are five different E-stops for the robot, you must perform five separate checks. Most industrial robots have at least one E-stop each on the teach pendant and controller. This may not be all the E-stops that control the robot, as it is common to place several more around where the operator works. Once you have confirmed that the E-stops are working, check the safety systems, such as door interlocks for the safety cage, sensing mats, light curtains, and any other devices that have a primary function of protecting your safety. Next, check all the various sensors to ensure that they have power and seem to be functioning properly. Even if it is not on your official daily checklist, a quick look at the indicators and LEDs of the sensors can prevent frustration during operation.

If you have an alarm condition, you must figure out what kind of alarm it is. Some alarms are simple notifications, such as the battery or tooling life has expired; in this case, you can reset the system and then run it. We commonly call these "reset and run" or "reset and forget" alarms. However, failing to deal with the root problem could cause broken tools, damaged parts, or a lost robot position down the road. A critical alarm prevents the machine from running (until it is corrected) when a situation such as an E-stop problem or a loss of positional data for an axis occurs. Complex industrial systems come with maintenance manuals that address the causes for specific alarms and often include a best-case corrective action as well. You can always search the Internet and contact the manufacturer if you need more information. (We will examine troubleshooting and maintenance in detail in Chapter 10.)

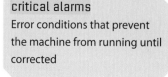

critical alarms
Error conditions that prevent the machine from running until corrected

After the safety checks are complete, examine the robot's hydraulic or pneumatic system for leaks, proper pressure, and clean filters. Most hydraulic and pneumatic systems have indicators that let you know when the filter needs to be changed. For hydraulic systems, check the tank for proper fluid level and make sure any cooling systems are functioning properly. Low or high fluid can cause pressure issues, while a clogged or faulty cooling system, sometimes called a heat exchanger, often leads to excessive heat in the system. There may be specific checks related to the robot you are working with, depending on the design of the system, and this is a good time to perform these checks as well.

The whole point of powered checks is to make sure the system is safe to run and ready to go before you engage the first program of the day. Many times, nothing changes from day to day, and this can lead to complacency, where operators feel that the checks are just a waste of time. If the choice is to run through the same set of checks a thousand times or suffer injury or death, which would you choose? I hope that your answer is a thousand checks, and that this sticks with you during your robotic career. (For those of you who think I am exaggerating, look at Figure 8-5.)

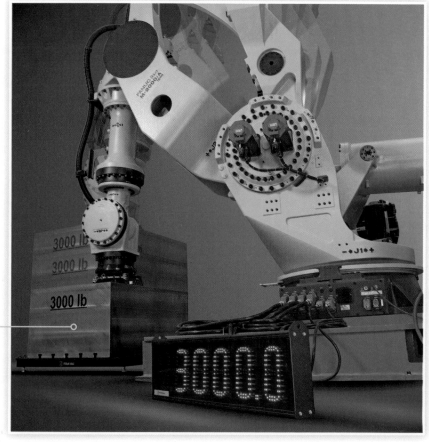

Figure 8-5   As you can see, this robot is working with more than a ton of weight. Going through the proper daily checks can mean the difference between normal operation and a 3,000-pound problem landing on your head!

safety note

Even though you have a daily checklist and safety systems in place for your protection, you are ultimately responsible for your own safety. Simple things, like taking the daily checks seriously and paying attention to what the robot is doing, can make all the difference when it comes to personal safety.

Moving the Robot Manually

Now that you have gone through the initial power up checks, put the robot in manual mode and move the various axes while you look for jerky motions, listen for odd sounds, and pay attention to any changes in operation (see Figure 8-6). This generally only takes a couple of minutes, but it can save time and money down the road in lost production time and repair parts. For this portion of the testing process, you will put the robot in the manual mode and you will use the teach pendant to control the operation of the system. Remember, if you enter the work envelope of the robot for any reason, you must take the teach pendant with you so that you have control of the system and access to an E-stop. While

figure 8-6 It is crucial to be careful when you move the robot while in close proximity!

performing movement checks, make sure that you do not run the robot into any stationary objects, as this could result in damage to the robot, object, or both. A good way to know how the robot will move in manual mode is to use the right-hand rule for robotics.

The right-hand rule is a way to determine how the robot moves while using a Cartesian-based frame motion with our thumb, forefinger, and middle finger. In this case, frames are a way to reference movements and points in the work envelope that control how the robot moves. A standard feature of modern industrial robots is a World frame, which is a Cartesian system based on a point in the work envelope where the robot base attaches. When the robot is in this frame of movement, it moves in straight-line motions that we can predict with the right-hand rule, which allows the user to avoid unexpected impacts or damage. While using a World frame, orientate yourself in the same direction that the robot is facing and, using your right hand (lending the rule its name), stick your thumb up, point your forefinger straight forward, and point your middle finger toward the left side of your body at a 90-degree angle to your forefinger (see Figure 8-7). Your thumb will be pointing in the positive direction for the Z-axis, your forefinger in the positive direction for the X-axis, and your middle finger in the positive direction for the Y-axis. Bear in mind that this only works with the World frame and you must orientate yourself in the same direction as the robot or look from the robot toward the work envelope. If you try to use the right-hand rule while you are facing the robot, you will actually be pointing in negative X and negative Y, which could cause a crash if you think of these as positive directions.

On the topic of frames, let us discuss the differences between Joint, World, and User frames, or coordinate systems. The term frames and coordinate systems are interchangeable when used in reference to robot movement, so do not be alarmed if you see frames in one place and coordinate system in another throughout this book or in the reference material for a robot. World frame, as denoted earlier, is a Cartesian-based system that uses the origin or zero point for the robot as the zero reference of the frame. The zero point for industrial robots is usually a point in the middle of the base that we bolt to the floor or other surface.

User frame is a specially defined Cartesian-based system where the user defines the zero point and how the positive directions of the axes lay, including the option of straight-line motions moving at an angle instead of the flat lines of the world frame. Because this is a user-defined orientation, the right-hand rule is useless with this frame, as the positive directions can be orientated however the programmer desires. We typically define User frames in relation to fixtures or other points of interest within the robot's work envelop to make setup and movement easier for the programmer. Most modern industrial robots allow for multiple User frames, with 10 or fewer the common option. Both World and User frames result in straight-line motion of the robot while moving the major axes in manual. This straight-line motion often requires multiple axes of the robot working in unison under the direction of the robot controller and the mathematical formulas therein.

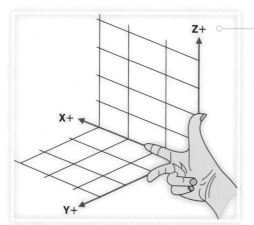

right-hand rule
When facing in the same direction as the robot, your thumb will be pointing in the positive direction for the Z-axis, your forefinger in the positive direction for the X-axis, and your middle finger in the positive direction for the Y-axis

frames
A way of referencing movements and points in the work envelop that controls how the robot moves

world frame
A Cartesian system based on a point in the work envelop where the robot base attaches

coordinate systems
Another term for frames

user frame
A specially defined Cartesian-based system where the user defines the zero point and how the positive directions of the axes lie

Figure 8-7 Remember to have your index finger pointing in the same direction the robot faces—in other words, in the same direction as the main part of the work envelope.

figure 8-8 Here is an example of the types of motion that you can expect while moving in the Joint frame or mode.

Based on: Miller Welding Automation

Joint frame
A reference system that moves one axis at a time, with the positive and negative directions determined by the setup of each axis's zero point

Unlike World and User frames, the Joint frame only moves one axis at a time while in manual mode with the positive and negative directions determined by the setup of each axis's zero point. This is the perfect mode to test each axis of the robot independently and, in the case of problems, determine which specific axis is making noise or moving oddly. The downside to the Joint frame is that the only linear movements of the robot are those axes that are linear in nature. Since most robots primarily have rotary axes, the robot will move in a circular fashion during manual motion (see Figure 8-8). On top of the nonlinear motion, the multiple axes' movements that the controller performs for the other frames becomes your responsibility in Joint mode. The result is a near inability to perform straight-line manual movements in Joint frame and the need to switch among multiple axes to get the robot to a desired point in the work envelope.

While we have the robot in manual, it is a good time to change out tooling, move the robot to inspect areas that we could not see previously, or move the robot out of the way so that we can reach other items in the work area. While working on anything in the robot area, especially those tasks requiring two hands, make sure you can reach the E-stop easily and quickly. If you moved the robot to gain access to cabling for tightening or repair, make sure you power down the robot before doing so, as this could result in unexpected movement or operation of the system. No matter why you move the robot manually, make sure you return it to a safe or home position before starting the system up. Leaving the robot in a random position could result in the robot alarming out at startup or worse, impacting fixturing or other items in the work envelope as it travels to the home position.

Home position is a point defined in the work envelope of the robot where the system and its tooling are clear of impact with any fixtures or other equipment in the area. Generally, this is an arbitrary point chosen during the initial setup of the robot that is used by all the programs to make sure that the robot starts clear of any impacts.

Nonindustrial systems should also have some form of manual movement; however, you will often only have one type of frame or motion available. Complex systems, such as a Quadcopter, may only offer you World frame motions, while a hobby robotic arm may only have Joint frame motions. When dealing with these systems, make sure that you take the time necessary to understand what your movement options are and how the robot responds to manual commands. When you first start experimenting with moving these systems, make sure that you have plenty of room around the robot and that everyone is a safe distance away to minimize any risks involved.

When it comes to manual motion, no matter what type of robot you are working with, you need to take the time to learn how your system moves and get a feel for what your options are. When you first start moving a robot in manual, set the movement speed at a very low rate if at all possible. This will give you a chance to respond to any issues that might arise and minimize the force involved should contact occur. Do not be afraid to create a cheat sheet or put some kind of a movement reference chart in a convenient place to help you remember how the robot moves. Taking the time to get to know your robot and its motion can make all the difference when it comes to avoiding or detecting motion problems.

What Do I Want the Robot to Do?

At this point, we have checked our robot without power. We have also powered up the robot and checked safety systems, sensors, and all the basic functions within reason. We have moved the robot to check proper movement as well as any other pre-startup tasks involving the physical condition of the robot. You might think, So we are done, and everything is cherries and roses, right? While it is true that the system is ready for duty, we have one very important task left to perform, determining what we want the robot to do! Without a mission, function, task, program, or however you wish to define it, the robot is at best an expensive toy and at worst a techno paperweight. So now, we need to give the robot something to do to make it a useful piece of equipment. This is where the rubber meets the road.

We often define the function of a robot by where and how we use it. In industrial settings, there are some general names for these tasks. Pick-and-place robots take parts from one place and put them in another (see Figure 8-9). Welding robots fuse pieces of metal together using a welding system for EOAT (see Figure 8-10). Inspection robots use some type of specialized sensor(s) to determine the quality of parts and then respond accordingly. Palletizing robots take boxes of product from conveyors or elsewhere and stack them neatly on pallets for transport. Lawnbots keep the grass to a set length while avoiding obstacles and living things. The list goes on and on…. I am sure you get the idea.

Figure 8-9 Here is a Panasonic welding robot in action.

Image courtesy of Miller Welding Automation

figure 8-10 A FANUC system palletizes bags of chocolate for future delights. As you can see, working with robots can be a sweet job.

Given that the task sets what the robot does and that the robot is already set up for that task, we should just be able to hit the go button and let the robot do the rest, right? Again, it is not that simple. Most industries create more than one part, and thus have the robot perform more than one task. During the process of turning raw materials into finished goods, the raw parts and materials tend to undergo many changes, which affect how the robot deals with those parts (see Figure 8-11). Sometimes you need to run a subroutine or mini program before starting normal operation, such as making sure that the welding gun tip is clean or that the right tool is in position for the process to begin. All of this variation causes situations in which the robot must react in specific manners to the situation at hand, which equals multiple programs to choose from. These multiple programs are part of the flexibility of the robot and its value to the world, but these multiple programs require us to *choose* the right program for what we want the robot to do. So, the first question is, "What do we want the robot to do?" Your answer will lie in the parts you are running, the operation you are staffing, or some other data that you receive from a source outside of the robotic system (see Figure 8-12).

Next, you have to decide how we will use the robot by picking the proper program or creating a new one to match the operating conditions. If you are familiar with all the programs loaded into the robot, you simply need to pick the correct one for the task. If you are not familiar with the programs, you can likely

Figure 8-11 This robot is performing both loading and unloading operations, which we commonly refer to as "pick and place."

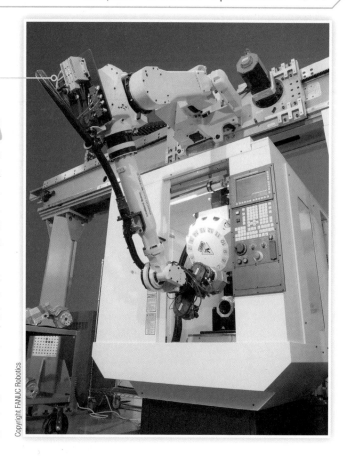

get the information you need from a work procedures book, the reference material that came with the robot, or someone with more experience. For cases where you have no reference material or help available, you will need to carefully test out each program so that you understand its operation and then choose accordingly. (See Chapter 9 to learn about programming a robot when you do not have the program you need and have to create a new one.)

Once you have chosen the program, make sure that the robot is running that program. In systems with multiple programs, there will be some manner of designating a program as the running or loaded program. This may require downloading the program from the robot controller, a computer, or a network, while others may simply require highlighting it in a list and pressing "enter." The ways in which to accomplish this are as varied as the manufacturers of robotic systems. Once you are familiar with a couple of different systems, you will start to see

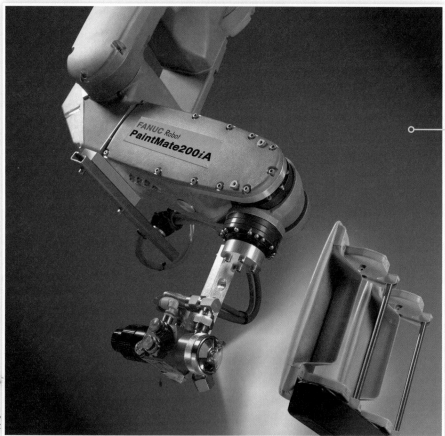

Figure 8-12 A FANUC PaintMate hard at work.

similarities; however, there is no universal procedure for program loading. The best advice I can give you is to make sure that you understand how the particular system you are working with loads programs.

Once you have reached this point, you can technically hit the green start button and it is off to the races. However, I would offer a few words of caution before doing so:

- Make sure the robot is at the home position or another safe position before startup.

- If the robot has experienced anything that may have moved it in relation to the base, check for proper tooling position before running it.

- Never assume a new program is ready to go without testing.

- Though rare, corruption of programs does happen. That said, never just hit the start button and walk off.

- I recommend watching at least one full cycle with your hand near the pause or stop button, just in case.

Failure to heed these warnings could result in damage to parts, equipment, or personnel. Ultimately, the better you understand how your robot and the process it performs work, the better the chances of successful operation.

How Do I Know What the Robot Will Do?

Given all the cautions at the end of the previous section, you may find yourself wondering, "How do I know what the robot will do?" The best and safest way to know is to run the system in manual mode before you turn it loose in automatic mode. Manual mode is where you maintain control of robot operations and can stop action as quickly as you can react. Automatic mode is where the robot runs the program without your continued input, based on the operating parameters of the program. The following is a closer look at each mode of operation.

In manual mode, you can move the robot to specific points or run a program while monitoring its operation. Most robots allow for operation in step or continuous modes during manual running. In step mode, the robot moves from line to line of code executing the commands while requiring some action from you to progress from the current line to the next. This allows you to verify each line of the program and make sure that things are working properly. In continuous mode, the program runs as it would normally, provided that you maintain the proper inputs, such as pressing the dead man switch and manual button at the same time or some similar arrangement of inputs (see Figure 8-13). Both of these manual modes allow you to check the program for functionality and give you a clear picture of the program's operation. Another bonus to manual operation is the fact that most systems automatically slow down the operating speed. Instead of having to match your reaction time to a full-speed robot move, most systems slow everything down to 50 or 60 percent of the maximum speed automatically, with the added option of slowing things down to as low as 10 percent. This gives you more time to react to problems in the program and prevent damage. When proving out new programs or testing the system after a crash, I always put the system in the slowest setting to give me maximum time to react to any problems during testing.

There are some downsides to manual operation. You have to hit buttons or maintain pressure on buttons to advance the system through the program. With most systems, if you release pressure on the dead man switch or any of the other buttons necessary for manual operation, the robot stops immediately. In situations where you are manually working through a welding program or other such

manual mode
A condition in which the operator maintains control of the robot and can stop its action as quickly as he or she can react

automatic mode
A condition where the robot runs the program without continued operator input based on the operating parameters of the program

step mode
A condition where the robot moves from line to line of code, requiring some action from the operator to progress from the current line to the next

continuous mode
A condition where the program runs as it would normally, provided that the proper inputs—such as the dead man switch and manual button or similar—are maintained

Figure 8-13   The whole time you are editing or creating a program in manual, you must keep the dead man switch engaged or the system will alarm out and stop moving.

operation, stopping in the middle of the program could cause defects in the parts or compromise part quality. Not all robot systems allow you to engage the tooling while in manual mode, and this could result in damage to tooling, parts, or fixtures depending on how the program works and the tooling involved. If you are unsure of how the program works or have an operation where the tooling removes material, I recommend running through the program without any parts and carefully monitoring for potential contact between the tooling and fixturing. Some moves require quick motion by the axes involved to reach a programmed position; in these cases, the robot will either alarm out or display a caution statement and then move suddenly, negating the slow safety speed or user-designated slower speed for that movement. If you happen to be in the work envelope at the time of a sudden move, there is the potential for injury, so make sure that you stay clear of the robot, especially near the fixtures or primary work area, during manual operation.

As long as you know the limitations of the robot and remain vigilant, manual mode testing is the best way to determine exactly what the robot will do during the operation of a program. This is also a great way to determine the performance of new programs and to check for variations in operating conditions at the beginning of the day. Some systems allow you to step through a program in manual to a desired point and then switch to automatic when you need to rework parts or restart an interrupted program. If you have a program for checking the position of a robot at startup or after crashes, I recommend running this program in manual instead of automatic, as you will have a better chance of stopping the robot before any impact due to misalignment. Manual mode is for those times when you need to slow things down or maintain control of the system.

Figure 8-14   This operator is monitoring the system in action. It takes a great deal of knowledge about what to expect to be that close to the robot during its normal operation.

Image courtesy of Miller Welding Automation

Automatic mode is where you hit the go button and the system runs the program. Once a program starts in automatic, the system continues until the logic of the program stops the system, the program is completed, alarm conditions occur, or the user stops the program through one of the various inputs for doing so (see Figure 8-14). Most of the time, this is what we want the robot to do, and this is why robots thrive in industry where there are many repetitive tasks. We set up programs to loop repeatedly and use various signals from sensors and other machines to pause or start the program as needed. The options for program parameters are near limitless; thus, there is a multitude of ways to set up a program for a particular task.

Just as manual mode has downsides, so too does automatic mode. While in manual, most systems limit the speed of the robot, but automatic mode has no restrictions other than the top speed of the robot or the speed designated in the programming. In many cases, the robot is capable of moving faster than we can react, making it difficult if not nearly impossible to minimize the damage when things go wrong. Another problem associated with automatic mode is unexpected startup. Since we often loop the program of the robot, a robot at rest is often waiting for a specific signal to start up once more. It does not matter if the signal comes in a few seconds, a few minutes, or a few hours, once the signal comes in, the robot is back in action, typically at full speed. The lulls in activity tend to give operators the false belief that they can do something in the work envelope and get clear again before the robot moves. Only the newer systems designed specifically to work side by side with humans provide the proper level of safety for humans near or in the work envelope during automatic operation. (See 8-1 Food for Thought to see just how dangerous this can be.)

You should only operate a robot in automatic when you feel confident that the system will perform as expected and you know what to expect. If any of the operating conditions have changed, if the robot receives damage, if there are alterations to the program, if you are unfamiliar with the program, or if anything else makes you nervous about starting the system, run through the program in manual. If for some reason you cannot check the system in manual first, make sure that you are in a safe location with a hand on the stop button during the initial pass through the program. A robot will only do what we program it to do; if we tell it to do something crazy, then it will do something crazy.

food for thought

8-1 It only takes a moment of inattention

While working in industry, I personally saw the aftermath of an individual trying to take care of something very quickly without shutting the system down and the price that he paid for his actions.

The incident in question involved a welding machine with a robotic feeding device. The part in this incident had become misaligned, and the technician in charge of keeping things running stepped into the work envelope to make a quick adjustment to the part before it caused problems in the machine. Normally, the technician would place the system in manual, enter the work envelope, correct the problem, exit, place the system back in auto, and start things up once more. This time, the technician assumed that opening the cage to the area would place the system in manual and it would be safe to adjust the part. He was half-correct: it placed the robot that fed the welder into manual, but did not place the welding machine into manual. While working on the part, he got it into position, and the welding machine, seeing all the proper sensors made, started the process of spot welding. The problem was that the technician's hand was in the way and thus welded.

The spot welder pinched the technician's hand between the thumb and forefinger, severely dislocating the thumb and pumping a large amount of amperage through the hand between the two points of the spot welder. By the time I reached the area to help, the individual was free of the welding machine and sitting on the floor. The technician's hand had third-degree burns and he was shaking with pain and shock from the incident. The whole time I helped this individual in prep for transport to the hospital, he kept saying how he knew better; he had not thought about how opening the cage would not stop the machine, just the robot.

I had a chance to talk with this individual a few years ago. While he recovered from the injury for the most part, there is some permanent loss of feeling and mobility in his hand. He was extremely lucky that the thumb was only dislocated, not crushed as it first appeared, and that the damage was limited to the hand. One moment, one instance, one time of trying to skirt the safety rules is all it takes to change your life forever or end it. Keep this in mind as you work with robots and equipment, no matter where your professional career takes you.

So, I Crashed the Robot... Now What?!

In the industrial setting, there is a high probability that eventually the robot will make unexpected contact, or crash. Often when this happens it is at full speed and the results, while spectacular, are devastating to the various parts and systems involved. The damage from a crash varies with the energy and type of contact involved: parts shatter, machine covers are bent or torn, tooling is destroyed, motors slip, belts break, gears lose teeth…. The list of possible damage goes on and on. No matter the type of robot involved in the crash, we can use the following basic guideline for dealing with the situation:

crash
The robot's unexpected contact with something in the work envelope

- First: Determine Why the Robot Crashed

- Second: Get the Robot Clear of the Crash or Impact Area

- Third: Determine how to Prevent Another Crash

- Fourth: Check the Alignment of All Equipment Involved in the Crash

- Fifth: Determine What to Do with the Parts Involved in the Crash

First: Determine Why the Robot Crashed

This sounds simple, but in the heat of the moment, many technicians and operators have overlooked this obvious first step. Before you move the robot, before you deal with the damage, before you do anything, determine what caused the problem to begin with. Once you move the robot and start doing things to react to the crash, there is a good chance you will lose important information about what specifically happened. Think of it as if it were a crime scene and you are a CSI tasked with determining what happened. Take pictures, look at the crash from different angles, and check programs, paying special attention to the line on which it stopped for each machine and system involved. Do whatever you need to gather enough information so that you understand exactly what happened.

In cases where you cannot gather enough data to draw a definitive conclusion about what happened, do the best you can. Sometimes there is no clear-cut explanation for why the robot hit something, such as intermittent problems that occur randomly with no discernible pattern, making it difficult to find the root cause of the problem. When you are unsure of the why, record as much information as possible and save it for future reference. Unfortunately, sometimes the system has to fail, and yes, even crash a few times, before we can get enough information to find the root cause. Just do the best you can with the information you have available. Do not forget to talk with other operators or technicians who have experience with the system, as they may be able to give you some insight or tips on possible causes.

Second: Get the Robot Clear of the Crash or Impact Area

Once you have gathered the necessary information, it is time to get the machine(s) untangled. Different systems will require different steps to accomplish this. Sometimes the robot becomes stuck or hooked on something in the crash area, and you have to use tools to free it. In these cases, try to minimize any additional damage by applying the right amount of force in the right ways instead of going straight to using a sledgehammer at full force! Some systems will let you reset the alarm and then move the system manually. In this case, make sure that you move the robot in the proper direction and make all initial movements small ones to minimize any potential secondary damage. This is where having a good working knowledge of the robot's movements comes in handy. You can also use the right-hand rule, as long as you are in the World frame and remember what each finger represents.

Some systems have manual-brake-release switches or buttons to allow free movement of the robot (see Figure 8-15). If you are going to use this to free the robot from the crash, there are some very important things to keep in mind. Once you release the brake there is nothing to hold the robot in place. For the minor axes of the robot, this may not be a major concern, but releasing the major axes of the robot can result in a sudden and dangerous collapse of the arm. Use the axes number from Chapter 3 to make sure that you are releasing the proper axis; always double check that you are using the switch for the axis you want and *not* the one next to it. Case in point, during a demonstration of the brake release method for some of my students in a class, I inadvertently pressed the brake release for the axis next to my intended axis. The robot arm suddenly fell downward under its own weight at the third axis. Luckily, I was off to the side a bit, but the arm caught the edge of my safety glasses and knocked them across the room. If I had been over a few inches, I would have had a nasty knock to the head and likely a trip to the emergency room. Yes, even instructors make mistakes; however, we share those mistakes so you do not have to re-create the learning event.

figure 8-15 Look closely on this Panasonic robot to see four white pushbuttons and a red one on the side, near the base, with a black background. These are break release switches used to help in crash recovery of the system.

Image courtesy of Miller Welding Automation

In a worst-case scenario, you may have to physically remove motors or remove power and turn gears by hand to move the robot. This should be your last resort, and not a first plan of attack. If you cannot clear the system alarms to move the robot manually, check the repair or maintenance manuals for a way to release the axes of the robot for movement. Often, clearing alarms that do no reset requires going to menus and settings rarely seen, but the majority of the systems out there provide some form of alarm override. The hobby and personal robotic systems tend to require the worst-case response due to the programming and level of operator input. The upside is that many of these systems are small or light enough so that you can physically move the whole system out of the crash area, thus negating the need to move the robot via controls.

To complete the second step of crash recovery, you have to understand the operation of the robot in question. This is yet another reason why taking the time to understand the robot you work with is time well spent. These machines are truly marvelous in what they can do and in their flexibility, but they require a great amount of knowledge on the user's side to work properly.

Third: Determine How to Prevent Another Crash

If step one was the CSI investigation, then step three is the arrest. This is when we make the necessary modifications to the programming or operation to prevent future crashes. In some instances, step three and four, checking alignment, combine into one mammoth step with the goal of preventing future damage to the robot system. This could require changing the position of a point, putting a pause in the timing of the program, adding an input to the robot system, replacing a worn out/faulty sensor, tightening bolts in a fixture, removing things that do not belong in the work envelop, and so on. Some of these fixes may fall into the range of repair that you can perform, while others may require maintenance personnel, engineers, programmers, tech support from the manufacturer, or other individuals. The main point of this step is to ensure that the problem will not happen again, causing even more damage.

In the case of the intermittent or unclear problems, it may be difficult to determine a course of corrective action to take. If you have to run the system without a corrective action, try to minimize any potential for damage. When you start it back up, watch the system run with your hand on the stop button instead of just letting it run while you do other things. Reduce the operating speed of the system, if possible. Do a few dry runs where no part is present, if that is an option. Do anything that you can to minimize the potential damage while checking the system for symptoms to help with the diagnosis. Unfortunately, you may simply have to let it run for a while before any new information surfaces. If this is the case, use your best judgment and do what you can within the constraints you have to work with.

Fourth: Check the Alignment of All Equipment Involved in the Crash

As mentioned previously, the unexpected contact of a crash can cause physical damage to the machine and cause the various axes of the robot to move. This creates a difference between where the robot actual is and where it thinks it is. When this happens, all the various points in a program become affected and the programs do not perform properly. In some cases, the difference may be small enough that there is no apparent change in the robot's operation until we examine the program as a whole or run the alignment check program. As a rule, the tighter the tolerances required of the robot, the less room there is for error or axial slip.

When you determine that there is error in the system, you need to take some form of corrective action. The action depends on the robot, the type of tooling it is using, and the amount of error. Many systems allow for corrective offsets as long as they are not over a set amount, usually somewhere under an inch. Some types of tooling slip within their holders to prevent damage, and in these cases, there is often a fairly quick and simple procedure for realigning the tooling to get back to business. This is another one of those instances where you need to know how your system works and how best to deal with positioning errors of that system. (Have you noticed a trend yet?)

When checking alignment, do not forget to check the fixture(s) as well. If the error is in the fixture and not the robot, you can cause yourself a lot of grief by changing the position of the robot instead of fixing the alignment of the fixture. Yes, once you align the robot to the new fixture position the system will function properly, but what happens if you change fixtures and go back to the proper fixture alignment? Suddenly the robot is out of position once more, but there is no crash to blame.

Another mistake to avoid is realigning the program instead of the robot. This is the quick fix and a common action for operators in a hurry. Yes, this will get the system turning and burning once more, but what happens when someone changes the program? Once again, we have positioning problems with no clear-cut cause. If this progresses for a long time and the operator adjusts multiple programs and then something happens that warrants the realignment of the robot that should have occurred in the first place, can you guess what happens next? Yep, the programs changed before now need changed back. This is another reason why we check programs when we first load them instead of just pressing the go button and assume that everything is all right.

Remember, we may combine steps three and four, but we must still meet all the criteria of each. The best hope for normal operation is to correct any damage or misalignments caused by the crash instead of just treating the symptoms. Any shortcuts taken at the time of the crash to get the robot up and running are likely to haunt the operation with points in the wrong position here and odd alarms or operation there until someone finds the original root cause and fixes it. Then we lose even more operation and production time removing all the various symptom fixes from the system to get it back to normal operation. This is why you need to take your time with these steps instead of rushing to get the system back up and running.

Fifth: Determine What to Do with the Parts Involved in the Crash

Before we start the robot back up and let it continue with the tasks at hand, we need to determine what to do with any parts involved in the crash. This may seem an odd step to some and not apply to every robot crash, but for industry, it is key. With lean manufacturing and the desire to reduce inventory, a company may produce just the number of parts needed by the customers, with no extras. If this is the case and the robot destroyed one of those parts, what happens to the customer's order? Well, he or she does not get the full order. Just as you would be mad if your favorite restaurant forgot part of your order, so too do customers become upset when they do not receive all the products requested. The fix for this is usually fairly simple: use the proper in-house procedure and let whoever know that you need another part (see Figure 8-16).

Lean manufacturing is a cost-saving set of operational procedures that has gained a great amount of popularity in industry. The principle is to remove all the wasted time, space, and motion possible by placing similar operations close together and streamlining the production process. Another piece of this is on time manufacturing, where industry produces parts that have been ordered instead of anticipating parts that might be ordered and stocking them in a warehouse or other storage area.

Sometimes we can save crash parts by finishing out the process. These cases often require the operator to manually step through the program up to the point of the crash and then start the system in automatic or finish the part in manual, depending on the process and robot in question. This is why we take care of the

Figure 8-16 These parts are produced by a robotic welding system. Parts of this size are often small runs, so any lost parts could represent a real problem if not handled properly.

Image courtesy of Miller Welding Automation

crash parts before beginning normal operation. It is easier to run the system manually before we start normal operation as opposed to stopping the system at some later point and trying to get everything back in position. If you are unsure how to save a salvageable part, ask someone with more experience for help.

Once you have determined if the part is salvageable, make sure you follow your employer's guidelines. If it is junk, report it, get a new one, and carry on as directed. Make sure to put junk parts in the proper area for disposal or recycling. I have seen many parts that were placed somewhere "handy" remain there for days, weeks, or longer. Worse yet, do not try to hide the damaged part. Eventually it will come to light that a crash or damage happened; if you try to hide it, you will look as if you know you did something wrong. This will likely lead to repercussions that you would rather not experience, such as a reprimand or loss of job. If you salvage the part, pay extra attention to the parts' quality (no one wants a part that looks good but is junk in disguise). Some parts seem salvageable, but in the process of reworking, become scrap with all the responsibility that entails.

Once you have completed the five steps of crash recovery, you are ready to fire up the system and let it run. I would recommend doing at least one manual run from start to finish, if possible, before switching back to automatic. This gives you a chance to ensure that everything is working right and that you did not miss any other problems with the system. Your employer may have some specific guidelines that apply to these five steps, and these guidelines may require some actions or steps that we did not cover. If you are unsure or have questions anywhere along the process of crash recovery, talk with someone who has more experience that can help you out. Mentoring, or the sharing of knowledge through teaching, coaching, and helping with experiences as they happen, is one of the primary ways that knowledge is passed in the industrial setting.

mentoring
Knowledge shared by those more experience with their trainees through teaching, coaching, and helping the trainees with job experiences as they happen

review

We have covered the basics of running robots, but this is by no means all the situations and concerns that you may encounter in the world of robotics. Just as each type of robot differs, so too will the specific requirements. Feel free to modify what you have learned in this chapter for the specific systems you work with and the specific ways in which you are using those systems. For many people, their primary interaction with robots comes at the operator level, so there is a good chance you will get to use the skills learned in this chapter out in the field. During the course of this chapter, we covered the following topics:

- **Before powering the robot up.** Here we covered some of the basic checks one should perform before turning on the power.

- **Powering the system up.** This section was about what to check after the system has power to ensure proper operation as well as user safety.

- **Moving the robot manually.** Whether it is the first checks of the day or just getting the robot into a handy position, this section provided valuable information to help.

- **What do I want the robot to do?** This section talked about some of the tasks robots perform and why it is important to know what you want the robot to do.

- **How do I know what the robot will do?** This section gave you some tips on how to know what to expect from the robot during operation.

- **So, I crashed the robot… now what?!** This section gave you the knowledge necessary to recover from the dreaded crash.

key terms

Automatic mode	Critical alarm	Mentoring	User frame
Continuous mode	Frames	Right-hand rule	World frame
Coordinate systems	Joint frame	Slag	
Crash	Manual mode	Step mode	

review questions

1. What are some things that we are look for when we check the robot for damage?

2. When is a good time to clean the robot, and what is a precaution to keep in mind while cleaning the robot?

3. What is the possible danger of intermittent communication between a motor encoder and the robot?

4. What are some of the pre-power checks that one might have to perform with a welding robot?

5. What is going on during the robot boot sequence?

6. What are some of the safety features that we check at startup?

7. What are some universal checks for hydraulic and pneumatic systems?

8. When we move the axes as part of our powered checks, what are we looking for?

9. Explain how to use the right-hand-rule to determine robot motion.

10. Describe the motion of the robot while moving it manually in the Joint frame.

11. How do you choose the right program if you are not familiar with all the programs loaded in the robot?

12. What are some precautions to keep in mind when you start the running program?

13. What are some downsides to manual operation?

14. What stops the program when the system is in automatic?

15. What are some downsides to automatic operation?

16. What are the five steps of crash recovery?

17. Why do we determine why the crash happened before we move the system or start to correct things?

18. If you can move the robot manually after a crash, what precautions should you take?

19. What are some things that may be required to correct situations that cause robots to crash?

20. What should you do if you could not determine the cause of the crash and are unable to take corrective actions?

21. How do we finish out the process on salvageable parts involved in a crash?

Programming and File Management

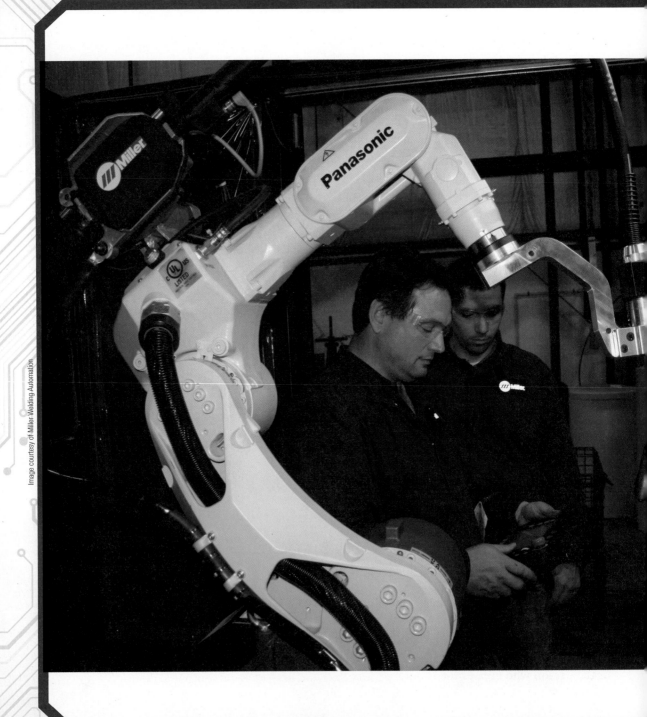

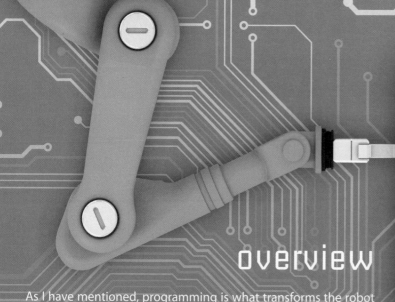

overview

As I have mentioned, programming is what transforms the robot from an expensive toy or technological paperweight into a useful device. Each robotic manufacturer has a different way of creating programs, with some having differences in program creation from model to model. Because of these differences, this chapter focuses on the process of developing programs and taking care of important data rather than the specifics of writing a program. The good news is that once you learn how to program one type of robot, most of you will find it easy to learn how to program others. The process remains basically the same; it is the syntax of the programming or how you enter and create programs that is different. In our journey through the world of programming, we will cover the following topics:

- Programming language evolution
- Planning
- Subroutines
- Writing the program
- Testing and verifying
- Normal operation
- File maintenance

What You Will Learn

- The basic evolution of the robot programming language
- The five different levels of programming languages
- The key points in planning your program and what each step entails
- What a singularity is
- What subroutines are and how they help the programming process
- The difference between local and global data
- The basics of writing a program and the main motion types available
- How to test a program once you have it written and what to look for
- What to do when the robot is ready for normal operation and some things to look out for
- How to manage the data of the robot to ensure proper operation and save yourself time and effort down the road

Programming Language Evolution

Before we get into how programming languages have evolved, let us take a moment to define exactly what a robot program is. A robot program is the list of commands that run within the software of the robot controller and dictate the actions of the system based on the logic sorting routine created (see Figure 9-1). You could think of the robot program as a list of instructions telling the robot what, when, and how to do whatever it is the robot does. The performance of a robot is only as good as the person who creates the program. If the program instructs the robot to add wasteful motions, the robot will be wasteful. If the program creates a near-perfect weld, the robot will weld near perfectly. The good thing about a program is that we

robot program
The list of commands that run within the software of the robot controller and dictate the actions of the system based on the logic sorting routine created therein

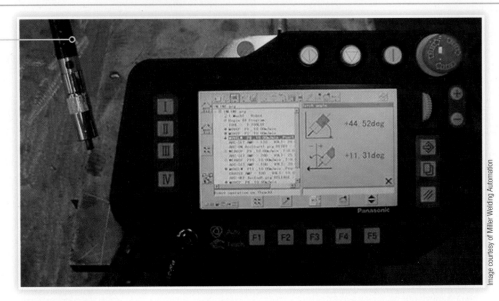

Figure 9.1 This is what a Panasonic Welding program looks like when viewed on the teach pendant.

Image courtesy of Miller Welding Automation

programming languages
The rules governing how we enter the program so that the robot controller can understand the commands

can continue to modify and change the program until we have the best program for the application. Moreover, with proper maintenance of the robot and its programs, once we have the ideal program, the robot can continue to perform its tasks in an ideal way for days, weeks, months, and even years. Programming languages are the rules governing how we enter the program so the robot controller can understand the commands.

Before the digital age, we controlled or programmed robots by using devices that opened and closed contacts in specific sequences. Punch cards, peg drums, relay logic, and other methods controlled these systems; these setups required creating punch cards with a different set of holes in them, moving the pegs on the drum, rewiring the relays, or some other physical alteration of the control structure, respectively, for a change in the operation of the system. As you can imagine, these changes often took a large amount of time to complete and were less than user friendly. Today it is rare to see this type of system; however, you may run across one of these older systems in the industrial world. If you do, there is a good chance that it is one major crash or breakdown away from replacement, as parts are likely either hard or impossible to find, leaving the options of either making a new part or getting a new robot.

In the early 1970s, computer technology changed the way we program robots and ushered in a new age. As with many major changes in the way we do things, there were some growing pains along the way. The first approach for computer programming of the robot involved using a language designed for the needs of the robot. The result was a programming language or technique that was great for controlling the robot, but difficult for users to work with. Programming these systems was unlike any other programming of the day, and this language was not well suited to data processing. Several companies produced systems with this type of programming, such as Cincinnati Milacron's T3 language, but complaints from users made this a short-lived method.

Since the customer is always right, robotic companies took users' complaints to heart and tried another method for robot programming. This time, the manufacturers started with a known computer language, such as BASIC or FORTRAN, and added commands to control the robot. The result was a language that computer programmers of the day understood with data processing capability. From the programmer's side of things, this was terrific, but it was not the optimal way to control the robot. Since these languages were for computers and not robots, there were problems with the motion commands, which translated into inefficiency of motion as well as other translational difficulties. Unimation was one of these companies: it developed the VAL language for its PUMA robots based on the computer language BASIC.

Over time, manufactures found a way to give users the best of both worlds by creating programming languages that combined the efficiency of those specifically designed for the robot with the established programming flow provided by common computer languages. The resulting languages were easy for those with a programming background to understand yet removed the inefficiency of using a computer language only. These hybrid languages continue to evolve; today, we have systems that no longer require the user to have a computer programming background. In fact, many users can successfully program modern systems with no more than 40 hours of training. Obviously, it takes more training and time to reach the expert level, but a week of specialized training or its equivalent of on-the-job experience allows most workers to edit existing programs or write new, effective programs for the system. These hybrid languages are responsible for the differences in programming methods that we discussed earlier between manufacturers and sometimes various models by the same manufacturer.

When looking at the evolution of robot programming, I like to break it down into five different levels:

- Level 1: No processor

- Level 2: Direct position control

- Level 3: Simple point-to-point

- Level 4: Advanced point-to-point

- Level 5: Point-to-point with AI

These levels start with the simplest systems as level 1 and end with the most complex as level 5. As we explore these five levels of robot programming, I encourage those of you that have experience with robots to classify the systems that you have worked with. Advancements in AI may lead to a sixth level of programming down the road, but at this point it is hard to predict what benefits and capabilities this level would include.

Level 1: No Processor

As the level name implies, these systems lack computer or processor control. They work with relay logic, mechanical drums with pegs to control the timing of actions, and any other system that works without the need for a computer or digital processing chip. To change the operation of these systems, one must physically change something in the system. Change the pegs on the rotating drum, create a new punch card to control which contacts are made and when, rewire the relay system, or some other physical manipulation of the system.

Though most of the modern robots do not fall into this category, there is one group that works purposely at this level, BEAM robotics. In the quest to build robots from the bottom up, the whole point of BEAM systems is to create robots with no advanced processor and see what we can learn from them. Like the insects that they work to emulate, BEAM robots seem to react to their environment instinctually instead of in a programmed manner. The information yielded from BEAM research is of special interest to the field of swarm robotics, which focuses on ways to use a large number of simple robots to perform complex tasks.

swarm robotics
The research field that focuses on ways to use a large number of simple robots to perform complex tasks

Figure 9.2   The spider robot I built used limit switches engaged by contact with the legs to trigger a specified turning response.

The hope is to some day use swarms of simple robots to combat things such as oil spills or diseased cells in the body, improving our quality of life with teams of robots that are simple and cheap to create.

Level 2: Direct Position Control

As the name implies, this system of control requires the programmer to enter the positional data for each axis as well as all the motion, processing, and data gathering commands required to create a program. This is the most basic level of processor control and the most labor-intensive for the programmer. Many of the hobby systems based on microcontrollers, like the Arduino, utilize this method of control. To simplify things, these systems often react in a predetermined manner to various sensor input. For example, one might have a robotic vehicle that drives forward until the ultrasonic sensor detects an object, at which point it will turn both wheels in opposite directions for one second to turn the vehicle before advancing once more (see Figures 9-2 and 9-3).

For industrial applications, this type of programming requires knowledge of the position of each axis when the robot is in the desired location. If there was no way to gather the information directly from the robot, the programmer must perform the complex math necessary to find the position for each axis of the robot. As you can imagine, this style of programming is very labor-intensive, with a good possibility of error, and is thus not a favorite among programmers. I have not run across any industrial systems still using this form of programming, and I am fairly certain any left in industry are just one loss of programming or serious maintenance event away from replacement.

Level 3: Simple Point-to-Point

This was a common way to program robots in the mid-to-late 1970s and throughout the 1980s; many of the older industrial systems used this method of programming. The programmer must still enter the motion type, data gathering, and other aspects of the program but does not manually enter the positions. When programming these systems, it is common practice to write the basic program offline, with each position having a label but no coordinate data. Once the programmer is happy with the program, he or she uploads it into the robot and then physically moves the robot to each point and records the positional data. Sometimes this requires moving through the program one line at a time, and sometimes it requires moving the robot to the desired position and then saving the data under the proper position label. The exact method depends on the model of robot, controller, teach pendant, and manufacturer of the robot (see Figures 9-4 and 9-5).

Some systems allow the programmer to create the program directly from the teach pendant using this method. Having programmed level 3 systems both by creating the program in advance and on the fly from the teach pendant, I can honestly say that I prefer to create the program offline and upload it. This is mainly because it is easier to type on a keyboard than a teach pendant. Teach pendants are

```
// These constants define which pins on the Ardweeny are connected to the pins on
// the motor controller.  If your robot isn't moving in the direction you expect it
// to, you might need to swap these!
const unsigned char leftMotorA = 7;
const unsigned char leftMotorB = 8;
const unsigned char rightMotorA = 9;
const unsigned char rightMotorB = 10;
const unsigned char leftswitch = 2;
const unsigned char rightswitch = 4;
boolean leftswitchclosed;
boolean rightswitchclosed;
// This function is run first when the microcontroller is turned on
void setup() {
   // Initialize the pins used to talk to the motors
   pinMode(leftMotorA, OUTPUT);
   pinMode(leftMotorB, OUTPUT);
   pinMode(rightMotorA, OUTPUT);
   pinMode(rightMotorB, OUTPUT);
   pinMode(leftswitch, INPUT);
   digitalWrite (leftswitch, HIGH);
   pinMode(rightswitch, INPUT);
   digitalWrite (rightswitch, HIGH);
      // Writing LOW to a motor pin instructs the L293D to connect its output to ground.
   digitalWrite(leftMotorA, LOW);
   digitalWrite(leftMotorB, LOW);
   digitalWrite(rightMotorA, LOW);
   digitalWrite(rightMotorB, LOW);
}
// This function gets called repeatedly while the microcontroller is on.
void loop() {
   if (digitalRead (leftswitch) == LOW)
     {leftswitchclosed=true;}
    else {leftswitchclosed=false;};
    if (digitalRead (rightswitch) == LOW)
     {rightswitchclosed=true;}
    else {rightswitchclosed=false;};
      // Turn both motors on, in the 'forward' direction
   if (!rightswitchclosed && !leftswitchclosed)
   {digitalWrite(leftMotorA, HIGH);
   digitalWrite(leftMotorB, LOW);
   digitalWrite(rightMotorA, HIGH);
   digitalWrite(rightMotorB, LOW);}
   // Turns robot right
   else if (!rightswitchclosed && leftswitchclosed)
   {digitalWrite(leftMotorA, HIGH);
   digitalWrite(leftMotorB, LOW);
   digitalWrite(rightMotorA, LOW);
   digitalWrite(rightMotorB, HIGH);}
   // Turn robot left
   else if (rightswitchclosed && !leftswitchclosed)
   {digitalWrite(leftMotorA, LOW);
   digitalWrite(leftMotorB, HIGH);
   digitalWrite(rightMotorA, HIGH);
   digitalWrite(rightMotorB, LOW);}
     // Robot goes in reverse
   else
   {digitalWrite(leftMotorA, LOW);
   digitalWrite(leftMotorB, HIGH);
   digitalWrite(rightMotorA, LOW);
   digitalWrite(rightMotorB, HIGH);};
   // Wait 1 second
   delay(500);
}
```

Figure 9.3 The program that runs my "Spider Bot."

Figure 9.4 A level 3 program with the positional data entered at the end. This program was written on a Mitsubishi five-axis robot.

```
10 'knm
20 '4-16-08
30 'box template
100 MOV P_SAFE
110 POFFZ.Z = 10
120 MOV P1 + POFFZ
130 MVS P1
140 MVS P3
150 MVS P3 + POFFZ
160 MOV P5 + POFFZ
170 MVS P5
180 MVS P2
190 MVS P4
200 MVS P4 + POFFZ
210 MOV P8 + POFFZ
220 MVS P8
230 MVS P6
240 MVR P6, P7, P8
250 MVR P8, P9, P10
260 MVS P10 + POFFZ
270 MOV P11 + POFFZ
280 MVS P11
290 MVS P11 + POFFZ
300 MOV P12 + POFFZ
310 MVS P12
320 MVS P13
330 MVS P13 + POFFZ
340 MOV P14 + POFFZ
350 MVS P14
360 MVS P15
370 MVS P15 + POFFZ
380 MOV P16 + POFFZ
390 MVS P16
400 MVS P17
410 MVS P17 + POFFZ
420 MOV P18 + POFFZ
430 MVS P18
435 MVS P22
440 MVS P19
450 MVR P19, P20, P21
460 MVS P23
470 MVS P23 + POFFZ
480 MOV P_SAFE
P1=(160.710,-89.430,502.100,111.960,110.010,0.000)(6,0)
P2=(221.920,-89.430,506.690,111.960,110.010,0.000)(6,0)
P3=(286.430,-97.050,511.610,111.960,110.010,0.000)(6,0)
P4=(288.360,-42.670,511.840,111.960,110.010,0.000)(6,0)
P5=(170.600,-34.020,501.570,111.960,110.010,0.000)(6,0)
P6=(268.230,29.240,511.020,111.960,110.010,0.000)(6,0)
P7=(256.280,19.570,509.430,111.960,110.010,0.000)(6,0)
P8=(268.450,4.780,511.020,111.960,110.010,0.000)(6,0)
P9=(283.020,1.360,511.740,111.960,110.010,0.000)(6,0)
P10=(288.140,25.940,511.760,111.960,110.010,0.000)(6,0)
P11=(247.520,72.240,509.920,111.960,110.010,0.000)(6,0)
P12=(261.970,70.880,511.830,111.960,110.000,0.000)(6,0)
P13=(289.730,65.980,514.110,111.960,110.000,0.000)(6,0)
P14=(252.980,110.240,511.950,111.960,110.000,0.000)(6,0)
P15=(292.690,101.820,515.120,111.960,110.000,0.000)(6,0)
P16=(269.820,95.340,512.500,111.960,110.000,0.000)(6,0)
P17=(269.820,119.110,513.100,111.960,110.000,0.000)(6,0)
P18=(189.260,156.660,508.670,111.960,110.000,0.000)(6,0)
P19=(258.110,139.580,513.790,111.960,110.000,0.000)(6,0)
P20=(249.680,160.070,512.990,111.960,110.000,0.000)(6,0)
P21=(257.540,171.670,514.020,111.960,110.000,0.000)(6,0)
P22=(293.830,127.760,515.400,111.960,110.000,0.000)(6,0)
P23=(291.210,166.100,515.970,111.960,110.000,0.000)(6,0)
POFFZ=(0.000,0.000,0.000,0.000,0.000,0.000,0.000,0.000)
```

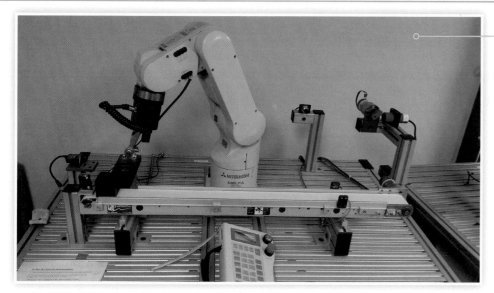

Figure 9.5   The program in Figure 9-4 was written for this robot.

like the older cellphones, where each key has three letters on it. To type out a word, you have to press a key one to three times for each letter. This takes more time and effort. For editing a line or two of code the teach pendant is fine, but to write a long program, I prefer to use a computer. Another drawback to the teach pendant is that the display screen usually only shows a few lines of code at any given time. This makes it hard to keep the flow of your program clearly in mind as you go along. If you do plan to enter the program from a teach pendant, I recommend having a paper copy of your program in advance so that you can enter it in without having to worry as much about the logical flow.

Level 4: Advanced Point-to-Point

Advanced point-to-point is an advancement of simple point-to-point that makes the process of writing a program many times simpler. With level 3 languages, the user bore the brunt of all the action commands and direction commands. With level 4 languages, writing a program really is as simple as creating a new program, entering a string of points with the proper motion label to reach those points, and testing out the program (see Figure 9-6). There is no need to memorize the larger number of movement and logic commands that are required by simpler programming methods. The programmer has to determine the key points of the program, how the robot moves between those key points, and any logic filters that might be necessary. The software of the robot handles the rest. (We will look closer at movement types and logic filters later in this chapter.)

Level 4 programming languages truly changed the world of the industrial robot. Suddenly, operators who had little or no experience with the system were able to

Figure 9.6   An example of a level 4 program on the FANUC controller written in my robotics class. In this program, students used linear and joint movements between the various points.

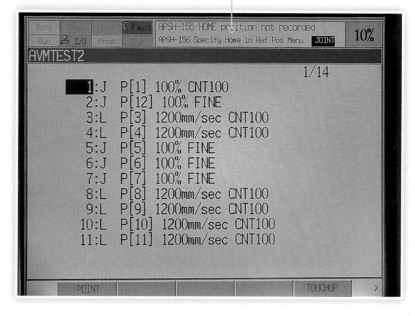

figure out how to create simple programs. Instead of needing a heavy background in programming and logic, a week's worth of training at the manufacturer's site was enough for most to begin writing working programs. Manufacturer training usually lasts 30 to 40 hours and is presented over the course of one week at a training facility, with a general cost of about 2,000 dollars, give or take a few hundred. Some manufacturers will waive the training cost if an industry buys a certain number of their robots, helping to sweeten the sales deal. Advanced programs still require the programmer to have a good understanding of programming as well as some form of advanced training. Many robot manufacturers offer this training as well, but it is another week or two of training beyond the initial 30 to 40 hours mentioned previously.

The lion's share of industrial systems fall into this category, as this level of programming makes it simple and easy to modify or write new programs. Another benefit is that the simplicity of this language makes it easy to write a program directly from the teach pendant. In fact, the teach pendant is the common way that programs are entered for these systems, as the variables and logic for programs are options selected from menus. Offline programming is still an option and involves the use of advanced software that can simulate the motions of the robot with accurate models of the work envelope, allowing the programmer to create viable programs with little need for adjustment. There is a high probability that you will work with this programming style if you spend any time with industrial robots. Unfortunately, there is no universal level 4 language, so you will still have to learn the specifics of the brand of robot that you are working with. On the plus side, those who learn a level 4 language tend to pick up other manufacturer's level 4 languages quickly and easily.

Level 5: Point-to-Point with AI

The newest advancement in computer programming is the addition of Artificial Intelligence (AI) to advanced point-to-point. This takes the convenience of level 4 languages and adds the ability of correction for error as well as advanced teaching methods. While it is easy to set points in a level 4 language, the only leeway for error is that provided by the tooling. Level 5 systems use vision or some other advanced sensing method to determine the difference between where the system should be versus the taught position and makes the offset. This adds a completely new level of flexibility to the industrial system and allows the robot to work in conditions that were impossible previously.

Another development in the level 5 programming styles is the way that we teach points. The standard way to teach a point is use the teach pendant and manual mode to move the robot into position via various motion types and a combination of movements. Some of the level 5 systems allow the user to physically grab the robot and move it into whatever position he or she desires. This turns teaching from an exercise resembling getting a stuffed animal out of a claw machine into something as simple as taking someone's hand and leading him or her wherever you want to go. This type of teaching is very organic and natural and makes it easier to position the robot as desired. The Baxter robot is a great example of this type of system, with advanced safety features, camera vision to make offsets on the fly, and manual positioning teach mode where you move the robot's arm instead of using a teach pendant.

Level 5 languages are new to the robot world and are still making inroads, but I fully expect these to be the norm in the next 5 to 10 years. As customers become familiar with all the options for operation and programming of these systems, the demand will increase, causing manufactures to respond by offering new models with this capacity or updating the models that they currently offer. At a recent trade show, I saw a vendor with a third-party (a system designed by someone other than the manufacturer or customer) hardware package that would take almost any level 4 system and advance it to a level 5 system. In my mind, this demonstrates the ease of upgrading the current designs to level 5; in many cases, it may be as simple as adding a vision system or other sensors and updating the operation software.

As AI software continues to advance and we develop new ways for robots to deal with tough decisions, there is a good chance that you will see an advanced programming language where all you have to do is tell the robot what you want it to do. Instead of setting points, we might simply take a picture with the vision system and use a special pen to mark where we want the weld. Instead of detailing pick-up and drop-off points, we may simply tell the robot to pick up the bottle and put it in the box. We have some systems that are close to this already, but they still need some basic points taught by the programmer as well as other bits of help from their human operators.

> **third-party**
> Systems, software, or other items designed by someone other than the manufacturer of the equipment or customer

Planning

Now that you know a bit about how programming evolved, it is time to learn the basic process of programming. This begins with planning (see Figure 9-7). Just as most things in life require a certain level of planning to achieve success; you need to have a game plan in place before you start writing a program. The level of planning depends on the complexity of the task that you have in mind for the robot. If you want the robot to move from point A to point B in a straight line, the plan is fairly straightforward. Make sure that the robot can reach both points and that there are no objects in the straight-line path for the robot to hit. If you want the robot to hit multiple points while welding two pieces of metal together and avoid impact with any of the fixture parts, then you will need a more complex plan. Breaking the planning process down into steps will help to ensure that you do not overlook anything.

Image courtesy of Miller Welding Automation

Figure 9.7   As you can see from this picture, not all robot tasks are simple, with easy avenues of approach. Many programs take a great deal of planning to avoid positioning problems, hitting objects such as clamps or fixtures, and other issues that might affect the quality of work.

Figure 9.8   If you look closely, you can see that this part has multiple welds positioned on top of each other to create the strength required. Without proper planning, these welds might look fine but actually have imperfections and other flaws that reduce their structural strength.

Image courtesy of Miller Welding Automation

Step one: What do you want the robot to do (see Figure 9-8)? Before you can start to plan, you have to know what the goal is. Remember, a program is a series of logical steps designed to control the actions of a robot. A robot cannot wing it or just make it happen, it can only perform within the confines and rules set by the program that you create. If you create a program with no clear-cut idea of what you desire from the robot, do not be surprised when the robot fails to meet your expectations or do anything of value. When you add in the fact that robots are often faster and stronger than we are, a poorly planned program could literally be dangerous to the people around the system. Take the time to figure out what you want the robot to do, from start to finish. I recommend writing down the tasks you have in mind or sketching out what you want the robot to do. During this step, it is important to make sure that the tasks you have in mind are something that the robot is capable of doing and not beyond the system's reach, strength, speed, or tooling type.

Step two: Task mapping. Now that you know what you want the robot to do, take the time to figure out how it will do it. Ask yourself these questions:

- What kind of tooling does the robot need?

- How do I want the robot to move between points?

- Are there any obstacles for the robot to avoid?

- What is the robot doing at each point?

- What is the robot doing between each point?

- Are there any conditions or other factors that I need to consider in the process?

- Is the process logical?

We will look at each of these questions in detail, as these steps can save you a lot of time, trouble, and frustration when you start writing the program.

What kind of tooling does the robot need? Before you begin mapping out any of the tasks, it is important to make sure that the robot has the necessary EOAT installed (or you have it available). If not, you will need to order the proper tooling or come up with a different plan. For instance, a pick-and-place system with a gripper installed that was not designed for welding would need new tooling, an extra

Figure 9.9 As you can see, some pick-and-place operations require something beyond the generic gripper.

power supply, and proper software to become a welding robot. Other times, it may be as simple as modifying the fingers on a gripper to hold the geometry of a new part. If the process requires more than one type of tooling, you will need to either allow for the swapping of tooling or install multiple tooling on the robot. Regardless of the need, I recommend addressing this concern right out of the gate because the wrong or improper tooling is a fast way to ensure failure (see Figure 9-9).

How do I want the robot to move between points? There are four primary motion types in robotics: joint, linear, circular, and weave. Joint motion is point-to-point, where all the axes involved move either as fast as they can or at the speed of the slowest axis, with no correlation between the separate axes involved. Because all the axes are moving independently, there is a good chance that the motion between the points will not be a straight line. The tooling may curve, dip, and/or move in odd ways due to the amount of distance each axis moves. I have seen several programs where a joint movement caused a crash with objects in the work envelope due to this unexpected motion. I recommend using joint motion only when the robot is away from fixed objects and there is no real chance for impact.

Linear motion is where the controller moves all the axes involved at set speeds to insure straight-line movement. Take a yardstick or other straight object and place it between your two points, and you will see the path of the robot. These motions do include angularity in cases where there is a difference in height between the two points, so keep that in mind as well. Be careful when you follow an object with elevation changes that are uneven. For instance, if point A on the part is a quarter-inch thick, and point B on the part is a half-inch thick, and there is a point between the two that is three-quarters of an inch thick, the robot will impact the high point between the two, as it is in the way of the straight-line motion. To avoid these kinds of issues, you may have to program extra points along the straight-line distance to help the robot work around changes in thickness or other obstructions, such as clamps or fixturing.

joint
Point-to-point motion where all the axes involved move either as fast as they can or at the speed of the slowest axis, which may result in nonlinear robot motion

linear
Motion where the controller moves all the axes involved at a set speed to ensure straight-line motion

figure 9.10 This is an example of how a circular program or portion of a program is set up in FANUC systems.

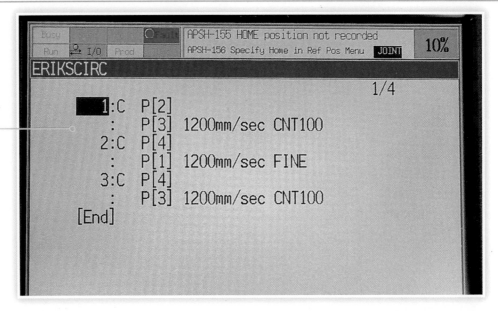

```
 Busy                        APSH-155 HOME position not recorded
 Run   I/O   Prod            APSH-156 Specify Home in Ref Pos Menu  JOINT   10%
ERIKSCIRC
                                                              1/4
        1:C   P[2]
         :    P[3]   1200mm/sec CNT100
        2:C   P[4]
         :    P[1]   1200mm/sec FINE
        3:C   P[4]
         :    P[3]   1200mm/sec CNT100
       [End]
```

circular
A motion described by no fewer than three points to create arcs or circles

arcs
A portion of a circle

weave
Straight line or circular motion that moves from side to side in an angular fashion while the whole unit moves from one point to another

Circular motion is the formation of arcs and full circles as described by no less than three points. For arcs or part of a circle, you need to teach at least three points: the start point, middle point, and end point. For a full circle, you will need at least four points (see Figure 9-10). However, five or more is better with a start point, a point every quarter of the circle (or 90 degrees), and an end point that is the same or just past the start point along the circle. For all languages level 3 and above, the robot controller handles the math necessary for linear or circular motions based on the points defined. Level 1 and 2 languages require you to figure out how to make this happen, which is another reason for their rarity in industry. The higher-level languages will calculate the best arc or circle based on the points you create to complete the motion. If you teach one of the points in an odd location, your arc or circle may end up a bit distorted, so make sure you test these motions thoroughly.

Weave motion is straight-line or circular motion that moves from side to side in an angular fashion while the whole unit moves from one point to another. Think of it as a stitching type motion across the normal line of motion (see Figure 9-11). To set up this kind of motion, you have to set two additional points besides the normal required points for linear or circular motion. These two points determine

figure 9.11 An example of what a weave weld can do. Notice how wide the weld path is and the almost wavy patter that it creates.

how far to each side of the normal motion the robot weaves as well as the distance it moves forward between weaving motions. We can program any application or robot system to use weave motion, but welding applications are the most common place that you will encounter this motion.

Are there any obstacles for the robot to avoid? Once you have decided how the robot moves between points, look for anything that the robot might hit along the way. The commonly overlooked item here is the fixturing. Fixtures hold parts in place for various industrial processes by clamping or holding them in some manner. The way we hold parts in place can become an obstacle to the robot, especially when the robot performs a task other than pick and place. I have seen several occasions where the robot ran into fixture clamps during operation, but did not alarm out and stop. Instead, the robot shifted out of its programmed path slightly and created a weld that did nothing to bind the two parts together (see Figure 9-12).

While on the topic, make sure not to leave tools and other objects in the work envelope, as they may be in the robot's path. Many machines that robots work with and around have a monitor and operator control system that is movable. I saw the results of one spectacular crash in which the operator had changed out a tool in a CNC machine, but forgot to push the monitor back into the safe area before starting the robot again. When the operator started the robot, it smacked and shattered the display while merrily performing its program. That several-hundred-dollar mistake shut the machine down for two days while we got a new monitor in. If the object struck is loose, such as a wrench or hammer left in the wrong spot, it has the potential to go flying, which could create added mayhem.

What is the robot doing at each point? Does the robot stop at that point? Does it open or close a gripper at that point? Do we need to turn on a welder, paint sprayer, glue dispenser, or other device? Do we need the robot to hit that point exactly or just get near it? You have to know why you set each point to make sure that the robot does what you want (see Figure 9-13). Some systems give you the

fixture
A device that holds parts in place for various industrial processes by clamping or securing them in some manner

Figure 9.12 Sometimes there are multiple clamps, hoses, and other problematic items that we have to navigate the robot around during programming.

Image courtesy of Miller Welding Automation

Figure 9.13   This example shows the importance of thinking about what the robot is doing at each point in the process. If the programmed point for pickup is to one side or the other, the metal prongs on the gripper will impact the rollers on the conveyor. If the gripper closes too early, the prongs could rip the bag open and spill the chocolate.

Copyright FANUC Robotics

cycle time
The time it takes to complete one cycle of the program

staging points
Positions that get the robot close to the desired point but are a safe distance away, allowing for clearance and rapid movement

singularity
A condition in robotics where there is no clear-cut way for the robot to move between two points that often results in unpredicted motions and operational speeds

option of hitting the point exactly or just getting close to decrease the time it takes to complete the program known as cycle time. You may have to pause at a point so that the gripper or glue applicator can do their job. The list is as varied as the reasons we teach robot points. The point is that if you do not know what the robot is doing at each point, how can you expect to program the robot properly?

What is the robot doing between each point? The answer often determines the motion type between points and other program specifics, such as whether the robot is welding and how fast can it go between the points. As a rule, movement between points for staging purposes is at full speed, while movement during tasks such as welding, painting, part pick up, etc. is slower. Staging points are positions that get the robot close to the desired point but are still a safe distance away, allowing for clearance and rapid movement. I highly recommend adding a staging point before and after any precision movements, as coming in too fast or leaving too abruptly can cause problems. You can always remove staging points later if you determine they are unnecessary.

Are there any conditions or other factors that I need to consider? This is the step in the planning process where you factor in such things such as sensors, waiting for machines to finish their part of the process, singularities, and anything else that might affect the program. Singularity is a condition in robotics where there is no clear-cut way for the robot to move between two points. This is the result of lining up two axes, such as four and five, in a straight line, where the robot could go two different ways to reach the programmed point. Figure 9-14 is an example of singularity. When this happens in your program, the robot may move erratically and faster than the designated speed. Some systems alarm out or warn the programmer when these conditions occur, others just do the best they can, which can result in damage to the robot, part, or both.

Is the process logical? This is a simple question, but it can cause you a great deal of trouble if you do not take the time to ask it. Remember, a program is a set of logical steps, so if you have an illogical plan of operation, your program is likely to fail. Make sure that your movements flow in a way that makes sense and not just whatever you happened to program first. Make sure you have the events sequenced properly. Make sure you have added in the necessary steps for any data collection or looping of sections of the program as needed. This is where you make sure your program is a viable set of commands to complete the tasks that you have in mind. For example, it makes no sense to start the movement to glue a part before you turn on the glue application process, but it does

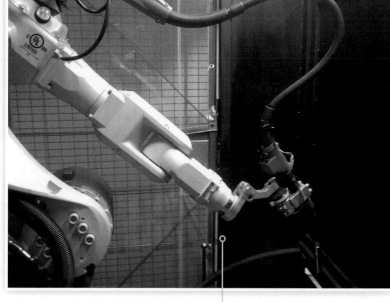

Figure 9.14 Because axes four and five are lined up, the robot can run into computational trouble during straight line motions that pass through this point. The first time that I encountered this in a welding program everything was going great until the robot made a fast, odd movement trying to navigate the singularity that caused the weld to skip and scrapped out the part.

make sense to unload the finished part from a machine before trying to load in a raw part. This is the step where you take a few minutes and look for anything that is out of order or other problems in the flow of the tasks. In other words, you need to channel your inner Vulcan, roboticist, programmer, engineer—or however you prefer to think of it—to make sure the process does not have illogical steps.

If you take the time and answer all of these questions, chances are you have a solid plan for your robot program. It may seem odd the first few times that you go through these steps, but as you write programs and work with robots, you will find that it will become a natural part of your programming process. You may find other questions that help you plan your program or that changing the order of the questions makes it easier for you. These are guidelines, not iron-clad rules, derived from my years of programming and teaching programming for robotics. Feel free to adapt them in a manner that works best for you.

Subroutines

Now that you have a plan in place and have mapped out all the various tasks involved, it is a good time to talk about subroutines. Subroutines are a sequence of instructions grouped together to perform an action that the main program accesses for repeated use. The purpose of using subroutines is to reduce the lines of code in a program and to make it easier to write programs. Some examples of subroutines are opening or closing grippers, a tooling change, alarm response actions, and other such actions that support the operation of the robot. There is not a specified limit on the lines of code in a subroutine, but often these programs are much smaller than the operating program and may have as few as three or four lines of code. Depending on the controller, you may need to add a specific line of code to return to the main program at the end of a subroutine, while others may require only an end statement. No matter how it is accomplished, there must be some way for the systems to return to the main program once the subroutine is complete, otherwise the system will not function properly.

By taking a few minutes to identify repetitive actions that can be turned into subroutines, you can save yourself a large amount of time when you are writing

subroutines
A sequence of instructions grouped together to perform an action that the main program accesses for repeated use

the program. For instance, if it takes three lines of code to open the gripper and another three lines of code to close the gripper, a program that requires 10 changes in gripper state would require 30 lines of instruction just for the gripper action. If we turn the opening of the gripper into a subroutine and the closing of the gripper into a second subroutine, we would only need 10 lines of instruction to complete the 10 changes in gripper state. This saves a lot of time when writing a program and reduces the chances of making a mistake. Once we have established a subroutine, it will continue to function properly until we modify the program or something changes with the basic operation of the robot.

Be sure you create subroutines as a global function instead of a local function. Global functions are variables, subroutines, and other code or data accessible by any program that you create on the robot. This is true of all systems that offer global data sorting and saves you from having to re-create subroutines or record key points when you create new programs. Over time, this translates into a huge savings in time and effort for the programmer. Some common global items are open and close subroutines for grippers, home position for the robot, tool change subroutines, and other positions or actions that are useful to multiple programs. Local function means that only one program can access the data. Most of the positions that you create while writing a program are local data. If you create a subroutine and make it local only, then only the program you created it in can use the routine. To determine if you are creating a local or global subroutine, you will have to understand how the controller sorts and stores data. You may have to create the subroutine as its own small program for global access. Some systems call global subroutines macros, which is an adaptation from the computer science world, where macro instructions invoke a macro definition to generate a sequence of instructions or other outputs. The way in which you establish local or global subroutines all depends on the system you are working with and the way its software is set up.

Again, we want to determine what tasks we want to turn into subroutines *before* we begin writing the program. The whole point is to save time and effort, so it only makes sense to figure this out during the planning phase. If by chance you forget this step or realize as you write the program there are things you could turn into subroutines, there is no rule that says you cannot do so mid program. In fact, if our program example from above used the three lines of code to change the gripper state five times and a subroutine to change the state the other five times, it would work just fine. The danger is that we may have made a mistake inputting one of the 15 lines of instruction we used for the first five changes and of course, we lost a bit of time putting in the extra lines. The only other possible concern is usage of memory, but in most modern systems this will be a rare problem.

global function
Variables, subroutines, and other code or data accessible by any program created on the robot

local function
Data accessible by only one program

macro
A single instruction or specified button used to generate a sequence of instructions or other outputs

Writing the Program

Now that you have your game plan, have mapped out the various tasks and their order, and have decided what to turn into a subroutine, it is time to write the program. This is when you must know how your particular robot works, how the controller organizes information, and most importantly, how to build a program for the specific robot that you are using. A lot of this portion depends on what base computer code or data management system the manufacturer used. Many of the microprocessors for hobby robots and the older industrial systems require programming that resembles CNC or one of the basic computer language codes. Panasonic robots use an operating system similar to Windows, so programming on

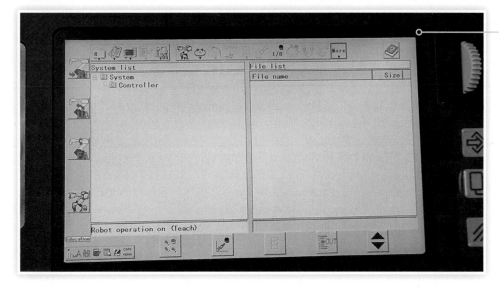

Figure 9.15 If you look close at this picture, you can see an icon in the top left corner looks like a file. It allows you to do the same things that the file tab does in most Word documents. The adjacent icon allows for edit functions, such as cut, paste, copy, etc. There are several icons that are related to the robot only, such as the robot icon and the one that looks like a circle with two arrows.

those systems is similar to creating a Word document (see Figure 9-15). FANUC uses its own software that has evolved with its robots, but it is a level 4 language and user friendly. The Baxter robot, due to its level 5 language, has you physically move the robot to create a program. The LEGO MINSTROMS NXT and NAO robot both use icon-based programming in which the blocks represent actions and the linking of the blocks determines program flow (see Figure 9-16). This is why writing a robotics program is model- and controller-specific and why we are not focusing on any specific system in this chapter.

On a positive note, it has been my experience that once a person learns how to program one type of robot, he or she tends to learn new styles of programming quickly. The process of writing a program is fundamentally the same from robot to robot; it is the specifics or syntax that changes from system to system. If you learn a level 3 language first, it is often easier to learn how to program a level 4 or 5 system. Going from a higher-level style of programming to a lower level may prove a bit more difficult, but having programming experience of any kind under your belt will be a definitive plus.

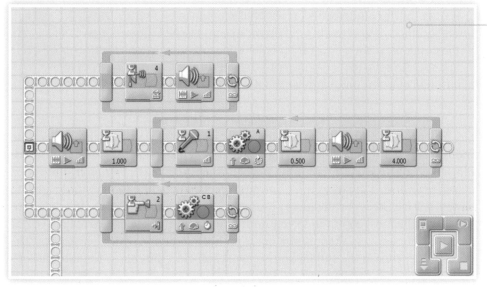

Figure 9.16 Here is an example of a LEGO NXT program. As you can see, it is a set of logic blocks linked to create the logical sorting of the program. Each block controls a specific function, such as sensing, logic filters, or motor operation, with certain variables that the program can adjust for desired operation.

Start with your global subroutines if your system allows for such. Since we are going to use these as part of the main program, it is easier to have them in place before we write the program. Oftentimes, this step has already been done as part of the initial robot package, especially in the case of tooling subroutines. In most cases, you will simply need to check the labeling of the subroutine in your system and then write that down to help with the programming process. If this is the initial setup of the robot, you are using new tooling, or you have a new function in mind, you will have to create these subroutines. When creating subroutines, make sure to test them out before you just assume they are good to go. (We will talk about this process shortly.)

Now it is time to create the program. The specifics of this depend on the system, but the common options that you need to be aware of at this point are naming the program, determining the program type, designating the frame, picking the proper tooling, and designating offset tables to reference. Again, not all systems will have these specific options, and some systems will have different options added to the mix as well; however, you should be able to find the basics in the programming manual or learn them via specialized training. When you name the program, make sure you pick a name that fits within the length constraints of the system and does not duplicate a name already used. Using a name that is too long may make your program unrecognizable to the controller or alarm out the system. Using a name already in use is likely to erase the previous program. You may not get a warning before this happens.

Picking the right reference frame and tooling for the program is crucial for operational success. Remember, the reference frame defines the zero point for the work envelope and is the point from which the robot defines all other positions. In many cases, the default frame is the World frame, which uses the robot base as the origin point. If you change the reference frame after creating the program and saving points, there is a high probability that your taught points will change: this could be a recipe for disaster. In systems that only use one type of tooling, you only need to verify that the tooling is present in the program parameters; this is often a default setting. When working with a system that has multiple tooling, make sure to select the proper tooling or tooling group. Changing the selected tooling after you have created a problem usually does not cause problems with positioning, providing you taught the points with the proper tooling in place.

We use tooling groups when a robot has more than one set of tool position data. This is how the robot keeps track of where the positional point or tool center point is for each unique tooling. You can set a tooling group as a default on most systems, ensuring that new programs use this data.

And

A logic function that requires two or more separate events or data states to occur before the output of the function occurs

Or

A logic filter where at least one, or more than one, of two or more input events must be true for the output to occur

The process outlined above to create a program is standard among modern industrial systems, but is not the way that all systems create programs. Some systems, the icon-based programming styles for instance, require you to simply open software and start creating the program (see Figure 9-17). Other systems try to simplify this process by creating programs in which the fields are already populated with the most common selections for your robot. Some robots only have one frame of reference, so there is obviously no need to define which one to use. The specifics depend on the system you are working with, the controller, and the software used by the manufacturer.

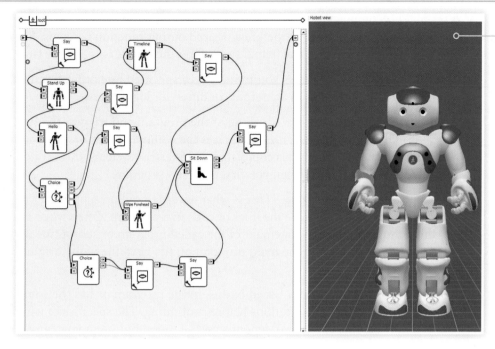

Figure 9.17   The NAO robot is a perfect example of a system in which you open the program, drag the elements you want into the work screen, link them to create the logic flow, and then upload them to the robot. You can also move the robot manually and record positions to create your own unique motions, allowing for the best of both worlds. Here is a program created by one of my students for the NAO robot.

Once you set the main parameters and double-check any other variables in the program creation process, such as offset tables, you are ready to finalize the creation of the program. Finalization saves the parameters set during program creation and allows you to start teaching the specified points and set the logic of the program. This is where you take all the information that you gathered and outlined during the planning process and start to create your program. When you set the points of your program, make sure that you set the proper motion type to reach that point. Pay attention to robot speeds and tooling conditions as you create the program. Make sure that you add in any subroutines or other logic functions at the appropriate points. In short, this is the main part of the programming task. In standard line-based programming, each line will have a unique number that dictates the flow of the program and allows you to reference specific parts of the program with various logic commands.

For those of you who are not familiar with basic programming logic, let us take a quick look at the common logic functions added to programs. The following is a list of common logic commands with a brief description of each:

- **And**—This logic function requires two or more separate events or data states to occur before the output of the function occurs.

- **Or**—If at least one of two or more events happens, then the output of the function occurs. If more than one condition is true, the output of this logic command engages as well.

- **Nor**—With this logic filter, all input conditions must be false before the output occurs.

- **XOr** or **Exclusive Or**—This works like the Or command mentioned above, with the exception that only one of the conditions can be true. If more than one condition is true, the output does not occur.

- **Not** or **NAnd**—This is the opposite of the And command, in that all the input conditions must be false before the output is triggered.

Nor
A logic sorting filter in which all input conditions must be false before the output occurs

XOr or Exclusive Or
A logic filter where the output is only activated when any one of the inputs is true, but not more than one

Not
The opposite of the "And" command, in that all the input conditions must be false before the output is triggered

NAnd
The opposite of the "And" command, in that all the inputs must be false before the output is true

If Then
An advanced logic filter that allows you to set a complex set of conditions to occur before a desired output function happens

Jump to
This command advances the reading of the program to the designated line, avoiding the execution of the lines skipped

End
This command stops the scanning of the program and triggers the system's normal end-of-program responses

Call program/subroutine
A command that calls up subroutines or other programs to help reduce the lines of code in a program

Wait
This command creates timed pauses in the program or has the robot wait for a specified set of conditions before continuing with programmed actions

Math functions
Commands that let you add, subtract, multiply, and divide in varying levels of complexity for data manipulation

- **If Then**—This command is an advanced logic filter that allows you to set a complex set of conditions to occur before a desired output function happens. For example, if input one is true and output two is false, then make output three true.

- **Jump or Jump to**—This allows you to jump to a specified line or label in the program. Often we use this as an output condition for logic filters.

- **End**—This command stops the scanning of the program and triggers the system's normal end of program responses. It is the command that lets the controller know that the program is completed; in industrial systems, this is often the only command present when you first finalize a program.

- **Call program or Call subroutine**—This is what we use to call up subroutines or other programs to help reduce the lines of code in a program. Often when we use this command, the end statement of the called program or subroutine directs the controller back to the main program on the next line of the original program's code.

- **Wait**—This command creates timed pauses in the program or has the robot wait for a specified set of conditions before continuing. The specifics of what you can do with this command depends on the robot, but often you can set up Wait commands that will pause the program until an input is turned on or some other specific event has happened.

- **Math functions**—With math functions, you can add, subtract, multiply, and divide in varying levels of complexity for data manipulation. This is commonly used to keep track of parts ran, create offsets in the system, or modify data in other useful ways. These are the same functions that the controller uses to compute straight-line motion when in the World frame and to create the arcs for circular movements.

You may encounter other logical filters within the robotic systems that you program, but those listed above are common to many industrial systems. Among the icon-based programming styles, you will often find these as prepackaged blocks that you simply need to connect in the proper order to get the necessary logic. Since most of the icons available are similar to subroutines in nature, you may not have the ability to call or build subroutines. If this is the case, try a copy-and-paste approach to save on programming time. No matter what system you are trying to program, it is time well spent learning where these commands are located, how they are inserted, and the specifics of how they are set up in the programming system of your robot.

Another crucial part of the program-creation process is saving your data. I always tell my students, "Save early, save often, save your data!" One of the most frustrating things about programming is spending a large chunk of time writing a program only to lose all your work due to momentary power loss, a missed action on your part, or a glitch in the system. I cannot count the number of times I have had students spend an entire lab working on a program only to lose their data at the end of class. Shutting down the robot before saving the program is a great way to lose your data and bring a tear of frustration to your eye. Until you actually save your program, what you have is a theoretical program. Sure, it has code, logic, and positions. You may be able to step through it for verification, which is a process we will cover soon. However, until you actually save it in memory, it is vulnerable to human and system error. For long programs, I recommend saving the program every 10 minutes or so when you are working to create the program. At the very least, you need to save the program once you have completed entering it and before testing. Many of the modern industrial systems save the program automatically before you enter auto mode, but I caution you against getting lazy and relying on this method for saving all your programs.

The specifics of creating a program, saving points, setting the motion types, creating the logic filters, and calling subroutines and all the other bits and pieces that make up a program depend on the system you are working with, which is why I am not trying to teach a specific style in this chapter. Chances are that you will have the chance to do some programming in the class that requires this textbook, and your instructor will provide the specifics that you need to write a program for those robots used in the classroom. If this is the case, learn all you can: this knowledge will greatly help you in the field when you encounter new systems and new programming styles.

Testing and Verifying

Once you complete your program, there is still one crucial step before placing the system in auto and giving it a go, and that is to test your program. Even programmers with years of experience need to verify the operation of their program before declaring that it is ready to go or viable (see Figure 9-18). Helmuth von Moltke, a German military strategist, said it best: "No battle plan survives contact with the enemy" (Levy 2010). While programming is not war, it can sure seem that way when things go wrong and the boss is breathing down your neck for results. The correlation here is that there is a good chance that you will need to make changes to your original plan once you start writing the program or when you test it out for the first time. This is a natural part of programming process. Do not hang on to your planned program so tightly that you cause yourself grief during the testing process, as there is a high probability you will need to make some modifications to tweak the program and get the desired results.

The system you are working with will determine what testing options you have at your disposal, but most industrial systems offer a manual step-by-step testing method and a continuous testing mode. For most systems, manual testing requires you to hold the dead man switch on the teach pendant as well as press and possibly

Figure 9.18 Welding systems often give you the option of welding in manual mode during the testing process. If you are working with a system that performs an operation during testing, remember to use all the proper safety gear.

Image courtesy of ABB Inc.

figure 9.19 As you can see, sometimes you will be close to the action during program verification. This is why the robot moves more slowly in manual mode and why you must always take care when proving out your programs.

Image courtesy of ABB Inc.

hold some of the keys on the pad. The good part about this type of testing is that if you release the dead man switch or the key you must hold down, the system stops immediately. This gives you the control you need to keep the robot from crashing into something or damaging parts. Another good thing about manual testing is that the system typically runs at about half of its normal speed. As you learned previously, robots can move faster than we can react, so this slower movement speed allows operators to keep up with everything that is going on and gives them the chance to respond before disaster strikes or get out of the way should they find themselves in the danger zone (see Figure 9-19).

In Step mode, the robot advances through the program one line at a time and requires you to press a button on the teach pendant keypad before it reads the next line of the program and responds. Most systems will let you step forward and backward in the program so that you can bounce between two lines of code if you need to watch the motion over again. One thing to keep in mind is that just because the robot is only executing one line of code at a time does not mean it will not travel a large distance. If you set two points 10 feet apart, the robot will travel that 10 feet as you execute the line of code with the movement to point B. You can stop the motion by releasing the dead man switch or, in the case of systems where you have to hold a button on the keypad, by releasing the button. When testing, it is my recommendation to start with Step mode, as this allows you to determine exactly what the program does line by line and makes it easier to correct the proper portion of the program as needed.

step mode
A manual mode in which the robot executes one line of program code each time a specific button or button combination is pressed

Continuous mode works as the name implies. Once you start the program in manual continuous mode, it runs through the program until you stop it by releasing a button or dead man switch, it reaches the end of the program, or something causes the system to alarm out. Many industrial systems let you run this method of testing backward as well, so make sure that you know which way the program will flow. I recommend running the program in this mode once you have verified and approve of the operation of the program in Step mode. This will give you a good idea of how the program looks as a whole and let you see the big picture.

Some motions in your program cannot happen in the space allotted if the robot moves slowly. If you run into this situation, most systems will give you a warning alarm; once you accept or hit whatever is required to authorize the movement, the next movement that the robot makes will be at or near full speed. This is often the case when you make a radical change in the tooling orientation between points or other moves where the axes involved have to turn a great distance (see Figure 9-20). If you just hit yes and do not keep in mind what is about to happen, you could get a nasty surprise, especially if you are close to the robot when this move happens. Anytime you get an alarm on the robot, take the time to read it and determine what the alarm is telling you *before* you just rest it and go full tilt ahead. I have had students crash the robot, damage parts, have near misses, and mess up their program by hurrying and ignoring what the alarm said. This is a common error that many experienced programs make as operators get used to seeing the warning messages and get into the habit of just resetting and going.

Step and continuous manual modes are also a great way to advance through an existing program for rework and post-crash operation. When used this way, make sure you take special care to insure the tooling is functioning as needed to finish the parts. For instance, welding guns have to strike an arc before they begin the weld and you may have to do something extra to make this happen mid-weld. You may have to open grippers before stepping through to the part pick up portion of a program. The point is to make sure you understand what is supposed to happen when using the manual modes for crash recovery or manual operation of the system.

When dealing with systems that do not offer manual testing modes, do the best you can to verify the operation of the program. You may need to hover near the E-stop, make sure the work envelope is as clear as possible, warn those nearby, or do other things to minimize the risks during testing. Many of the hobby- or entertainment-type robots lack testing modes, so with these systems, you will just have to do your best to double check the program before you enter it and minimize the risks when you run the system. Luckily, most of these systems are low power and the potential for injury is minimal. Make sure you have a controlled environment when testing these systems, as the last thing you want is to unleash an out-of-control robot on a group of unwary people.

Once you have tested the program in both modes and are happy with the operation, make sure to save the program once more. This ensures that the program in memory is the one you want and not the pretesting program. Some programs turn out to be so laden with problems in the testing phase that the programmer is better off to start over than try to fix what they have. This may sound crazy, but sometimes it really is easier to start over than to fix a hot mess. Some of the warning signs to look for are multiple-alarm conditions, problems that get worse instead of better after several attempts to correct, code that causes the program to stop for unknown reasons, and/or a program that has so many corrections to it you no longer understand the logical flow. Some programmers also find it easier to create a completely new program than to make changes to an existing program written by someone

continuous mode
The manual mode in which a program runs until completed or a specific button is released

Figure 9.20 Look at these two pictures to see the types of movements that often require a rapid move in manual mode. Switching the orientation of the tooling in such a radical manner over a small distance usually requires a quick movement with a break in the welding, which can be detrimental to weld quality.

else. Just as battle may require generals to scrub their strategy and start over, testing may send the programmer back to the planning stage to create a new program.

Remember that the whole point of testing your program is to get the bugs out of it before you put it in automatic and let it run. The more effort you put into this step, the greater the chances of successful robot operation in automatic mode. Rushing through this step may cost you a lot more time repairing damage than you would have lost doing it right, so keep that in mind. When it comes to programming, quality is much better than quantity!

Normal Operation

Once you have completed writing and testing the robot program, the next step is to hit the green start button and let the robot get to work. Even though you are now ready to begin normal operation, this does not mean that you can just walk away with no concerns. Most systems with a manual mode run at a slower speed than the robot is capable of in automatic mode as a safety feature. This feature that improves safety for you when working with the robot may mask certain problems when the robot is running at full speed. On top of this, the changes that naturally occur in the world around the robot as well as changes in the robot's internal systems can also cause problems with normal operation. It is the job of the programmers and operators to monitor the robot's behavior, make the necessary adjustments as needed, and in some cases prevent disaster before it happens!

When running a new or edited program for the first time in automatic mode, make sure that you watch the robot go through the program a couple of times before you consider it ready to run with limited supervision. Many programmers have skipped this step only to discover problems with the program after the robot has ruined parts or damaged equipment. By verifying that the program is working correctly during a couple of cycles, you can avoid the expense and frustration of downtime as well as having the boss asking uncomfortable questions such as, "Didn't you test this thing?" or "What do you mean it crashed?" I would also recommend that you keep your hand near the E-stop or some other stop button during that first run in automatic, in case something does go wrong. You cannot always stop the robot in time, given that robots move very fast and we mortals are sometimes left with little time to react, but it will give you a fighting chance. After you have verified the operation by watching it run through the program a couple of times or cycles, then you can truly declare it ready to run.

Once you are running a verified program, it does not mean that the robot no longer needs human help. Variances in parts can cause problems with pickup and placement. As tooling and fixtures wear with use, positioning can change, parts can move, and the quality of work may decrease. As the robot wears out due to normal use, the positioning of the various axes may change. These and many other things may make enough of a difference in the working conditions of the robot that it needs human intervention; it is common in industrial applications for operators or programmers to tweak robotic programs from time to time to deal with these changes. This procedure often only takes a few minutes and can be as simple as adjusting one or two points in the program and then saving the changes. In these cases, there is no need to reinvent the wheel and write a new program; most times, the testing is minimal, such as stepping through a couple lines of code to verify the new positions' work (see Figure 9-21).

Whether the robot has an operator watching it or a complex sensor array that looks for problems, there must be some system for monitoring the robot. Remember, the robot can only do what we program it to do. It has no intuition. It has no feeling. It has no problem-solving skills other than those we give it through programming. As we figure out new and inventive ways for robots to solve tough problems, it is easier for systems to respond to unexpected circumstances; however, these responses are still program driven. Eventually, the robot needs a human to solve the tough problems, find better ways to perform tasks, and make determinations that are simply beyond the scope of its programming. This is part of the partnership between people and robots; each has its strengths and weaknesses. We capitalize on the strengths of both parties to compensate for the weaknesses of the other and thus perform tasks that would be difficult if not impossible for either party to do alone.

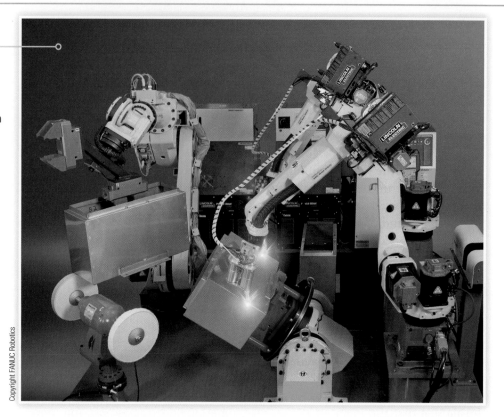

Figure 9.21   A properly designed and tested program allows multiple robot systems like this one to work together in fluid elegance.

File Maintenance

In most cases, robot controllers are specialized computer systems, and like a computer, you need to perform data or file maintenance from time to time. In many ways, this is similar to managing the various documents that you might have in your word processor program or the music you have on your MP3 player. You want to save the data you like, make sure you can find and use it when you need it, and delete the data that you no longer want. Let us take a closer look at what each of these functions entail in the world of the robot.

Earlier I mentioned to save early, save often, save your data. This goes beyond just saving the program into the teach pendant or controller of the robot. Most robot controllers do not write programs to a storage device unless specifically instructed to, which means that once the power is gone, so are all your programs! While most have battery backup, batteries only last so long and people are notorious for forgetting to change these when prompted or per the manufacturer's recommendations. If you do not have your programs written to some form of backup, such as a Flash Drive, SD card, hard drive, or other storage medium, all your programs are in peril of deletion! To avoid this, most manufacturers provide some way to back up data and programs from the robot. Often, you can set the time interval for this to occur. For instance, the Panasonic welding robots use SD cards as their backup media; these plug into a slot inside the teach pendant for easy access. There is a setting in the controller that allows you to set how often the system backs up all the programs as well as how many backups to put on the SD card before overwriting the oldest backup. This allows you to go back to previous saves in case you edit and save a program that is not to your liking. The process varies from robot to robot, with some giving you automatic options while others require you to manually perform each save. Make sure that you understand how your robot works and that you save all your programs and data.

Another data trick you might have available for use is sorting. Imagine for a moment all your documents or music thrown into one folder with only the title of the document or song to sort things. This would make it difficult to find specific items and place a large amount of data in one spot. To get around this in the computer world, we create folders within folders and use these to sort data into groups that make sense to us. Depending on the system, you may have this option with your robot as well. This is a handy feature, especially if you are looking for something specific in a backup file. If you have this option with your robot, make sure to create folders that are meaningful and sort the data appropriately. Failure to do so could lead to problems with both you and the robot finding needed data. Many industrial systems create specific folders automatically during the backup process; in those cases, you simply need to learn how the robot organizes the data for efficient navigation.

Do you have everything you have ever written saved on the hard drive of your computer? The likely answer is no. Just as we discard documents that are no longer useful or relevant, so too must we delete programs that are no longer beneficial. With modern hard drives, computers have massive amounts of storage, and deleting a Word document has very little impact on the storage capacity of the system. Unfortunately, the world of robotics is different. Robots often have a small amount of memory, and if you have several complex programs stored in the robot, you may not have enough storage space for another large program. To get around this, we save programs to external storage devices, as mentioned above, or delete programs that are no longer of use. Before you delete a program, I would offer one caution: make sure to save an external copy if possible. Obviously, if we are deleting a program to make space on the robot's memory, we cannot save the program on the robot. If you can save the program externally, make sure that you label the storage device properly or in some way document where to find the program so that you can retrieve it if needed. If saving the program externally is not an option, you will have to live with the fact that deleting a program now may require you to rewrite it later.

While on the subject of deleting files, make sure you keep track of what automated backup systems are doing. You may have to manually go in from time to time and delete old backups so there is room for new/current backups. In the fast-paced modern world, it is easy to ignore the automatic backups, as they are out of sight and out of mind. You many need to create a checklist or some form of automatic reminder to help you keep up with this if there is no alarm message (or the like) from the robot to warn you that the backup system has used all the available space.

You may run into tasks specific to your robot when it comes to file maintenance, so make sure that you understand how your robot deals with data and your responsibilities in this respect. If you run into a data snag, make sure that you take the time to come up with a corrective action and then implement your plan. Experience is a great teacher, and there is no greater teacher than hours of tedious work that you could have avoided with a few minutes of proper file management. When dealing with files and data, do not forget about things such as home positions, user frames, offsets, data tables, and other information that affects the operation of the robot. Just because they are not a program does not mean they are not important. In fact, loss of the zero point for an axis or axes can cause you a great amount of downtime and frustration as you correct the situation; where saving this data might make correcting the problem as simple as uploading the zero point. File management is another place where the how and what varies from robot to robot and sometimes model to model, so make sure that you take the time to learn your robot.

review

While we did not cover the specifics of programming any one system, the information from this chapter will help you in programming the robots you encounter during your exploration of the field. Remember, once you learn the basics of programming one system, it will make it much easier to learn how to program other robots. Whether you are working with a VEX, LEGO NXT, FANUC, Panasonic, ABB, MOTOMAN, or other robotic system, the process is the same as what you have learned in this chapter with merely differences in the syntax or how you create the program. The language level does have an impact on writing the program. Remember that the lower the level, the more work required on your part.

During our exploration of programming and file management, we covered the following topics:

- **Programming language evolution.** We looked at the five different levels of programming languages and how each of them works as well as the evolution of digital programming.

- **Planning.** We discussed the importance of making sure that we know what we are doing before we write the program.

- **Subroutines.** This section covered those small programs and sections of code that we use repeatedly during the course of operation.

- **Writing the program.** We looked at the process of writing a program and the types of motions that you can expect to use.

- **Testing and verifying.** We looked at double-checking your program for proper operation during and after the process of writing the program but before running the system in automatic.

- **Normal operation.** This section covered normal operation and the human responsibilities during such.

- **File maintenance.** Just as we manage the data on our computers, so must we manage the data and files in the robot.

key terms

And	Fixtures	Math Functions	Staging points
Arcs	Global function	NAnd	Step mode
Call program/subroutine	If Then	Nor	Subroutines
Circular	Joint	Not	Swarm robotics
Continuous mode	Jump to	Or	Third party
Cycle time	Linear	Programming language	Wait
End	Local function	Robot program	Weave
Exclusive Or	Macros	Singularity	XOr

review questions

1. How do we alter the programs on robots that pre-date or do not take advantage of digital technology?

2. Describe the operation of the first digital robot programs.

3. What was the second step in the evolution of digital programming?

4. What was the third step in digital robot programming and how does it affect the modern world of robotics?

5. What are the five different levels of robot programming?

6. Describe programming in a level 2 language.

7. Describe programming a level 3 language robot.

8. How does one write a program in a level 4 robotic system?

9. What are the benefits of a level 4 programming language?

10. What is the difference between a level 4 and a level 5 programming system?

11. What are the two main parts of program planning?

12. What are the seven questions you should answer during the task-mapping phase of planning?

13. What are the four main motion types and how does the robot move during each?

14. What types of motion can joint movement generate between points, and what is a potential danger from this motion?

15. How do you create arcs and circles using circular motion commands?

16. What are some of the questions you may need to ask yourself to determine the robot's operation at each point?

17. How does motion between staging points differ from motion during work operations?

18. What causes singularity?

19. What is the purpose behind using subroutines, and what are some examples?

20. What are the common options that you will need to consider when creating a program?

21. What is the potential problem of changing the frame of a program after saving points?

22. What is a good rule of thumb for saving long programs?

23. What is the difference between step mode testing and continuous mode testing?

24. What typically happens when you are testing in manual mode and the robot has to make a large axial movement, and what is the danger associated with this?

25. What should you do when running a program in automatic for the first time?

26. What is likely to happen to your data and programs in the robot should you forget to back them up to a permanent storage device?

Troubleshooting

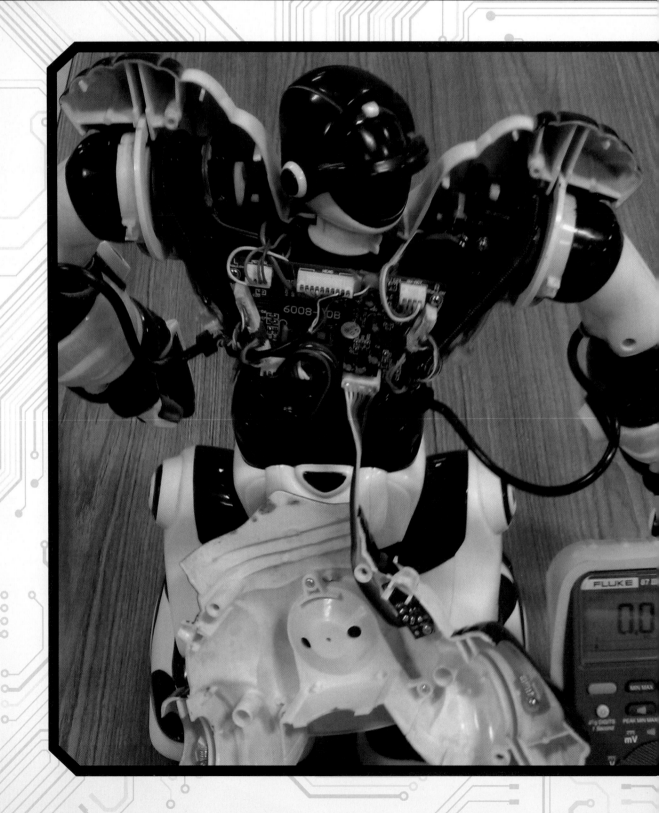

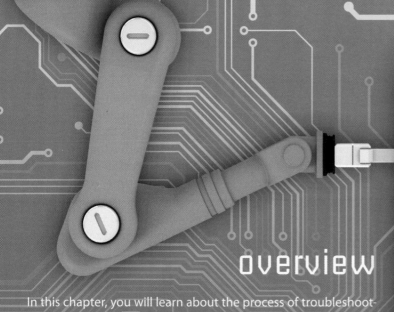

overview

In this chapter, you will learn about the process of troubleshooting. It is a simple, unpleasant fact that eventually every mechanical system or process will fail to perform as desired. When this happens, someone will use the troubleshooting process to determine what the problem is and how to get things functioning properly once more. In the course of learning about troubleshooting, we will cover the following items:

- What is troubleshooting?
- Analyzing the problem
- Gathering information
- Additional resources for information
- Finding a solution
- If at first you do not succeed…

What Is Troubleshooting?

Before we discuss the process of troubleshooting, we must first get some definitions out of the way.

Troubleshooting—The logical process of determining and correcting faults in a system or process

Troubleshooter—A skilled person who determines the cause of a fault and makes the corrections necessary to get the system or process working properly once more

You can find many variations of these definitions on the web, in dictionaries, and scattered through technical manuals, but we will use the definitions above

troubleshooting
The logical process of determining and correcting faults in a system or process

troubleshooter
A skilled person who determines the cause of a fault and makes the corrections necessary to get the system or process working properly once more

in this chapter, as they embody what we want to accomplish. With that in mind, let us begin to explore what troubleshooting really is. The definition begins with, "The logical process," and that is a key point to understand. Troubleshooting is not just some magical thing that happens, but a logical process that starts with the problem or condition and proceeds to the information-gathering phase. An action plan is formed and carried out, and the system or process is finally tested for proper operation. Sometimes this process works the first time; other times, it takes many cycles to find the correct course of action. That said, do not get discouraged if your first plan does not work. There will be times when you try an action plan and actually make the problem worse. Do not panic if this happens, as this too can be a part of the troubleshooting process. One of the greatest troubleshooters I have had the pleasure to work with told me, "If you can affect the problem for better or worse, you are on the right track." The important thing is to think about what you did and how it affected the problem. This information can give you great insight into what is causing the problem, and that is when you can truly fix the issue at hand.

The second part of our troubleshooting definition is "determining and correcting faults in a system or process." We will look at ways to organize the "logical process," but first let us spend some time talking about the system or process mentioned in the definition. Before you can fix a fault, you have to know how the system normally works. For example, imagine that you are standing in line for the new ride on its opening day at your favorite amusement park. When it is your turn, the operator tells you that there is a problem with the system and asks you to help fix it. You have never ridden this ride nor have you had the chance to talk to anyone who has. You only have a general idea of what the ride does and know nothing about how the ride is controlled. At this point, all you really know is that the ride is not working properly. What would you do to fix it? The answer is, without detailed information about how the ride works, how it is controlled, and what it is specifically not doing or doing wrong, your chances of success are near zero. The same is true of troubleshooting any system or process; you have to understand how it normally works and what is going wrong before you can fix the problem. A good troubleshooter may spend a large amount of time just learning the various processes and systems that he or she takes care of. Some of the best ways to accomplish this is to run the piece of equipment for a while or watch it operate under normal conditions until you understand how it works.

Let us revisit our previous example to show how knowledge of normal operation can help. You are back at the same amusement park. The ride has been in operation for several months, and you have been working as an operator. You know how long each run should take, how many people can ride at one time, the sequence of events, and, most importantly, you now know how all the controls for the ride work under normal conditions. You know its quirks, too. It is your day off, and you have brought your friends/family to the park to have fun and try out the ride that you work. As you are about to get on, the operator says the ride is not functioning properly. He asks you to help fix it because of your knowledge of the system. Do you think you could be helpful in the troubleshooting process under these conditions? The knowledge that you now have would give you a basis to ask helpful questions. You know how the system works and know its quirks, and there is a chance you have seen this problem sometime during your operation of the ride. All of this would greatly increase your ability to determine what the problem is and find a solution.

This brings us to another important aspect of troubleshooting: those who forget the repairs of the past are doomed to repeat the work. I recommend that everyone who is going into any technical repair field, or plans to do any trouble-shooting, keep what I call a repair journal. This is a place where you make notes about all the repairs you perform and anything that can help you in the future. For example, when you are working with a problem you have not seen before, make notes about machine conditions, alarms, indicator lights, and other information you find during the problem assessment phase. Do not forget to add notes on what you did to fix the problem, as this is just as important as the symptoms that you record. This will help you down the road should the same problem occur again and with problems that are similar to ones you have corrected. During my time turning a wrench, I have seen the same problem on the same piece of equipment separated by years of run time. It is hard to remember what fixed the problem two or three years ago, but a good set of notes can save you the time previously invested into a particular problem. Moreover, it is common for similar problems to have similar solutions. Even though the original solution may not fix the new problem, it may give you a very good idea of where to start, thereby saving a large amount of time in the information-gathering phase.

We began the chapter with another definition, that of a troubleshooter. A troubleshooter looks at a problem, finds a solution, and makes things work again. We must all become troubleshooters at times, but some of us take it to the next level and become professional troubleshooters. These are the maintenance tech-nicians, engineers, programmers, electricians, mechanics, machinists, managers, and installers, among others, who turn troubleshooting into a professional career. People who troubleshoot for a living have a set of specific skills and knowledge that allow them to solve problems effectively and relatively quickly in their chosen field. This specialized skill allows them to command a higher wage and gives them a greater level of job satisfaction than most entry-level workers experience. You may be surprised to learn that even when the economy is in a slump and unem-ployment is high, highly skilled troubleshooters can still find jobs and wages that match their skills. No matter what field you choose to go into, no matter which way you decide to specialize, you can be sure that you will have to perform some kind of troubleshooting.

After talking about troubleshooting and troubleshooters in general, you may feel that this is something beyond your skills or experience. However, that is far from the truth. Have you ever changed a flat tire or wedged a piece of paper under a table to stop it from rocking? Have you ever altered a recipe on the fly when you discovered that you did not have all the ingredients or tracked down a location after receiving bad directions? Have you ever fixed your favorite gaming system by checking the cables or with a DVD cleaning disk? If you can answer yes to any of these questions, or similar ones, then you have worked through a troubleshooting problem. If you take what you already know about troubleshooting and refine it with the tips in this chapter, you will be well on your way to becoming a profes-sional troubleshooter, with all the benefits that entails.

Analyzing the Problem

Before we can fix a problem, we have to know what the problem is. Can you imag-ine a doctor handing you a prescription without taking the time to determine what is going on with you? Would you have any confidence in this treatment plan? No. You want the doctor to listen to you, perform an examination or run tests, and then determine the best treatment. It is the same when we troubleshoot a sys-tem or process. Before we can prescribe a corrective action to the problem, we have to figure out what the problem is and gather enough information to make an informed decision. If we fail to do this, our treatment of the problem has about the same success chance as the doctor's walk-in prescription. There are several different areas we can draw information from during the process of determining what is wrong with the system in the analyzing phase.

Operator(s)

There are many great resources for information about the problem. One that is often overlooked is the person or people who were operating the equipment just before the fault. They can tell you valuable information, such as, "Just before it quit there was an odd grinding noise"; "I put in a small offset, it ran three parts fine and then bottomed out"; or perhaps "I have noticed that the machine has been acting odd all day, seeming to take longer than normal to run the sequence and hanging up." The operator(s) generally have a front row seat when something goes wrong as well as a great understanding of how the system normally works. I have seen this resource ignored in the troubleshooting process and it has cost more than one technician many extra hours of work.

Sometimes operators have seen the problem before and have a general idea of what fixed it last time or who worked on it last. While the operator may not know the specifics, this information could point the troubleshooter in the right direction. Similarly, if you were the one running the equipment when the fault happened, think back to what you noticed or what you just did. Did you make any changes to the program, did you make any offsets, did you change tools, did you notice anything odd about the parts, did you hear or smell anything unusual, did you feel any excessive vibrations, or anything else that caught your attention? Troubleshooting often involves many (if not all) of your senses and requires you to communicate with others as well. If you find yourself tempted to ignore the operator(s), think back to a time when you noticed a problem with a piece of equipment (like your car) and the troubleshooter (automotive mechanic) ignored what you had to say. If you are like most people, you found this aggravating to some degree and likely did not appreciate your input being ignored or swept aside. You may have lost some of your confidence in the troubleshooter because of this or even your desire to work with that person again in the future, given a choice. You do not want to become the troubleshooter no one wants to work with, do you? Keep this in mind for the day when you are the one trying to fix something and an operator is trying to tell you what he or she knows about the problem.

The Equipment/System

After talking with the operator(s) to learn all that you can, the next step is to see what the equipment or system can tell you. Some systems use lights to indicate what portions of the machine are running or alarmed out while others display messages and/or information on teach pendants or user screens. It is common to find both information systems together in one piece of equipment, providing a better overall picture of what is going on. This is similar to vehicles that have lights, messages, and/or gauges to let you know that the vehicle is low on gas, that the tire pressure is low, what the temperature of the engine is, that the door is open, that the high beams are on, etc. No matter what equipment or system you are working on, it is important to get as much of this information as possible *before* you begin to fix the problem. Many systems will store alarm data, but this is only useful if you know where it is stored and how to retrieve and read it. You may have to refer to

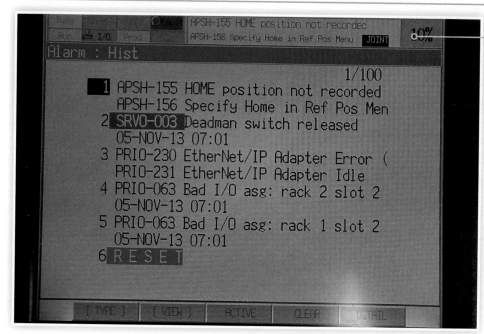

Figure 10-1 A sample of the information provided in the alarm history on a FANUC teach pendant. The alarm highlighted in red is of higher importance than the other alarms; it indicates that the system will not run until someone addresses the issue. Can you guess what mode the system was in when this alarm occurred?

the equipment's manuals or operating procedures to interpret this data, as many alarms display as a combination of letters and numbers.

For example, I was working with a FANUC robot system and pulled up the alarm screen on the teach pendant to see why it would not run (see a similar scenario in Figure 10-1). There were 23 alarms, each offering a piece of information about what was going on with the system. Ultimately, it turned out that the backup batteries had failed and the robot had lost its mastering, or zero position, data. Of the 23 alarms, 6 were telling me that an axis had lost its mastering data while the rest were lower level alarms related to the six axes no longer having a zero position.

This is what led me to the battery backup issue, which was the heart of the problem. The alarm data saved me hours of work tracking down the core problem and made fixing the system much easier, though it still took about an hour to get new batteries in the system, home out all the axes manually, and get the system ready to go.

As mentioned previously, some error codes are in an alphanumerical format (letters and numbers) and may need to be converted to a regular mathematical number before you can find the error information.

When checking alarms, you will often find a string of numbers, or an alphanumeric formatted code, that consists of a mixture of letters and numbers along with the axis involved or some other short message. Getting the full alarm information may require using a code list from the maintenance manual(s) with the pertinent information or converting the number given back into the base-10 numbering system we know and love before you can find more information. Regardless of how we get it, the information associated with an error code often gives us a great amount of information about not only what has happened, but also some of the steps to troubleshooting and repairing the system. This information helps us focus on where to begin the troubleshooting process during the analyzing phase and can save hours of work. (See Food for Thought 10-1 to learn more about numbering systems and how to convert them quickly using a computer running windows.)

alphanumerical format
Information in the form of letters and numbers

food for thought

10-1 Converting Binary, Octal, or Hexadecimal Numbers

Decimal Numbers

This base-10 numbering system is one that you likely have the most experience with. When you look at gas prices, work simple math problems, determine if you have enough money to make a purchase, or make other normal calculations, you are using decimal numbers. The 10 digits for any position are 0 through 9.

Binary Numbers

Binary numbers have a base-2 system with 0 or 1 for each position. These 0s and 1s represent decimal numbers in computing systems. Because there are only two states for each position, the controller of the system can process millions of these states per second and thus perform many calculations quickly.

Octal Numbers

Octal numbers have a base-8 numbering system that uses the numbers 0 through 7 to represent each position. This base makes it easier to represent large numbers with fewer bits of data and thus takes up less memory. It also helps with the computational speed when processing large numbers.

Hexadecimal Numbers

Hexadecimal numbers have a base-16 numbering system that uses digits 0 through 9 and letters A through F for each position. Because the base is so large, it takes very few bits to represent a very large decimal number. Many industrial systems, including robots, use this numbering system for their error codes, so expect to see these in the field.

Bit

This is the smallest unit into which data can be broken. Each digit of a number is a bit, no matter which numbering system is used. Sometimes the conversion from one format to another will actually require more bits of data. For instance, it is common for a binary number to have more bits of data than the decimal number it represents. This seems counterproductive, until we realize that the simplicity of the binary code allows the processor to make large calculations at astonishing speeds.

Conversion Steps

No matter what system you are converting to or from, follow this procedure:

Open the start menu in windows.

Select All Programs.

decimal numbers
The base-10 numbering system that most of us are familiar with

binary
A base-2 number system with 0 or 1 for each position

octal
A base-8 numbering system that uses 0–7 to represent each position

Personal Observations

Another way we gather information is with our own observations. By now, you should know how the equipment or process normally works, talked with the operator, and gathered any error message data or other indicators from the equipment. Now it is time to see what your senses can tell you about the situation. There will be times when just a few moments of looking at the equipment will give you all the clues you need. For instance, if you observe that the robotic arm in a malfunctioning piece of equipment has collided with and critically damaged the machine's controls, you have a good idea of the problem. The point in the process where the equipment stops can also provide information. A robotic arm that stops in the middle of loading/unloading could indicate a misalignment, an unexpected contact, or I/O that is not working properly. You may smell burning plastic or rubber, indicating a problem with the wiring, electronics, or a drive belt. In the process of

Select Accessories.

Select Calculator.

Once the Calculator is open, click on the View tab.

From this dropdown list, select Programmer.

You should now see a row of options to the left that say Hex, Dec, Oct, Bin: these are your hexadecimal (base 16), decimal (normal numbers/base 10), octal (base 8), and binary (base 2) numbering systems (see Figure 10-2). With Dec highlighted, you can input any number. For fun, try inputting the year that you were born. Now, click the button next to Hex to see your birth year in hexadecimal format. Then click the buttons next to Oct and Bin to see your birth year in octal and binary formats. You may have noticed that as the base decreases, the number of bits used to represent your birth year increases. The more bits it takes to represent a mathematical number, the more memory needed to store that information. This is why octal and hexadecimal take up less memory space when dealing with large numbers.

If you need to convert a number back to a decimal, follow these simple steps:

Select the proper numbering system: Hex, Oct, or Bin.

Enter the number as displayed on the equipment in the alarm code.

Press the button next to Dec and read the decimal number.

For those who have access to and like using smartphones or tablets, there are apps out that will

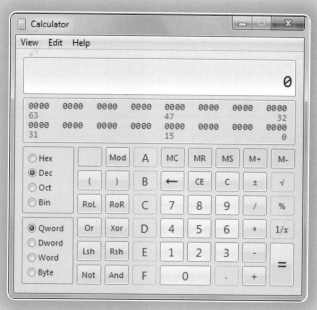

Figure 10-2   The calculator should look like this once you have selected the programmer option.

convert numbers among the various bases in a similar manner to the Windows calculator. The main point is to extract the information you need for troubleshooting purposes and save yourself some time and trouble.

If you are interested in seeing the math behind the conversions, check out **http://www.purplemath.com/modules/numbbase.htm**. You can also find this information in certain Algebra books and various technical manuals as well.

inspecting the equipment, you might feel a hot spot, indicating a possible component failure. You may hear a motor buzzing or whining because it cannot move as directed or you may do a test run of the equipment and hear something amiss, indicating a possible mechanical failure of gears or bearings. Again, the more you can learn about the problem in the information-gathering phase, the better your chances of correcting the problem.

Just like a CSI at work, the troubleshooter has to use all his or her senses to determine what happened. Many times, it is tempting to look at the alarms or indicator lights and jump right into trying to fix the system without taking a few moments to examine the situation for more clues. Just like many of the popular crime dramas, it is often the little details that help us find the culprit. More than one troubleshooter has ignored this step of analyzing the problem and ended up losing hours of time by going down the wrong path. This step relies heavily on the troubleshooter knowing how the system normally works and what is normal

hexadecimal
A base-16 numbering system that uses digits 0–9 and letters A–F for each position

bit
The smallest unit data can be broken into

Figure 10-3 When using your sense to find clues to machine problems, remember to factor in the normal operation of the system. For instance, the welding operation pictured here produces enough heat to melt metal as well as atomize some of the materials fused in the form of smoke. When troubleshooting this robot, the smell of smoke in the air or the welding gun being hot would be normal for the system and *not* a clue.

Image courtesy of Miller Welding Automation

for the various systems and components as well (see Figure 10-3). If it is common for the servomotor to run at 140°F, then the motor being hot to the touch is normal. If the robotic arm always slows down for loading and unloading procedures, then this not a symptom. If the system is in an environment where molten plastic is present, then the hot plastic smell is not necessarily a symptom. This is where the troubleshooter uses logic filtered through knowledge of the system to determine what is and is not important. Just like a CSI, the troubleshooter has to decide which of the clues are relevant and which are circumstantial.

Diagnostic Tools

Another way we can gather information about the system is to perform tests using specialized equipment. There is a wide variety of testing equipment available, but most technicians agree that the multimeter is a necessity for troubleshooting equipment. No matter which brand you choose, make sure that the meter has all the functions you need, is rated for the conditions of voltage and amperage you will expose it to, is dependable, and, most importantly, that you know how to use it safely. Often the checks performed using a meter put you in harm's way, so make sure that you are aware of the potential dangers before taking a reading. One common way in which we use multimeters is to track the power, where you know which components should have power and verify if they do. When you find the point where there should be power but there is not, you have a vital piece of information for the troubleshooting process. During this process, you are working around and with electrically powered equipment, which can be a very dangerous situation if something goes wrong. You also need to make sure you have the proper voltage type selected (AC or DC) and that you have selected a voltage level higher than what you expect to detect. Here's the rule of thumb: If you are not sure of the voltage level present, select the highest level on the meter and work down as needed. When taking voltage readings, make sure that you do not touch any metal components, including the tips of the meter probes. (See 10-2 Food for Thought for more on the dangers of electricity and Figure 10-4 for a diagram of a common multimeter. See Chapter 3 for more on AC and DC power.)

tracking the power
The process of verifying that the components that should have power do indeed have power

1- This plug is for taking amperage readings of up to 10 amps.
2- This plug is for small amperage readings of under 1 amp.
3- This plug is the common connection and where the black lead is plugged in.
4- This plug is for voltage, resistance, and electronic component testing and is where the red lead goes when not testing amperage.
5- This setting turns the meter off.
6- This setting is used to check AC voltage (note the wavy line to represent AC).
7- This setting is used to check DC voltage (note the solid line with the dashes below it to represent DC).
8- This setting is used to check very small DC voltages (less than 1 volt).
9- This setting is used to check capacitors, resistance, and continuity.
10- This setting is used to check electronic components.
11- This setting is used to check amperage in the milli amp to full amp range.
12- This setting is used to check micro amperage.
13- This is the selector switch used to designate what the meter checks.

Figure 10-4 If you are not familiar with multimeters, take a couple of moments to look over this image of a Fluke multimeter settings dial with explanations.

live
Electrically charged equipment

food for thought

10-2 Electrical Shock

Remember that the severity of a shock depends on the amount of current that flows through the body, how long it flows through the body, and the path it takes. You should always work to avoid electrical shock and take as many precautions as you can to prevent electricity from passing through your body. One simple precaution is to remove all jewelry and metallic objects that may make it easier for electricity to enter the body. Another precaution when working around/on live electrical equipment is to use properly insulated tools, thus reducing the hazard of electrical shock and arcing. You can place thick rubber mats on the floor to prevent grounding of your body, and in some cases place thin rubber mats over portions of the equipment to cover live (electrically charged) components that you are working near,

preventing contact with those parts. You should always be careful working with live electrical equipment, no matter the voltage level of the system or the task you are performing.

Below is the chart from the chapter on safety as a reminder of the dangers of electrical shock. It takes a very small amount of current to equal big danger for the person shocked; this is why you must be cautious at all times when working with electricity. One mistake, one moment of doing without thinking, one careless action is all it takes to become a part of the circuit, with all the dire consequences that could bring.

OSHA requires those working near live electrical equipment to wear proper protective gear based on the level of voltage present and how quickly the fusing systems can stop power flow during a short circuit situation. NFPA 70E covers the arc flash standard, which is how one works safely with live electrical systems, especially those in industry. This standard tells you what to wear, what distances different boundaries must be from the live electrical equipment, who can be within those boundaries, and how to determine the threat level of an electrical system. Before you work with an industrial-grade electrical system, you must review or be trained on the NFPA 70E standards to work safely.

Current in Milliamperes	Effect on the Body
1 to 3	Ranges from unnoticed to mild sensation
3 to 10	Painful shock
10 to 30	Muscle contractions and breathing difficulty begins, with loss of muscle control possible
30 to 100	Severe shock, with high possibility of respiratory paralysis
100 to 200	Ventricular fibrillation highly possible
200 to 300	Severe burns and breathing stops
2,000 to 4,000 (2 amps to 4 amps)	Heart stops beating, internal organ damage occurs, irreversible bodily damage possible

(*Note*: 1 milliamp equals .001 amps.)

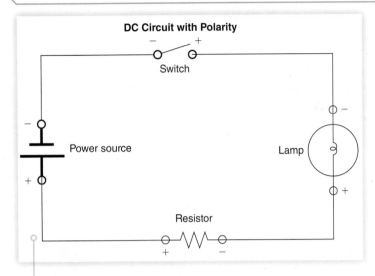

figure 10-5   A simple DC circuit, with the polarity of each component added in. Notice how the negative side is connected to the line that comes from the negative side of the power source, and how the positive of one component is connect to the negative of another.

In Chapter 3, you learned that DC (direct current) moves in only one direction through the system and has a defined polarity (positive and negative side) for everything in the system. The positive side of any component is electrically closest to and wired to the positive side of the power source; the negative side is the opposite side, or electrically closest to and wired to the negative side of the power source. (See Figure 10-5 for more information.) When making DC voltage checks, touch the black, or common, lead to the negative (−) side of the circuit or component and the red, or hot, lead to the positive (+) side. In many DC circuits, the polarity is labeled or indicated in some fashion; however, if you happen to get them reversed, do not panic; most modern multimeters will simply put a minus sign in front of your reading to indicate that you have the leads reversed. In some cases, this provides vital information about the circuit, the power supply, or a component as it indicates reversed polarity—providing you have the correct orientation of your black and red test leads. Reversing the polarity of DC circuits or components has effects ranging from circuits simply not working or working erratically all the way up to damaged or destroyed components and systems!

When taking AC voltage readings, probe placement is less of a concern, as we do not worry about polarity in AC systems. AC (alternating current) changes direction and magnitude many times per second and is thus going back and forth in the circuit; therefore, there is no defined positive or negative side for any component because the voltage is constantly changing its direction of flow. While we are not worried about polarity in our AC measurements, we still have to make sure we are testing at the appropriate places in the circuit and understand how the circuit works. The reading between two of the hot connectors in a three-phase system will be higher than between either of those line connections and ground. When taking readings using ground or a neutral wire, the standard procedure is to touch the black or common lead to the ground/neutral portion of the circuit. If you switch the leads from the meter, there is no change in your readings and, as long as you have selected the correct voltage type and range, there will be no damage to the meter. (See Figure 10-6 for examples of how to take power readings.)

If you are having trouble remembering how electrical systems work or skipped Chapter 3, I recommend you return to that chapter. Pay special attention to the sections on electrical systems that discuss the reason that systems need a neutral wire for 120V power while many three-phase systems only have three line connections and a ground.

tech note

continuity
Unbroken electrical connection between two points

Another great use of the multimeter is to check **continuity**, which is a term used to describe something that is unbroken or continuous over a given area or time. For our purposes, when we check continuity, we are looking for an unbroken electrical connection between two points. If there is an unbroken path, we get a

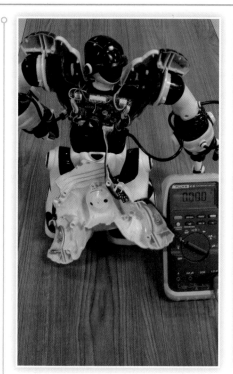

Robosapien used for DC voltage checks, shown with back removed.

DC power check from pin one to pin two on switch, no power flow.

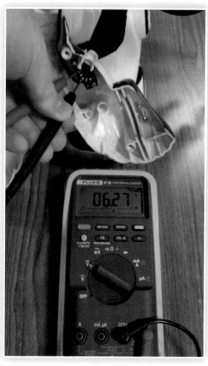

DC power check from pin one to pin three on switch, 6.27 volts DC present with correct polarity.

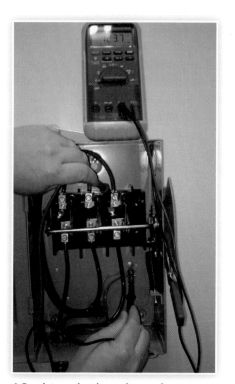

AC voltage check on three-phase system, hot leg to ground.

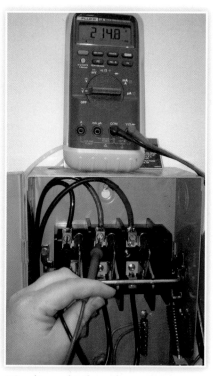

AC voltage check on three-phase system, hot leg to hot leg.

Figure 10-6

resistance reading in ohms telling us how hard it is for electricity to pass between the two points, with lower value readings indicating an easier path for electricity to pass through. If there is a broken path or no path for electricity to flow between the two points, the meter should show either infinite resistance or OL to indicate there is no path. This test is a great way to check for blown fuses, malfunctioning components, bad connections, and damaged wires.

Often continuity checks can give you deeper information about the equipment's state and really help you focus your troubleshooting efforts. You need to have a good understanding of the systems and components you are testing in order to properly use the data the meter gives you; otherwise, you may misinterpret the problem. For instance, relays, a common control component, generally have normally open (NO) and normally closed contacts (NC). If you perform a continuity test on a relay and check the NO contacts instead of the NC contacts, you may believe that a functional relay is the problem instead of finding the real cause. A misconception of this nature could end up costing you time and money as well as getting you to think on a track that is completely unrelated to the actual problem.

> Remember that the normally open (NO) portion does not conduct power when the relay is in the neural or de-energized state while normally closed (NC) portions will allow power to pass through. See Chapter 3 for more information on relays.

tech note

When performing continuity checks, the meter uses its internal battery to send a small amount of electricity through the system, determining the resistance between the two probe tips by measuring the potential difference. Because of this, the tested system *must be powered down*! If you do a continuity check on a system that has power supplied to it, there is a high probability that you will damage the meter, the battery, or both as system power feeds back into the meter! Depending on the type of meter used, there may be a couple of options for checking continuity. Either you can select the Ω symbol on the meter to get an ohms reading on the tested portion of the circuit or use the setting that looks like sound waves, which will generate a tone if there is a low-resistance circuit present. Some meters may combine the two, like the one in Figure 10-4, so be sure you know how your meter works. The tone setting is useful when you are checking several connections at once or making quick general checks, as you do not have to look at the meter to know if you have a complete circuit or not. In many meters, this setting is meant to check solid-state components instead of general resistance checks and thus may not tone out a circuit that does indeed have a complete connection. The Ω setting is useful for gathering deeper information on what is actually going on with the circuit and is more sensitive than the tone setting. Personally, I prefer the Ω setting as I like to know the specific resistance and I have been burned a time or two by no tone for a working circuit on the other setting.

A good example of the benefit of the ohm reading is when you are checking fuses in a system without removing them (see Figure 10-7). If you remove a fuse from the system, you will get an accurate reading as long as you are not touching both of the metal ends at once. However, if you leave the fuse in the system and do a continuity check, there will be times when you will read the resistance of the circuit and not the bad fuse. If there is a circuit or path for current around the fuse and the resistance is low enough, your meter would give you a tone on the sound setting. If you check the same fuse with the Ω setting, you will see the higher resistance of the circuit and be able to tell you have a problem. Most fuses have a resistance of a few ohms, while a circuit will often give you a reading in double or even triple digits. Fuses of the same amperage rating, basic size, and construction should be within 0.1Ω to 1.5Ω of each other. When checking a fuse bank, you can

determine which fuses need closer examination by looking for wide variances or differences in meter readings between similar fuses.

Fuse bank—Multiple fuses concentrated in one area for ease of wiring, checking, and replacement. These fuses are often of similar amperage, design, and operation, but there may be a mix of fuses in the fuse bank. Different fuse characteristics can vary the ohm readings, so when in doubt, remove the fuse and check it.

tech note

fuse bank
Multiple fuses concentrated in one area for ease of wiring, checking, and replacement

As mentioned earlier, the multimeter is not the only piece of equipment that you can use to gather information about the system. There are tools for checking signal strength through data cables, light transmission through fiber-optic cables, oscilloscopes for specific information about the power in the system, various temperature testing instruments, vibration analysis equipment, laser alignment and level checking devices, and other devices to help you gather information. The main thing is to make sure that you understand how the testing equipment works, how to use it safely, and how to use the data it gives you. It is all fine and good to know that the servomotor on a robotic arm has a case (outer surface) temperature of 140°F, but if you have no idea if this is a low, normal, or high temperature, then the data is of little use. One place you can go to give data context, or meaning, is a working piece of equipment. Many companies will have multiple pieces of the same equipment; if this is your situation, you can compare data from the broken machine to a working machine. Suddenly that 140°F reading you took earlier gives you valuable information when you discover that the servomotor in question seems to run at 110°F on another robot of same make and model that is working fine.

The point of using diagnostic tools is to gather useful information about the system, not tons of raw data unrelated to the problem. You could spend hours gathering data on the system, but if the data does not really tell you anything new about the system or help to point you in the direction of the problem, then it falls into the realm of busywork. As your troubleshooting skill grows, you will get a feel for when you need to gather deeper information versus just the basics with the multimeter. A good approach is to gather all the basic facts you can and then dig more deeply with more complex devices if you are having trouble finding the root problem or when directed to do so as part of an alarm response.

Gathering Information

Sometimes in the analyzing process, you discover that you still do not have enough data to properly understand and diagnose the problem with the system. When you find yourself in this situation, you often need to access additional sources of information to find a solution. A good place to look is in the various pieces of literature and materials you received with the system as well as the input/output tables. As with all the data that we have previously talked about, the information you get from these sources will only be helpful if you know what to look for and how to use the data. We will now look at how to use inputs/outputs, manuals, schematics, and wiring diagrams to glean more information about the system.

Inputs/Outputs

Inputs and outputs are signals that the machine or system uses to monitor and interact with the world around them (see Figure 10-8). Inputs, as the name implies, are signals coming into the system and are generally informational in nature. Outputs are signals that the system generates; these can be either data, which we can use as an input for another machine, or power, which runs something. For example, an input might be the signal from a prox switch indicating that there is a part or a signal from the machine that the process is complete. A part-loading robot might have a program that waits until several inputs, such as the raw part prox, the gripper open prox, and the input from the machine indicating the end of cycle, are all true or set to 1 before it begins the next cycle of loading a machine. During the loading/unloading sequence, there may be a need to blow coolant or chips off the finished part, buff the part, ream a hole, or carry out some other function that requires an external device to operate. We could control these with an output from the robot arm that would either power the device or send the data signal to initiate the process as needed. Outputs can also be simple elements, such as indicator

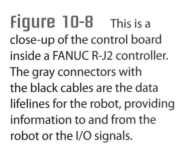

Figure 10-8   This is a close-up of the control board inside a FANUC R-J2 controller. The gray connectors with the black cables are the data lifelines for the robot, providing information to and from the robot or the I/O signals.

lights or data for the purposes of quality, order fulfillment tracking, or any other data function required in modern manufacturing. For instance, you may have your program set up to turn on an indicator light to warn those nearby that the robot is in auto mode or set an output to turn on a tone and flashing light in the case of an error with the system. The only limit to the use of inputs and outputs are the number that the system supports, the voltage and amperage level they are set up for, and your ingenuity. Many systems will have unused inputs and outputs built in or slots for expansion that offer you options in how you utilize I/O.

When you want to check the status of the inputs and outputs for a system, you have to know where and how to access the information. Oftentimes there will be a screen or data table accessible through the user interface that gives you this information; unfortunately, what you normally get is an address for the I/O and its current state with very little additional information. The data for the state of an I/O depends on whether it is digital, which has a value of either 0 or 1, or analog, which has a range of values based on the sensor or device. Often analog inputs and outputs range from 0 to 20 milliamps, 4 to 20 milliamps, or 0 to 10 volts with the value scaled in the program or system to have specific meaning, such as 0 volts could equal 0 degrees and 10 volts could equal 500 degrees with the values in between scaled appropriately. The address is simply some way that the system designates inputs and outputs so that it can process the data, often these consist of numbers or alphanumeric combinations. To determine what the I/O does, you need something that correlates the address in the table to the device it is tied to, or, at the very least, the terminal in the controller so that you can trace it manually. There may be a sheet among the materials that came with the equipment listing the I/O used and the data address for such, or you may have to look in one of the manuals for this information. In some cases, especially when the I/O was added once the machine arrived, you might find this data in the cabinet where the I/O is located or labeled on the wiring diagrams. For the extra inputs and outputs used, the best-case scenario is that the technician who added the I/O took the time to document it somewhere. If it not, you may have to ask around to see if anyone knows what it is connected to or do some testing of your own to determine functionality. Some programmers will put notes in their programs about the I/O, giving you another place to look for information. In short, expect to spend some time tracking this information down; once you find it, it will be another great addition to your technician's journal.

Once you know what the I/O data represents, you must next understand how the system or program uses the data. The use is as varied as the robots themselves. One system may require five separate inputs to be made or 1 before the system can run in auto mode, while the robot next to it needs only two or three to start turning and burning. When looking to see how I/O affects the program, you can usually gather this information by studying the lines of code in the program. Depending on the system, you may be able to watch indicator lights on the input and output modules that will tell you when they are active. If this is the case, watching a working system run under normal conditions and noting which inputs/outputs are active along with what is happening at that time can give you a wealth of knowledge. For the rest, you need reference manuals, training, or experience to guide you. If all else fails, see if anyone else knows how the I/O affects the system or try contacting tech support at the company that made the robot.

Often, a system will seem to have faulted out when a program is waiting for an input or will work intermittently because an input is blinking on and off instead of being either fully on or fully off. Intermittent operation is where the system will work perfectly for a while then fault out without any clear-cut reason why.

digital
Data with a value of 0 or 1

analog
Data with a range of values based on the sensor or device

intermittent
A condition where a system will work fine for a while and then fault out without any clear-cut answer as to why

These systems may work fine when reset for a cycle or several cycles before faulting out again. Because of this, they are some of the more frustrating faults to try to correct. Luckily, many modern robot manufacturers provide a way to track the state of inputs and store this data. If your robot has this option, I highly recommend learning how this works; it could save you hours of time tracking and troubleshooting intermittent problems.

Whether the robot has a few or hundreds of different inputs and outputs, knowing how they work and how they affect the system will serve you well in the troubleshooting process. Ignoring them or always trying to work around your lack of knowledge in this area is a great way to add hours of work (and frustration) to the troubleshooting process.

Manuals

You have read bits and pieces about manuals in this chapter, so let us take a closer look at this information resource. A manual is the book that comes with a piece of equipment or system that gives you more detailed information on how it works. Some manuals are rudimentary while others give you a wealth of information about how everything works. Complex pieces of equipment and systems may come with multiple manuals covering a variety of topics or specific subsystems. It is common for the variable frequency drive (VFD) to come with a dedicated manual, and the same may be true of an internal programmable logic controller (PLC) if it is present. We have already talked about error manuals or the error section of the manual and the input/output information in manuals, but most have much more information than that. You may find general diagrams of portions of the equipment, exploded views of assemblies, the technical instructions necessary to install or remove sensitive parts of the equipment, troubleshooting tests, preventative maintenance recommendations, and other information relevant to the piece of equipment. It is worth our time to explore some of the specifics found in the manuals.

Parameters are specific pieces of data needed to run equipment. Often, these are numerical in nature and detail such things as how many pulses an encoder has, set operational limits for speed, let the system know what type of motor is being used, limit axis movement, or set other numerical data for the equipment. If there are any parameters that need to be set during equipment installation or adjusted during operation, you should find this information in the manual. Many times, you have to go through a very specific procedure to get to a point where you can enter or change parameters. If that is the case for the robot in question, this procedure should be included in the parameters section of the appropriate manual.

Mastering is the process of setting or defining the home/zero position for each moveable axis. Without this information, it is impossible for most modern systems to move to defined positions or function properly. Because of this, many systems will simply not move in automatic mode, and all but the joint movement in manual mode may give you an alarm message. The mastering procedure for robots varies greatly from manufacturer to manufacturer and often somewhat from model to model. Many times when mastering a robot, the new defined zero position(s) will differ slightly from those previous set, necessitating the need to either adjust each programs point individually or offset the base robot positioning in some manner. If possible, the preferred method is to offset the robot, as this is a faster process than reteaching hundreds if not thousands of points.

Another useful tidbit is the parts section of the manual or parts manual for those systems that have enough pieces to warrant a dedicated book on the topic.

manual
The books that come with a piece of equipment or system that give you deeper details on how it works

variable frequency drive (VFD)
A control system that takes AC power and converts it to pulsing DC power to control the torque and speed of motors

programmable logic controller (PLC)
A specialized industrial computer used to control the operation of equipment based off inputs, data, and the entered program

parameters
Data that designates the type of equipment, signals, and information that the controller needs to have for proper communication and control of various devices

mastering
The process of setting or defining the home/zero position for each moveable axis

It is a common practice to show complex assemblies in this portion of the manual as an exploded view. An **exploded view** is an assembly of complex systems drawn in such a fashion as to show all the parts and how they fit together. These drawing often show the parts in the sequence of assembly, or there will be a line showing you how they go together, giving troubleshooters insight into how they come apart and go back together. If you are lucky, there will be tech tips to help you with tricky steps along the way, specific tightening torques, notes on special tools needed to work on the equipment, and other information that is not obvious from the drawing. If the company also sells replacement parts, you can expect part numbers included in this section of the book to make it easier to order replacement parts from the manufacturer. If you order replacements from someone else, you may have to work harder to find the right parts; however, in some cases, the cost savings is worth the effort.

There may be advanced troubleshooting procedures in a manual that tell you the what, where, and how of testing along with an explanation of the gained data. Some manuals have this in the error code section and some near the exploded views, while others may have separate sections of the manual or entire manuals on specific portions of the robot. Regardless of where you find this information the first time, make sure you know how to find it quickly the next time you need it. There may be programming data in the manuals that show you how to program specific operations or allow you to use the advanced functions of the unit. These sections can be quite helpful if you have a background in robot programming; if you do not, you may find them more frustrating than useful. You may find notes scribbled along the side of a section from previous technicians that could prove helpful. I remember several times in industry when I looked up an exploded view and found some notes from the last technician to work on the system that saved me time and frustration. Learning what information the set of manuals has for your system is just as important as learning the basic operation of the robot and its limitations.

The downside to a manual is that there is no set standard for what it should contain or how the information should be presented. You may find manuals that are similar to one another, which is often true of equipment from the same manufacturer, but there are no guarantees. Some companies provide a wealth of information but it is oddly organized or written in such a complex manner that it is very difficult to use. (I joking call these "stereo instructions," as the old stereo systems came with horridly written, overtechnical, jargon-laden thick instruction sets.) Other sets may contain so little information that they are barely of any use. The bottom line is that you should see what manuals you have access to and what information they contain *before* you are in a troubleshooting situation. The added stress of trying to fix the equipment may keep you from fully realizing what information is present and how it could help you. Time is often of the essence in troubleshooting situations, so you do not want to spend a lot of time flipping back and forth in the manual trying to find the section or information needed.

> **exploded view**
> Assemblies of complex systems drawn in such a fashion as to show all the parts that make up the assembly and how they fit together

Schematics and Wiring Diagrams

Another place where troubleshooters look for information and clues is the equipment's schematics and/or wiring diagrams. Schematics and wiring diagrams are two different methods to imparting information about the electrical side of the equipment. For anyone who is planning to go into a technical troubleshooting field, understanding how to read and use schematics and wiring diagrams is a crucial skill. Let us take a closer look at both.

schematics
Drawing of the electrical flow or logic of a system with no regard to the component placement or design in an effort to make understanding easier

legend
Pictures of the symbols used in a drawing with details of what each symbol represents

Schematics are drawings of the electrical flow or logic of a system with no regard to component placement or design. They illustrate what is happening in the system and how it works in a clean, concise manner using symbols that represent the various components instead of pictures of the components. When you first start looking at schematics, it will take a while to become familiar with the common symbols and thus figure out how the circuit works; however, over time, you will find it easier to understand the logic of the system's operation and find faults quicker and easier. The main benefit of schematics is that they show the logic of an electrical system in an manner that is easy to read and understand. The downside is that there is no set standard for symbols used in a schematic. A good schematic will have a legend that shows pictures of the symbols used and details what each symbol represents. If the schematic you are looking at does not have this, you will have to rely on what you know about the system and your experience with similar schematics to determine what unrecognized symbols represent. Keep in mind that there is no reason why you cannot modify the legend for a schematic or create a legend if there is not one. This information would make a nice addition to your technician's journal as well. Figure 10-9 shows some common schematic symbols. Bear in mind that this is by no means a complete list.

When you read a schematic, there are only three rules to remember:

RULE 1: Read a schematic from left to right, just as you would read a book. Your schematic tells the story of how the equipment works. When you read from left to right, the schematic indicates what must happen and the sequence in which it must happen. This can be very helpful when you have tracked the power and are trying to figure out why power is missing from a component. It also shows the components in a system that must work together to make something happen. For instance, the overloads that stop a motor from running if it pulls too many amps for too long are placed in series with the motor to shut down the circuit if something goes wrong. On the schematic, it is easy to see this connection and operation; in the field, it may be hard to equate loss of power at a contact in the control panel as to why the motor on the other side of the machine is no longer working.

Figure 10-9 Some of the common electrical symbols you may find on a schematic.

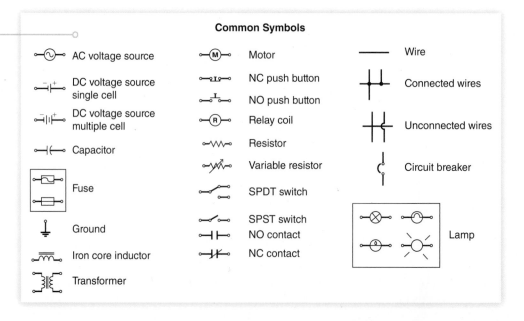

Common Symbols

AC voltage source		Motor		Wire	
DC voltage source single cell		NC push button		Connected wires	
DC voltage source multiple cell		NO push button			
Capacitor		Relay coil		Unconnected wires	
Fuse		Resistor			
		Variable resistor		Circuit breaker	
Ground		SPDT switch			
		SPST switch		Lamp	
Iron core inductor		NO contact			
Transformer		NC contact			

RULE 2: **Components are drawn with no regard to location or design.** The point of the schematic is to show the logic of the circuit, not where it is in the machine. You may have components side by side on the schematic that are located at two separate ends of the equipment or even in different buildings of the facility. In addition, many of the symbols show the logic of the system, not what the part looks like. For example, you will often find contacts from the motor starter, which is square or rectangular part on the machine, on multiple lines of the schematic represented as two lines with an open space between them and some form of label above it. This symbol gives you zero information about what the motor starter looks like.

RULE 3: **All components are drawn in the de-energized or off-state.** If you were to check the system before you ever even hooked it to power, the paths for the flow of electricity would match the schematic. When you apply power to the system, there is the potential for change in the electrical paths, and this is actually the principle behind making equipment perform as desired. For instance, when you engage the control coil of a relay, all the contacts that it controls change state. The normally closed (NC) contacts will open and stop conducting electricity while all the normally open (NO) contacts will close and allow current to flow. Another example is the start and stop push buttons. They are in the unengaged condition on the schematic and change state when you push the button.

Look at Figure 10-10 to see the schematic for a simple control circuit. Refer to the schematic as you learn how the circuit operates. This system runs in a forward direction when the switch is on and then turns when the limit switch encounters something. If you look at the schematic, you can see that the two lines running vertically along the sides are labeled L1 + and L2 –. These are where the components draw their power and how the circuit is completed. You will also see three lines, called rungs, going horizontally across the page. The main power switch is in the positive power line to kill voltage to the whole system when turned off.

rungs
The lines going horizontally across the page of a schematic

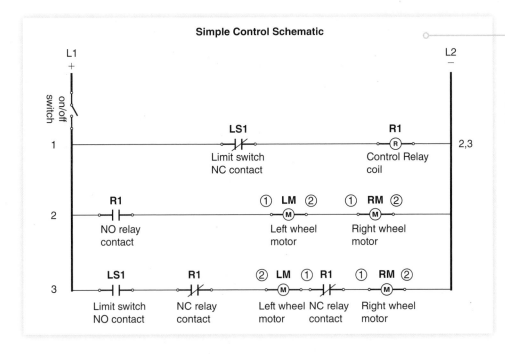

Figure 10-10 This is the schematic for a simple control circuit. Refer to this circuit as you read about its operation in the text.

The first line, rung 1, has the limit switch NC contact and the coil for the relay. The second line, rung 2, has a NO contact for the relay, the left motor, and the right motor. The third line, rung 3, has the limit switch NO contact, two NC contacts from the relay, the left motor, and the right motor. When the switch is turned on, power is allowed to flow to the NC limit switch contact and then through to the coil of the control relay. This causes the relay to engage and change the state of all of its contacts. This means the relay contact on line 2 closes and the relay contacts on line 3 open. When the contact on line 2 closes, this allows power to flow into the left and right motors and starts the motors turning. The open contacts in line 3 ensure that we do not double-feed the power to motors or feed the left motor negative voltage to both sides. The system will operate in this fashion until something triggers the limit switch. When this happens, the limit switch contact on line 1 opens and the limit switch contact on line 3 closes. In turn, the relay coil no longer has power (line 1) and all its contacts return to their de-energized or normal states. Also, there is now a complete path for power on line 3 as the limit switch contact is closed and the relay contacts are closed as well (due to returning to their normal state), allowing power to the left and right motors. At first glance, it seems that there is no difference between lines 2 and 3 as far as running the motors is concerned; however, if you look at the numbers in the circles near the motors, you will see them switched on the left motor. These circled numbers represent the motor terminals and designate how we connect the circuit. When we reverse the direction of power flow through the left motor, we reverse the direction of the motor. This means that when line 3 has power flowing through it, the left motor turns backward and the right motor spins forward, turning the system. The purpose of the two NC relay contacts in line 3 is to ensure that at no time will the left motor receive power at both contacts simultaneously. The first NC relay contact ensures that power is only supplied to contact 2 on the motor after the relay has returned to the de-energized state while the second NC relay contact in the line is to isolate the connection from contact 1 of the left motor to contact 1 of the right motor. If we removed the first NC relay contact from line 3, the circuit would function, but if the relay was slow to return to its de-energized stat, the left motor could become double fed and thus damaged. You may have also noticed that there are numbers to the right of line 1. These are the lines with contacts controlled by the relay in line 1. This makes it easier to troubleshoot systems by quickly locating related lines of control. Not all schematics have the lines numbered and the relay contacts referenced, but you can always add these in as you see fit.

A wiring diagram is like a schematic in that it shows how the power moves through the system; however, it represents this information in a different way. A wiring diagram shows components, to some degree, as they actually appear. You do not read a wiring diagram from left to right, so you have to trace the wires to determine functionality; thus, the logic of the circuit may not be clear when you first look at the diagram. A wiring diagram is very helpful when it comes to tracking down components in the system and checking connections, especially when it includes color-coded wires or the wire numbering. It also provides insight on where to find the various contacts and components that you wish to test. Ideally, you would have both schematics and wiring diagrams for a system, but many times, we do not get the ideal situation. You may even find a blending of the two in your equipment documentation. If this is the case, you may have some of the logical line elements of a schematic mixed with pictorial representations of switches, relays, motor starters, etc. When reading a mixed diagram, remember that all the components are shown in their de-energized state; the components may or may not be shown in pictorial format so may sure you understand any symbols used;

wiring diagrams
Drawings that show how the power moves through the system with components drawn to some degree as they actually appear

and you must start out trying to read the diagram from left to right. If this is not possible, then you will have to trace it out.

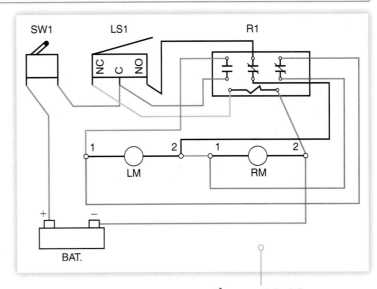

Figure 10-11 shows the wiring diagram equivalent of the simple control circuit shown in Figure 10-10. At first glance, the wiring diagram may look like a completely different circuit, but I promise you it is the same one using rough sketches of the parts and their connections with colored lines to show the circuit. Refer back to the wiring diagram as we cover the various connections. The red wire connects to the positive post of our battery power source and from there it runs to one side of our switch. This is so that we can kill all power to the system when it is not in use. From the switch, we go to the common contact on the limit switch. The common connection on devices like this are used to create a path for power or signal to the contacts of the unit. This is how the NC and NO contact of the limit switch access the power of the system. Next, we have the red wire running from the common contact on the limit switch to one side of the NO contact on the relay. With this setup, anytime the switch is on, we have power at the common of the limit switch and one side of the NO relay contact. The gold wire connects the NC limit switch contact to one side of the relay coil, providing power to engage the relay until something triggers the limit switch. The green wire finishes the circuit for the relay coil and the motor circuit by connecting back to the negative terminal of the battery. The blue wire connects the other side of our NO relay contact to terminal 1 of the left motor and also makes the connection between terminal 2 of the left motor and terminal 1 of the right motor. This is the power circuit for forward rotation. The black wire connects the NO limit switch contact to one of the NC relay contacts and the other side of the same NC contact to terminal 2 of the left motor. This is how the left motor receives the reversed power needed for turning. The orange wire connects terminal 1 of the left motor with terminal 1 of the right motor through the other NC relay contact. This provides the electrical isolation necessary to prevent power from bypassing the left motor all together and creating issues. If we had a wire connected directly from terminal 1 of the left motor to terminal 1 of the right motor, there would be a low-resistance path for electricity around the left motor. If we left this connecting in the circuit when we tried to go in the forward direction, the right motor would most likely turn fine, but the left would barely turn or not turn at all—creating a system that literally runs in circles.

You can create a wiring diagram from a schematic. To do this, we first need to number the schematic. We use each number *once* to prevent confusion; the number will change each time the wire passes through a component. All things connected to the same wire will get the same number. Once you have a numbered schematic, match the numbers to your component diagrams. Then it is simply a matter of connecting the dots. See Figure 10-12 for an example of the process.

When gathering additional information about the system, you may use some or all of the resources we have talked about; however, remember that the whole point is to find information that helps you fix the problem. Inputs/outputs, manuals, schematics, and wiring diagrams are all great sources of information, but you could spend hours checking all of this information. When you access these resources, make sure to keep the problem in mind and only look for information

Figure 10-11 The same circuit we saw in Figure 10-10 shown as a wiring diagram. If you are like most people, it will take you longer to figure out the operation of the circuit when it is presented in this way.

common connection
Connections used to provide power or signal to contacts controlled by switches or relays

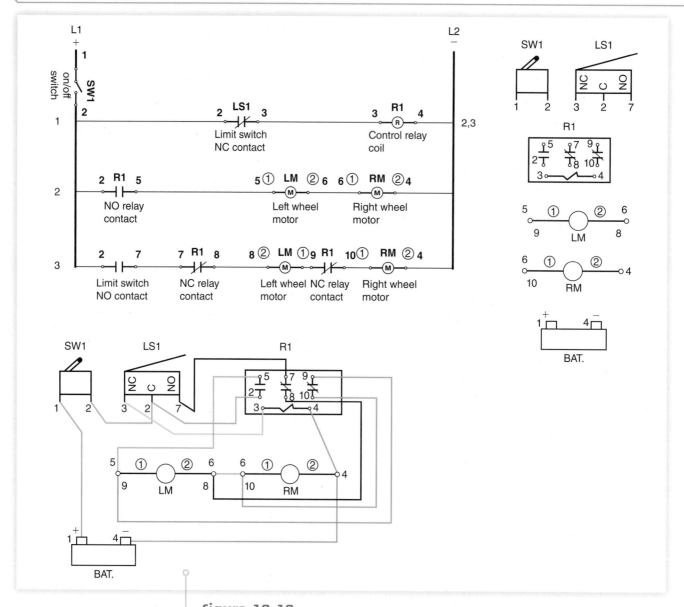

Figure 10-12 Make sure you pay close attention to the component numbering as it is transferred from the schematic to the components. This becomes the roadmap for wiring the system when we start to connect the numbers together.

related to what you are trying to fix. For instance, it is great to know how to master the axis of a robotic arm, but it is unlikely to help in fixing a problem with the gripper. Another example is to be looking at the wiring diagram for the robot arm when the controller is where the problem is. You must always filter the data you have and the data you are looking through with your understanding of the system and the problem you are working to correct.

Additional Resources for Information

We have discussed ways to analyze what is wrong with a system or piece of equipment when it faults out and some common resources that you can use, but where do you look when you still do not have enough information? In this section, we will discuss additional resources that we often overlook but that can be of great help.

Two Heads Are Better than One

A common pitfall of troubleshooting is tunnel vision, where a person focuses on a specific way of thinking or seeing the problem and ignores everything else. This is especially problematic when the troubleshooter focuses on an incorrect diagnosis of the problem or the wrong path to fixing the machine. Sometimes this happens when the symptom or symptoms of a problem becomes the focus of the troubleshooting process instead of the root cause. Hurrying through the troubleshooting process or not having a firm grasp of all the elements of the system could also lead to tunnel vision. For example, a robotic arm is alarming out during the loading of parts into the fixture. The troubleshooters comes over, analyze the system to determine the error, and decide that it is the fixture because the side of the part is hitting the edge of the fixture. They spend hours trying various things such as replacing portions of the fixture, installing a new fixture, adjusting the position of the fixture, and so on, without fixing the problem. They do not go back and look at what else may be causing the fault, such as a bad program position, or reevaluating the data, such as the machine had crashed the day before.

When a troubleshooter runs out of ideas or thinks that he or she may be a victim of tunnel vision, a good resource is the other people around who have experience with the system. Everybody views the world in a slightly different fashion as they interpret information filtered by their own experiences, understandings, perceptions, and the unique way they process information. Because of this, other people can offer a different point of view and help the troubleshooter to work on the problem in a new way, breaking out of the trap of tunnel vision. They may have experience with the problem you are working on and be able to offer advice or tips that can save you large amounts of time. Even if they do not know the solution to the problem, they may tell you things that they have tried in the past that did not work, saving you time that way.

Another technique that involves other technicians is brainstorming, where a group of people try to come up with a variety of ideas about a topic or problem. This is great for new problems that no one has seen before, tough problems that offer no clear-cut course of action, intermittent problems that appear randomly with few symptoms, or troubleshooting problems for which you have run out of ideas. Brainstorming does not have to be a large production or a formal meeting, it can be as simple as getting two or three other people to gather around the equipment, or wherever is convenient, and talking about the problem. During the brainstorming process, it is important to let ideas flow freely and avoid ridiculing any offered, as this will stifle the sharing atmosphere and may suppress the idea that in fact fixes the problem. It is fine to work out the logic behind an idea to check its feasibility as long as it is *not* done in a condemning or attacking fashion. It would be okay to ask, "How do you think the servo alarm might be related to the fixture?" as opposed to, "There is no way the fixture could affect the servo, what are you thinking?" The first question invites the person to share what he or she sees as the cause of the problem; the second attacks the idea and may very well keep the person from sharing any more ideas. When asking for input, it is important to be open to input and not just look for the ideas that match your own. The whole point of brainstorming is to get new ideas, new perspectives, and new ways to try to solve the problem.

Regardless of how or why you are asking for input from others, remember that you did ask for help. I have seen more than one rift created when someone asks for help and then rejects all suggestions with little thought or investigation. Many times when a technician gets to the point of asking for help he or she is frustrated

tunnel vision
When a person focuses on a specific way of thinking or seeing the problem and ignores everything else

brainstorming
A group of people getting together to come up with a variety of ideas about a topic or problem

with the troubleshooting process. The person you ask for help is not responsible for what happened to the equipment—so do not take out your frustration or anger on him or her. This will only add new problems to your day. Many people will only offer or try to help once or twice if their efforts to assist are meet with dismissal, ridicule, or anger. When working with others, the golden rule is the best approach: "Do unto others as you would have done unto you."

Technical Support, Please

A wonderful, but often overlooked resource is the manufacturer of the equipment. While you may have one (or even several) of the machines in your facility, the manufacture has likely built thousands and received feedback on that equipment from a multitude of customers. What this means is that even though the problem may be a new one for you or your facility, it has probably occurred before, and the manufacturer most likely has information about the issue. A few minutes with the manufacturer's tech support or a repair technician could save you hours or days of work trying to find the solution. The downside to manufacturer tech support is that there may be a cost involved. This will often depend on how many pieces of their equipment you have, who you bought it from, how old it is, and if you purchased any kind of a tech support or warranty package with the equipment. Many manufactures will give you a year or two of free tech support when you purchase new systems and then offer you a reduced rate service after that.

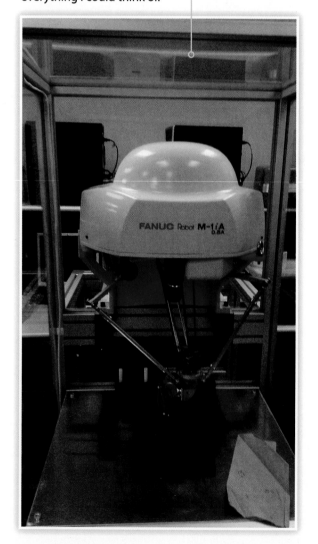

Figure 10-13   I ended up calling tech support for this robot after I had tried everything I could think of.

The best course of action is to call tech support and ask if there is a fee for the service and, if so, if your company has a service agreement. Make sure you ask how much the service costs and if the company offers any package deals. It could be that buying a year of tech support is the better deal than paying for an hour here and there. Even if the company charges for tech support, they may be willing to give you some information on the problem or send you any manuals that you are missing. It is better to call then and end up no worse off than you started than to assume it will cost money and miss a valuable resource.

Here's an experience I had: I contacted FANUC about some issues with our robots (see Figure 10-13). I reloaded the operating software that came with the system to try to correct multiple issues and get a start fresh. The process seemed to be working correctly, but it kept giving me an alarm when I tried to boot the system after the software install. I checked the alarm manual, but that specific alarm was not there. I tried reloading the software to make sure it was not just a glitch . . . same issue. There were two load profiles for the system, so I tried both. I did everything I thought of, so I decided to talk with a technician I know at FANUC. I told him everything that was happening, he asked me some pointed questions, and we determined that somehow my system software file was corrupted and that I needed a new set of core software. He also told me how to make backups so that once I had the new core software, this would not happen again. This saved me countless hours of trying to find information on an alarm not covered in the error manual and looking through all the manuals for some kernel of knowledge that would help. Best of all, there was no cost for this service, and FANUC was great to work with.

Another service that many companies offer is technician service. This is where someone specifically trained in the equipment comes out to make the repairs. This can be a rather expensive route as there is often an hourly charge, travel, and expenses for the technician not to mention the time spent waiting for them to arrive (they are often out of state). Some companies have a minimum charge as well—say, assistance for up to eight hours will cost $1,000 plus travel and expenses, with time over eight hours charged at $125 per hour. This explains why many manufacturers have an in-house maintenance department, as the cost is considerably less. If the maintenance department has tried everything it knows to get the equipment running and it still not working, then the company is faced with the choice of continued downtime until someone finally finds the solution (which may involve a lot of changed parts), buying a new machine, or calling in a factory technician. Calling in a technician from the company is generally a last resort due to the costs and time involved.

Another problem you may have with contacting tech support is availability. If the company only has a few specialized technicians tasked with fixing equipment and helping customers, it may be difficult to arrange a time to talk with them. If you need a technician to come to your site, it may be a few days or even a few weeks before someone is free. Another issue may be your facility's hours of operation: if the equipment faults out just after the manufacturer's tech support line closes for the day, it is highly likely that management will want you to try to fix the problem instead of waiting until the morning. When you run into these kinds of situations, you will just have to use your best judgment on how to proceed.

No matter which route you go regarding contacting tech support, you need to have some key piece of information ready. One of the first questions tech support will ask is the make and model of the equipment. Tech support may also want to know such things as serial numbers, build numbers, controller types, user interface type, teach pendant type, etc. The more information you can gather about the specific piece of equipment before you call, the better. Another common question is what alarms or error messages is the machine displaying. You should already have this information, as that was one of the crucial analysis steps we discovered earlier, but remember to have a copy of the specifics with you when you call. Telling tech support that "it has a servo alarm" is not sufficient; what they need is something similar to "it has a 411 servo alarm: 5th axis—excess error." You need to have this information for *all* the alarms that the unit is displaying, not just one or two from the list. Remember, in some cases you will have to go to the alarm screen or page to find all the alarms. You also need a list of things you have tried to correct the problem. Often you will have already tried several of the steps tech support suggests, so having this information will ensure that you get to ideas you have not tried, thus saving time. Make sure you write down how each attempt to fix the problem affected the machine. The more information you have on the system, problem, symptoms, and attempted solutions, the better your chances of getting the help you need. If you only have vague information and answers for tech support, then you can expect vague information and suggestions in return.

Before you call for help, I would recommend that you know what your options are: Do you need tech support or do you want a technician to come to your location? This way you can be straight up with the company should the question of payment arise. Do not lead with, "We can pay you $1,000, can you help me?" Instead, know your budget so you can haggle.

WWW....

Another great resource is the Internet. There is a vast wealth of information online that has been posted by companies, professionals, technicians, and others that you can access quickly and easily. You can look up error codes, search for schematics or wiring diagrams, find exploded views of assemblies, track down tips and tricks, and many other useful bits of knowledge. Many companies have started to upload all the documentation they have on their equipment with large amounts of free information for consumers to use or download. You can sometimes find chat rooms or blogs about specific issues with a piece of equipment along with corrective actions. You may find training on the equipment or system to get a better understanding of how it works. The best part about the Internet is that every day, the amount of information available is growing, which means that a search today that was not helpful might turn up just the information you need tomorrow.

No matter what you are looking for, there are a few things to remember. First, if you do a vague search, your results are likely to be vague as well. For example, a search on "error code 411" netted over 2 million results; a search on "error code 411 servo" returned 359,000 results. While both numbers are too large for you to able to look at everything, you can imagine that the first search has many results that are not beneficial in any way, whereas the second offers more relevant results. It is also common for most people to check the first two or three pages before they give up on a search. Because of this, search engines tend to put the best results in those first few pages. This does not mean that what you are looking for is not on page 10 of the search results, but it does mean that the results in the first few pages are more likely to be helpful. A way to narrow the search is to be a specific as possible. When you do a specific search, you may get a very few results or "no matches found"; however, you can always broaden your search from there to get more results. If you get an exact match for the error code you are researching, those results will be more helpful than one that contains a few of the words. You may also try doing an exact phrase match or must contain all words search. Usually, these options are available as part of an advanced search and it may take a little bit of work on your part to use these filters. An exact phrase match will find sites that include that exact phrase in them. This will often get no results, but it is worth a try. Must contain all words searches will likely net more results, as the site can have the words anywhere on the site and not just side-by-side in a phrase. There is a fair chance that sites that mention everything you are looking for have something of value, but you may have to wade through a large amount of data to find it.

Do not be afraid to try different searches and search engines. There are several different mainstream search engines out there, and not every site registers with all of them. This means that a search on Bing might give you different results from the same search on Google. Many of us get in the habit of using a search engine we are familiar with and do not check to see what the others might find for us. If searching for the specific alarm does not get you the information you need, try searching on the problem. Instead of looking for "error code 411 servo" try something more general, such as servo error on manufacturer, style, model, robot, etc. The results will vary, and you will get a lot of junk, but if the specific search did not return what you needed it is time to broaden the search parameters. Do not forget that you can specify what type of data the search engine looks through. A shopping search is going to give different results as compared to an image search. The more you think about what it is you need to know, the better your chances of finding it on the Internet. We want to work smarter, not harder!

The additional resources mentioned here are by no means every source available, but they are some commonly overlooked ones. As always, you want to approach these sources with the problem you are trying to correct in mind. This will help you process and filter the information provided to find what you need. Save the useful information you get from these sources somewhere safe, like your technician's journal, in case you need it again. I would also recommend sharing what you find with your colleagues. This way, if the problem should happen again later, you are not the only one trying to remember the information!

Finding a Solution

By this point, you should have plenty of information on the problem and are ready to actually fix the issue. The first thing you must now do is find some way to take all the data gathered and process it so that you can fix the root cause of the problem. This part of the troubleshooting process is as varied as the people involved, and if you were to compare two different troubleshooters' methods, they may be similar or as different as night and day. As long as they find the solution in an accurate, timely manner, the difference in how technicians get there truly does not matter. The main thing to remember is that often there are several ways to fix a problem: there may not be a "right" way, per se.

Dividing Up the System

One of the ways we can begin to sort the gathered data is by attaching it to the part(s) of the system it relates to. To do this, the troubleshooter must decide how to group the components of the equipment or various aspects of the system. The groupings could be broad, like signal, power supply, mechanical parts, electrical parts, and control, or something more specific, such as motors, relays, motor starters, fuses, inputs, outputs, etc. Complex systems will likely require more groupings than simple systems. The key to this step is to divide the equipment or system into something that makes sense to you and helps in the information-sorting process. If you are having trouble deciding how to sort the various systems and components, you may want to ask someone else with more experience for help.

Once you figure out the groupings, you must figure out how they mesh. You may want to draw a diagram or chart that shows the relationship of the divisions for a clear picture of those systems that work together and those that are part of other systems versus stand-alone operations. This is how you figure out where to focus, given a specific set of symptoms and information. Let us say that we have the three divisions in Figure 10-14 and we are using them to determine what is wrong with the system. The problem is that the robot arm does not move, and when you looked up the alarm, you find that it is a servo drive alarm. From our example, we can see that servo drives relate directly to the servomotors. So, until we fix the problem with the servo drive, the motors are not going to work and the arm will not move. At this point, the data would tell us to look closer at the servo drives and see why they faulted out. This would be our preliminary course of action.

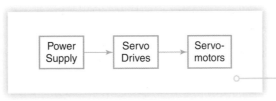

Figure 10-14 Notice how the division chart has the system broken down into broad categories such as power supply or control device. This is a great way to start to sort or group portions of the equipment.

Figure 10-15 Once we have narrowed down our search to one part of the machine, it is time to start looking at specifics such as indicator lights or fuses.

Power to servo drives → Alarm lights on servo drives → Fuses leading to servomotors

subchart

A chart that takes a portion of a larger chart and breaks it into greater detail

If you are using this method and find that your preliminary course of action is too broad, you may need to create a more complex division chart or a specific division chart for an area of the system. Let us continue with the previous example. We narrowed our search to the servo drives, but we really do not have a clear-cut course of action on what to check or how to fix the servo drives from our simple division chart. At this point, we could either create a more complex overall division chart or make a subchart for the servo drives. Since we are going to have to create a division chart anyway, we might as well create a subchart or a chart that takes a portion of the previous division chart and breaks it into greater detail. (See Figure 10-15 for a potential second level or subchart for the servo drives.) We would now refer this subdivision chart and filter our information once again to get closer to the root cause of our problem. If we wanted to, or if there were a need, we could create another division chart for the "power to the drives" box. We can repeat this process over and over again until we finally find the problem or run out of way to divide the equipment. You will want to keep any charts you create for future use, keeping in mind that the broad initial charts might actually work for more than the piece of equipment or system. This is another great addition to your technician's journal!

Power Failure

Another way we find the root cause of the problem is to track the power and find where it disappears. There are many instances where this technique can determine which component(s) failed, giving you the necessary information to fix the equipment when added to the information you gathered. To use this method of troubleshooting you have to know how the system works, how the various components work, and understand how they are all interconnected. This often requires the use of the schematics and wiring diagrams that came with the equipment so that you can understand how the systems feed one another and when you should have power as well as where. If you do not have the necessary documentation, you can also trace the physical wiring of the system. The downside to doing it this way is that it will likely require a lot of time to trace all the connections, especially with complex control systems, and there is a good chance that some of the wiring runs through conduit, making the process all the more difficult. In these instances, you will likely need a second person to help with the tracing process and may need special leads or test equipment as well so that you can check continuity between the wire ends to determine which wire goes to what.

Conduit is metal or flexible piping that we use to protect wires run over a distance. Conduit may run through control panel wires, around various parts of machines, through the middle of robot arms, and in any other place where there is clearance for the pipe.

tech note

These guidelines can help you use the power tracking method of information sorting:

GUIDELINE 1: **You can check power at various points.** This means that if you are checking a string of several components that should have power, you do not have to check them all. In many complex systems, there are components that are hard to access or require a fair amount of disassembly to reach. In these cases, you start by checking the areas of the circuit you can get to easily and then infer what is going on from this. For example, if you know that you have power coming out of the fuse, you can infer that there is power coming into the parts that feed that fuse. If you have power coming to the motor starter, you can infer that everything leading to the motor starter is working. Just make sure that you are checking components that are all interconnected and not two different current paths!

GUIDELINE 2: **Check component by component once you have found the trouble area.** After you have taken your general power checks and defined the area where you believe the problem to be, you need to check each component. These checks are filtered by two conditions: Does it have power coming into it, and should it have power flowing out of it? If there is no power coming into the component, you can move to where the power for that part should come from. With this in mind, it is faster to check the power on the input side or where the power comes in for various parts than the output side or where power goes out. Once you find where the power is coming in, but not flowing back out, you have found your focal point for guideline 3.

GUIDELINE 3: **Determine why the component is not working.** Even if it is not always apparent, there is always a reason why the component or assembly is not working. Sometimes it is the various connections of the part, sometimes it is outside inputs, sometimes you will see physical damage, and on occasion, there will be no clear-cut reason why it failed. Some technicians like to swap parts at this point and see what happens, but if you do not know why it failed, you risk damaging the new part as well. Be sure you do everything you can to figure out why the component failed *before* you toss in a new one and see what happens.

GUIDELINE 4: **Fuses blow and circuit breakers trip for a reason.** I have known many technicians who thought that a blown fuse was the problem, not a symptom. Fuses blow and circuit breakers trip because there is an excessive amount of either heat or magnetic field passing through them, caused by the amount of current in the circuit they protect. The whole point of these devices is to protect the circuit from shorts and other dangerous situations so that when one faults out it is a symptom rather than, the cause of the problem. A blown fuse or tripped circuit breaker is telling you that somewhere past that point in the circuit there is a short or excessive amperage draw in need of repair. A short circuit (or "short") is when there is a direct path between a charged wire/component and a ground or neutral point or between two powered connections. Because there are no resistances or components to slow the flow of current, the system allows nearly the maximum current to flow and thus the amperage spike. If you reset the circuit breaker or replace the fuse without looking for the reason why, there is a fair chance that the circuit will fault out again and possibly do more damage to the affected components.

short circuit
A direct path between a charged wire/component and a ground, neutral, or another powered wire/component

When replacing fuses or circuit breakers, always use the same size and type as the original. Fuses and circuit breakers often have characteristics designed for specific operating systems, so using the wrong fuse or circuit breaker can cause you a great amount of trouble and mask the true problem with a system. A good example of this is replacing a slow blow fuse with a fast-acting fuse in a motor circuit. The slow blow fuse withstands the amperage spike necessary to start the motor without opening while still protecting the system from overloads and shorts. When you use a fast-acting fuse in this type of circuit, it opens with any spikes in amperage and will often blow when the motor is starting up. This could happen the first time or after several times of starting the motor, causing an intermittent fault that may have nothing to do with the core problem you were trying to fix. Whatever you do, *do not oversize fuses and circuit breakers*! Over the years, technicians have used oversized fuses, circuit breakers, or in some way forced the circuit to remain on to find the problem or simply keep the circuit running. This is very dangerous, as there is a high probability of component damage, wiring damage, fire, and events that are hazardous to technicians and operators alike—such as arc flash. `Arc flash` is an electrical catastrophe where two main legs of a system short out and draw enough amperage to create an electrical explosion complete with pressure waves, shrapnel, vaporized metal, and enough heat to instantly cause second- and third-degree burns!

arc flash
An electrical catastrophe where two main legs of a system short out and draw enough amperage to create an electrical explosion complete with pressure waves, shrapnel, vaporized metal, and enough heat to instantly cause second- and third-degree burns

When performing the voltage checks, remember that you are working with live circuits. You need to take all necessary precautions to protect yourself from becoming a part of the circuit as well as insuring you do not short out anything with tools or other metallic objects. Individuals who are familiar with electrical circuits in general and the specifics of the piece of equipment they are working with should be the ones to perform this type of troubleshooting. If you are new to troubleshooting, work with someone more experienced so that person can guide you and help you gain the experience you need. For more on working with live circuits, read NFPA's *Guideline 70E: Standard for Electrical Safety in the Workplace*. This guideline will give you information on the dangers of arc flash and the requirements and restrictions on working with live circuits. It is part of the OSHA regulations by reference.

Signal Failure

The signal failure method is similar to the power failure method, except that now you are looking at inputs, outputs, and data paths as opposed to voltage flow alone. A good example of this is fiber-optic cables that transmit light, not power, over a distance to communicate data. While electricity is often a part of the signal system, it is usually at a much lower voltage and amperage than the running system. Some common voltage levels are 5 V, 12 V, and 24V, with amperage in milliamp range. Generally, the signal system detects the presence or absence of a condition as opposed to directly running a system or output. As with the power failure method, signal failure tracking will require a firm understanding of how the system works, which signals should be present, and the components role in that signal.

Remember, there are two basic types of signal: digital and analog. Digital signals are either present or absent, with no in-between state. These are great for applications where either something is or is not in position or present, but are a poor choice for conditions that vary. These variable conditions are where the analog signal comes in, as it is a range instead of on or off. With analog signals, we get a range of signal from the source that we then scale for use in program or the controller. For instance, we might have a 4- to 20-milliamp thermocouple that monitors a robot's servomotor, scaled so that 4 milliamps equals 0°F and 20 milliamps equals 500°F. With this scale, 10.4 milliamps would equate to a temperature of 200°F. The controller in turn could monitor this value, and if it exceeds a set value, trigger an alarm to let the operator know that one of the servos is hot.

To get the value of each degree, take 20 milliamps, subtract 4 milliamps, and divide by the total it represents (500°): the result is 0.032 milliamp change per degree. By taking the 0.032 multiplied by 200°F of the system and adding the 4-milliamp start point back in, we can determine the amperage generated by this temperature.

Here are some guidelines for use with the signal failure method:

GUIDELINE 1: **Check the input and output tables**. You may have already done this during the analyzing phase. If you have not, you will need to do it now. You also need to know which inputs and outputs should be present or on and which ones should be off or absent. This is another place where knowing how the system normally works comes in handy. If you are unsure of the proper state of the inputs and outputs for the given situation, you can refer to the documentation of the equipment, the program, or other working equipment if available.

GUIDELINE 2: **Match the input or output to the component/function it represents**. Once you have discovered which signal(s) is missing or present when it should not be, you have to match it to something in the system. This may be harder that it sounds, as some inputs and outputs are just for data purposes with no physical component tied to them. The robot program may turn on a specific output bit to let the machine it feeds know that the loading and unloading cycle has finished, or there may be an input bit tied to the robot to let it know when the machine has finished with the part cycle. With luck, there will be good documentation to help with this portion of the process; otherwise, you will have to figure out some way to find the information you need.

GUIDELINE 3: **Find the cause**. Once you complete the first two steps of the process, you then have to determine *why* the signal is wrong. If the signal is from a physical component like a prox switch or some kind of probe, you should check the component directly. In cases where the component is working properly, check the wires or the system transmitting the signal to the controller. When checking the data transmission line, look for bad connections, damage, and corrosion that may be causing problems. In cases where it is information sent from one system to another, instead of a physical component, you want to confirm that the sending system is in fact sending out the data and that the transmission line is working properly.

GUIDELINE 4: **If everything else is fine, check the signal interpretation device**. Many times signals, both inputs and outputs, go through cards on the controller that specifically manages data. If everything else seems to be working properly (the components, transmission sources, and the transmission lines), then it is fair to think that there may be a problem with the portion of the system that interprets or sends the signals. You can often remove a card and replace it in just a few minutes without having to unwire and rewire the system. If you must unwire all the connections, make sure that you know where everything goes and how they are connected. You may want to take a picture (perhaps with your cell phone) to help you remember where the wires go or sketch a quick wiring diagram. This would be a great addition to your technician's journal for future reference.

Even though signal systems are usually low voltage and amperage, you should still treat them with care. These systems are often right next to high-voltage systems, which present a danger of bodily harm. In addition, you never want to take any kind of a shock on purpose just to see if voltage is present! You may be tempted because you are dealing with a low-voltage system, but the moment you stop respecting electricity could be your last.

Many technicians favor a two-prong approach to tracking problems and use both power and signal to find the issue. If you combine these two as your method of choice, make sure that you keep in mind what you should have at each point. For instance, you may pick up 1 or 2 volts leaking through from some other portion of the system and think you have a signal, when in fact you need 120V power. The combined power and signal method is my personal favorite, but I have caused myself issues in the past by hurrying and not thinking about what should be there instead of just seeing if I could find anything at the measured point. Remember, the goal of troubleshooting is to fix the problem and get the equipment running once more, not to set a new Guinness record!

Flowchart

flowchart
A group of boxes with either a question or a task to perform with directions on where to go next depending on the answer or result

Flowcharting is a great way to help organize and process data to determine a course of action. The traditional flowchart has boxes with either a question or a task to perform, with directions on where to go next. For questions, the answer determines the direction, and tasks usually loop you to another question once completed. This process continues until you have either done all you can with the flowchart or found what you needed. This process can help you work through the data in order to find a course of action. Figure 10-16 provides an example of a flowchart based on the simple control circuit that we used earlier in the chapter.

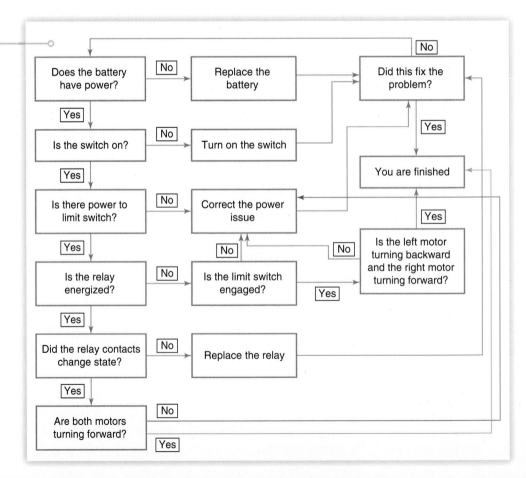

Figure 10-16 When working with a large/complex flowchart like the one pictured, you may need to run through the chart multiple times before you fully fix the problem.

Follow along with the flowchart in Figure 10-16 as we go through the process of reading the flowchart. Usually we start reading a flowchart in the same manner we read a book, with the upper left corner being the first place we look. You can see that the first thing the chart asks is "Does the battery have power?" If the answer is "Yes," you go to the switch question; if the answer is "No," you are directed to replace the battery. If you look at the box that says to "Replace the battery," you will notice that there is not a "Yes" or "No" option: it simply leads to another box. With this kind of format, you would replace the battery and then proceed to the directed box, which asks if this fixed the problem. From there you have the "Yes" or "No" option; yes takes you to the box that says you are finished, and no takes you back to the beginning of the process. With this kind of looping, you keep going through the flowchart until you fix all the problems or find an issue not addressed by the flowchart. The second box down on the left is the switch box: "Yes" leads to the next box under it, and "No" tells you to turn on the switch, followed by our looping statement. The third box down on the left asks if you have power to the limit switch: "Yes" leads to a relay question, and "No" leads to another box that is used multiple times. The "Correct the power" box is referenced by several boxes in the flowchart, thus making it a common corrective action. Because this is a generic statement, it does not tell you what to do specifically but rather gives you an idea of how to proceed. Under this directive, you may end up changing out parts, replacing wires, or checking connections—depending on what you found in the assessment phase. The last box on the left is the only one that has different-colored arrow lines. This was to prevent confusion, as the lines from the "Are both motors turning forward?" box ended up crossing other lines and might have been difficult to trace otherwise.

You can create your own flowcharts like Figure 10-16 for the system you are trying to fix or portions of the system you are working on. You would not have to put the chart on paper if you can follow the logic of the circuit and filter the information as you go. However, writing down the information will help to prevent skipped steps and allow you to save the procedure for later. (Plus, it is another great addition to your technician's journal.) The whole point of using the flowchart method is to help you process the information you have in order to find a corrective action for the fault you are working on. You can filter by major topics or really drill down into specifics. You can write a simple one-page chart or create one that references other charts and has many pages in the process. As long as you can use it efficiently and it is beneficial, the sky is the limit.

Finding Your Path

We all process information in different ways. Because of this, there is no one method of organizing information that works for everyone. You may find that you prefer to use bits and pieces from the various methods discussed to create a troubleshooting technique that works for you. You may find that you tailor your technique to the type of problem that you are working to fix. You may find a path completely different from anything we have looked at. It does not matter how you find the answer as long as you keep the following factors in mind when troubleshooting.

EFFICIENCY: When you have a problem to fix, you generally do not have huge amounts of time. Because of this, you need a method that works in the first few attempts—not eventually or after a large amount of time. If it takes you hours and hours to come up with a solution each time, you need to revise the method or try something altogether different. Remember that in industry and business, "Time is money." Employers want troubleshooters who can fix problems quickly and accurately without wasting time or resources in the process. Those who are the best at this will demand the top pay and positions. An example of wasted time would be taking several hours to troubleshoot an indicator light that has burned out due to three years of continuous use.

ACCURACY: Whatever method you end up using, it must enable you to find the root cause of the problem. You may take a wrong path from time to time or it may take a couple of attempts to find the problem, but a method that rarely gets you to the right answer is of little to no value. The whole point of troubleshooting is to fix the core problem and get things going once more, not to spend all your time troubleshooting. Troubleshooting is all about results. With this in mind, you need to use a method that will get you the *desired* results and not just keep you busy. For instance, it is much better to spend a half hour with the schematics and wiring diagrams to understand the system you are working with than to take a full workday tracing out all the connections by hand.

UNDERSTANDABLE: When you create your troubleshooting style, it must be something you understand and be easy to work with. Sometimes we come up with systems that are overly complex or blended in such a way that it is near impossible to analyze the data in order to find the solution. This often happens when troubleshooters first start out and emulate the styles of those around them or try a new style of troubleshooting. If you find yourself getting frustrated when you try to filter the information or get completely lost halfway through the process, you need to tweak your information filtering method. You may find that as you gain experience and confidence, your style will change and grow. Most people must learn the basics and build on them before they are ready for the higher difficulty, complex methods. Think of it as you would think of learning to drive. How many people do you know who started driving with the skill level of a racecar driver? For most of us, the answer would be zero. The same is true for troubleshooters; they have to master all the "driving" basics of troubleshooting before they learn to "race" with the complex methods.

START SIMPLE: Sometimes the simple answer is the best answer. One common pitfall for the experienced technician is to dig too deep into a problem without checking for simple solutions first. Many technicians have spent hours on a fault only to discover that they overlooked a simple answer that had been staring them in the face. As your knowledge and skill increase, so does the depth of your analysis. The downside to this is that many technicians will start with the complex analysis and ignore all the simple steps that served them when they first started. To prevent this, you may want to put a step in your method that reminds you to check the simple stuff; for example, does the machine have power or is the switch on?

In this section, we have discussed several methods of filtering information to find a solution for the core problem. However, do not be afraid to use or create a method not mentioned here. Keep in mind that the main purpose of filtering the information associated with a fault is to find the root cause of the problem and correct it in a timely, accurate, manner. You will often find that the ends justify the means in troubleshooting. This means that in most cases, as long as your method meets all the parameters we have set forth in this section, no one will question how you solved the problem. The how questions usually come up when a large amount of time and/or money is spent solving a problem with no positive results. If you have trouble defining your style or using one of the styles we have talked about, do not be afraid to ask for help from those with more troubleshooting experience. Moreover, do not be afraid to try new ways of processing information that you think of or see someone else using. One of my favorite quotes is "Nothing ventured, nothing gained" (which dates back to Chaucer, c. 1374) (Titelman 2000); in this case, it means that if you do not take the risk of trying anything new, you will not gain the benefit of a better way of troubleshooting.

If at First You Do Not Succeed ...

Comedian Steven Wright is famous for this statement: "If at first you don't succeed, then skydiving definitely isn't for you" (Goodreads 2013). Luckily, most of the instances we deal with are not life and death (like skydiving). Most of the troubleshooting situations in which you will find yourself allow you to try and try again as needed. There are a few things to keep in mind when your first attempt at troubleshooting does not succeed:

GUIDELINE 1: **Do not get frustrated**. If you allow yourself to become aggravated or upset that your first attempt failed, there is a good chance you will not be thinking clearly when you try again. This can lead to missed steps, ignored information, and illogical thought processes that will only impede troubleshooting. If you find yourself getting frustrated, take a short break from the problem or get some help. Another time that frustration and anger can get in the way is when you injure yourself while fixing the equipment. Most people eventually end up getting hurt in the process of repairing a problem. It may be a smashed finger, a cut, a pulled muscle, a tool dropped on the foot, hitting your head on something solid, or some other painful event. No matter what the cause of the injury, many of us react to these injuries with anger. This can lead to a missed step or steps in the process of the repair and either cause you more problems or at the very least prevent you from fixing the problem. If you find yourself slipping into this kind of negative cycle, you definitely want to do something to break the chain.

GUIDELINE 2: **Reevaluate the information, adding in what you have learned**. When you try something, no matter the outcome, it provides new information. Earlier in the chapter, we broached the idea that as long as you affect the problem, for better or worse, you are on the right track. With this line of thought, you must first confirm that you have not created a new problem that is masking the original problem. As long as this is not the case, you have valuable information to use in your next attempt at troubleshooting. If you made it worse, then you are on the right track, but likely need to do the opposite of what you just tried. If you make it better, but did not completely solve it there are two main possibilities. One, you are on the right track but simply did not make enough of a correction; or two, you have a core problem caused by multiple failures. Multiple failures happen when more than one component lies at the core cause of a problem. In some instances, multiple failures happen when one component goes bad and in the process damages other components connected to it. Last but not least, if your efforts had no effect on the problem, then you are likely on the wrong track altogether. No matter what the outcome, there is some new information gained from what you tried. The following are examples of the various situations mentioned:

> **multiple failures**
> When more than one component is at the core cause of a machine breakdown

- You determine that a soft stop needs to be adjusted to allow the robot to move appropriately in the work envelope. After your adjustment, the robot will not move as far as it did previously. This is an example of going the wrong direction with the adjustment and making the problem worse.

- We have the same situation as above. After we make our adjustment, the robot can move farther in the work envelope but still not far enough to correct the problem. This is an example of going in the right direction, but not far enough.

- The robot is giving a servo alarm. In the process of gathering data, you discovered that one of the motor leads had come loose and damaged the motor. You replace the motor and get a new servo alarm. This one leads you to discover that the system has a damaged servo drive. This is an example of a multiple failure and how it might progress in the troubleshooting process.

- The robot stops during the loading process, and you discover a part that was not properly picked up and is slightly malformed. You remove the bad part and restart the system, believing that this will correct the situation. However, the robot alarms out in the same spot on the next cycle. When you look at the system, you see that the new part improperly picked up, even though it seems to be a good part. This is a case of a corrective action that had no effect on the core problem.

GUIDELINE 3: **Try, try again.** If your first attempt at fixing the problem does not work, then you still have a problem that needs to be solved. Good troubleshooters realize that sometimes perseverance is the only way to find the core problem and correct it. This is especially true if you take an erroneous path or if you missed a step in the process on your first try. You may find that when you are working with something new, whether it is a problem, a piece of equipment, or a system you have little experience with, it may take you more tries than normal to find the root of the issue. Often, regardless of your experience level with the system, you will need several tries to eliminate all the possible variables in a complex problem. This leads us to guideline 4.

GUIDELINE 4: **"Once you eliminate the impossible, whatever remains, no matter how improbable, must be the truth"** (Doyle 2012). This famous quote, from Arthur Conan Doyle's character Sherlock Holmes, should be something that every troubleshooter remembers. This approach is especially beneficial when there are several possible causes for a problem. There are often simple things we can try first before we spend large amounts of time and money on more complex solution. The best way to organize how you work through your options is to use availability, time involved, and cost. We start with availability, for simplicity's sake. If you have to choose between three different things to try and one of them requires a part you do not have access to, then you would obviously want to try one of the other two. We also use availability to make sure that we exhaust all our available options before we order a part that may be costly or take a long time to arrive. Once we have decided what is available, we next look at the time involved. If checking one possible solution would take a half hour while another would take a day, we start with the quicker of the two. The last way we filter is by cost. On occasion, we may have to factor cost in before time involved, especially if there is a risk of damaging a high priced item; however, we generally use this as tiebreaker when we do not have a clear-cut path. We would continue to filter our options using this method and trying again until we have found the problem or have no other options left.

GUIDELINE 5: **Back to the drawing board.** Sometimes you will try everything that you can think of, and nothing seems to have any effect on the problem. In this case, you may very well have followed a flawed troubleshooting path.

Do not worry; this happens to every troubleshooter eventually, no matter what they claim. When you find yourself in this situation, take all the data you have and filter it in a new way. Look for things such as other systems or components that could affect the problem. See if there is data you missed in the analyzing and gathering phases. Get a fresh set of eyes to look at the problem and give you some suggestions. Look into what additional resources you have access to. The simple fact is that once you have tried everything you can think of, you will need to look at the problem in a new way. Albert Einstein said it best, "Insanity: doing the same thing over and over again and expecting different results" (Einstein 2012). If as troubleshooters we continue to try what we know does not work, it would be insane to think it will suddenly work on the third or fifth try.

Troubleshooting can be a very complex process, especially when we are working with multiple systems, problems with multiple possible causes, or multiple system and/or component failures. When working with these systems, there is always the possibility that your first solution will not work. What separates the troubleshooter from others is what they do next. Troubleshooters persevere, getting help as needed, and solve the problem. It is that simple. It will take time and effort to become an experienced troubleshooter, but it can be a very rewarding and profitable career path. Most technicians take pride in the repairs they perform and the problems they solve, making this a rewarding career choice. Those who have honed their skill are sought out by the industries and businesses that stand to benefit from their knowledge, even in times of recession. As long as we have complex systems and equipment in our lives, there will be a need for the troubleshooter.

review

In this chapter, we covered a wide variety of topics on troubleshooting. At this point, you should have a better idea of what troubleshooting entails and how one goes about correcting problems. As you have the chance to apply the principles in this chapter, feel free to look over the material again as your new frame of reference may lead to deeper understanding. Remember, everyone starts out as a novice troubleshooter and it is only through experience that one becomes an expert. The key topics covered in this chapter were:

- **What is troubleshooting?** This section discussed the difference between troubleshooting and the troubleshooter and gave a frame of reference to begin our talk on the troubleshooting process.

- **Analyzing the problem.** Here we talked about ways in which to get information about the problem from operators, the equipment/system, personal observations, and diagnostic tools.

- **Gathering information.** This section discussed how we use inputs/outputs, manual, schematics, and wiring diagrams to get deeper information about the equipment and the problem.

- **Additional resources.** This section covered how we could use the thoughts of others, technical support, and the Internet to find more information about a problem.

- **Finding a solution.** Here we talked about filtering information using methods such as dividing up the system, following power/signal, and flowcharting in order to find a solution.

- **If at first you do not succeed...** Here, we discussed persevering and gave general guidelines to help the troubleshooter when his or her first solution does not work.

key terms

Alphanumerical format	Digital	Mastering	Subchart
Analog	Exploded view	Multiple failures	Track the power
Arc flash	Fuse bank	Octal	Troubleshooter
Binary	Flowchart	Parameters	Troubleshooting
Bit	Hexadecimal	Programmable logic	Tunnel vision
Brainstorming	Intermittent	controller (PLC)	Variable frequency drive
Common connection	Legend	Rungs	(VFD)
Continuity	Live	Schematics	Wiring diagram
Decimal numbers	Manual	Short circuit	

review questions

1. What are the steps of the logical process of troubleshooting and in what order do they normally occur?

2. How does knowing the normal operation of a piece of equipment or a system help in the process of troubleshooting?

3. What goes in your repair journal, and what is its purpose?

4. What is the benefit of talking to the operator or the last person to run the system/equipment before it faulted out?

5. Why is it important to check the error codes and alarm messages associated with a fault?

6. What are some of the things you are looking for during the personal observation phase of troubleshooting?

7. What are the guidelines for selecting a meter, regardless of the brand?

8. When taking DC readings, which lead do we connect to the negative side and which to the positive side? What happens if we reverse the two?

9. Where do you commonly place the leads for 120V AC readings, and what happens if you switch the two?

10. When taking continuity readings, what must you make sure of before taking the reading, and what can happen if you do not?

11. What is the difference between input and output signals?

12. What is the difference between a digital input and an analog input?

13. What might you find in a robot's manuals?

14. How might mastering affect program positions and what is the best method to deal with this?

15. What are the three rules for reading a schematic?

16. What do wiring diagrams show, and how do they convey this information?

17. Draw a simple schematic for a SPST switch that turns on a light and engages the coil of a relay using the component symbols from the Schematics and Wiring Diagrams section of the chapter.

18. What are some things that cause tunnel vision, and how can we avoid this?

19. What do you want to avoid when brainstorming, and why?

20. When calling tech support, what pieces of information should you have available?

21. What kinds of information can you find on the Internet?

22. What are some of the ways we might group parts of a system?

23. What are the guidelines for tracking the power?

24. What are the guidelines for following the signal?

25. What factors should you keep in mind, regardless of which method of troubleshooting you chose to use, and why are they important?

26. What are the guidelines to keep in mind if your first attempt at troubleshooting does not work?

27. How do we filter the information when we try something and make the problem either worse or better?

28. What can happen if you become frustrated in the troubleshooting process, and what can you do to combat this?

troubleshooting questions

Read the following scenario. Use what you learned from this chapter to answer the questions below.

Troubleshooting Scenario, Robot Servo Failure

You are working on a robot that has a servo drive alarm. You find blown fuses on the drive. In the course of gathering information, you discover that the motor has a loose connection, which caused it to overheat, damaging the internal windings and the motor encoder. Once you replace the motor and encoder, you get another servo drive alarm that indicates there is a bad card in the drive.

1. What kind of a fault would this be?

2. After you replace the servo drive card, you find that the system will not run and is giving you alarms indicating that it is does not have information on the type of motor and encoder being used. What would be the likely corrective action to this problem?

3. Once you have the system recognizing the motor and encoder, you get an alarm indicating that it does not know where the zero or home position for that axis is located. What would be the likely corrective action to this problem?

4. Given the information above, what do you think was the root cause of all the problems, and why?

5. What information-gathering techniques do you think would work best with the scenario and why?

Repairing the Robot

overview

What You Will Learn

- The definition of preventative maintenance

- Some common preventative maintenance tasks for electrical, hydraulic, pneumatic, and mechanical systems

- What lockout/tagout (LOTO) is, how to perform it, and why it is important

- Some good tips to help with the repair process

- Why swapping parts is not necessarily fixing the robot

- Things to look for before starting a robot up after repairs

- How to tell if you have actually fixed the system

- A series of questions to ask if the robot is still having problems after a repair to get to the root cause

- Three key tasks to perform after repairs are complete and everything is running correctly

In the previous chapter, we talked about how to determine the root cause of problems or to troubleshoot the robot. Bear in mind that once you know what is wrong, you have to do something to correct the problem. In this chapter, I will share some of my knowledge about repairing equipment along with tips and tricks to help you in those endeavors. While the specifics of a repair depend on the specifics of the problem and the equipment involved, I have found that certain principles and practices apply to many situations and offer that information here. During the course of this chapter, we will cover the following:

- Preventative maintenance (PM)

- Precautions to take before you begin repairs

- Repair tips

- Part swapping versus fixing the problem

- Precautions before running the robot

- What to do if the robot is still broken

- The robot is running, now what?

Preventative Maintenance (PM)

De Legibus (c. 1240) perhaps said it best, "An ounce of prevention is worth a pound of cure" (Titelman 2000). When it comes to repairs, proper preventative maintenance can equate to a savings in both cost of parts and time involved. Preventative maintenance (PM) is the practice of changing out parts and doing repairs in a scheduled fashion *before* the equipment breaks down or quits working. You could think of PMs in the same way as you think of a yearly checkup at the doctor's office or changing the oil in a vehicle. The whole point is to find problems in the early stages or to prevent the problems from occurring in the first place. Moreover, since we schedule preventative maintenance before the machine breaks down, we

> **preventative maintenance (PM)**
> The practice of changing out parts and doing repairs in a scheduled fashion (based on previous equipment and component performance) before the equipment breaks down or quits working

can prevent the loss of production or functionality of the robot by doing the work at a time that is convenient instead of in a reactionary mode when it finally breaks.

The literature for many commercial robots, especially the industrial varieties, includes a schedule of PMs. This schedule will tell you how often to check or replace various components and what needs to be done daily, monthly, quarterly or every three months, semiannually or every six months, yearly, and over a span of years. For instance, checking the level of a hydraulic tank is a daily task; tightening all the electrical connections would be a yearly task, in most cases. Greasing the bearings and tasks of this nature are usually required every three or six months, while replacement of the wiring tends to fall in the three- to five-year category (see Figure 11-1). Because of the way we divide up the preventative maintenance tasks, you may only perform a few, taking a small chunk of time, while at other times you may need a couple of days of downtime to complete a long list. For instance, you may have monthly, quarterly, semiannual, and annual tasks due at the same time, giving you a lengthy list to work through. If a three- to five-year task happens to be due at the same time, you may end up adding a day or two to an already busy schedule. Good thing we can schedule when we do the PMs!

At this point, you may be wondering, "How do they determine when tasks are due?" We schedule PMs based on the previous performance of equipment. At some point, the equipment runs under normal operating conditions and someone keeps careful records of part failures. Each time something breaks, wears out, loosens up, or ceases to function properly, it is recorded somewhere. Over weeks, months, and years, this creates a pool of data that can be analyzed for trends or patterns of failure. For instance, if 5 out of 10 robots need to have their wiring harness replaced after five years of normal use, then the manufacturer will recommend that all robots have their wiring replaced every five years. If the grease in the bearings tends to sling out or become contaminated after three months, then the recommendation is to grease all the bearings every three months. These recommendations become the preventative maintenance schedule issued by the manufacturer. The base information sometimes comes from other equipment that uses the same components in a similar fashion. For instance, if the same motor used for an axis of a robot is used by a CNC machine and proven to fail after 10,000 hours of use, then it is reasonable to assume that the same motor should be replaced in the robot after 10,000 hours of use.

The systems of the robot determine the types of preventative maintenance required. Electrical systems need all their connections tightened yearly; if the wires

Figure 11-1 The gears that turn the minor axes of the FANUC Delta robot. If you look closely, you can see the grease that reduces friction in the system; this grease needs to be added to or changed from time to time.

run through any of the moving parts, say, the center or outside of a robot arm, then you will have to replace them at some point (see Figure 11-2). Electric motors wear out and generally have a use life of only so many running hours. You may need to clean or replace encoders based on where and how they are used. Another common task is checking the cooling fans for functionality and dirt buildup. You may have to pull the electronic cards from the controller and put them back in or **reseat** them to insure proper connection. There will often be a series of voltage and amperage checks to perform at various points. The verification of electrical safety systems and other sensors is another common task in the electrical area. If the robot you are working with has battery backup, then expect to replace the batteries on an annual or lengthier basis (see Figure 11-3). Failure to comply with the battery change PM

reseat
The process of removing electronic cards from the controller and putting them back in to insure proper connection

Figure 11-2 Here is a close-up of the internal components of a FANUC R-30iA Mate controller. A common practice is to tighten all electrical connections, like the ones in this controller, each year to prevent loose connections and all the havoc that could cause.

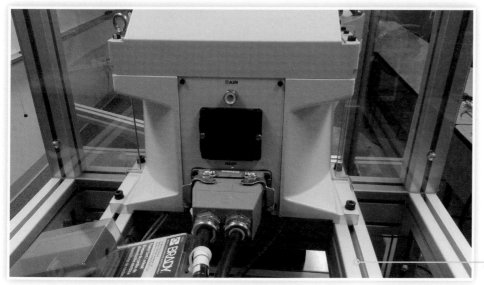

Figure 11-3 Under the black cover in the picture is where you'll find the backup batteries for the Delta robots in my classroom. Make sure you know where the batteries for your robot are and how often they need to be changed.

is a good way to lose robot data and create hours of work. The full list of electrical tasks depends on the robot you are working with and its equipment.

Hydraulic systems require periodic monitoring of the fluid for contaminants such as dirt, water, and metal fragments as well as replacement of the oil (commonly at the one-year mark). Periodic testing of the oil for the proper consistency and chemical makeup is another oil-related PM task. Cursory checks for leaks are a daily task, where pulling off all the covers and looking deeply into the system may be on the semiannual or annual list. You need to change out the filter for the system whenever it shows a clogged condition or periodically, as directed, again typically at the one-year mark (see Figure 11-4). If your system uses accumulators, which are devices used to store hydraulic pressure and then release it back into the system as needed, you might have to check the gas charge, depending on the type. (See Food for Thought 11-1 for more about accumulators.) Some of the valves may require removal and cleaning, which involves taking the valve apart and cleaning the internal components. Checking the hoses for leaks and signs of wear is another common PM task for hydraulic systems. These checks are common to most hydraulic systems, not just those used with robots. So, do not be surprised if you see a list like this for other equipment that you encounter in industry.

Pneumatic (or air) systems have their own checks as well. Of course, any fluid power requires checks for leaks, damage to hoses/piping, and the changing of filters, so you can expect to do this on a predetermined basis (see Figure 11-5). Additionally, many pneumatic systems have lubricators that add oil into the compressed air as it passes to lubricate moving parts in the system. Over time, the system depletes the oil and must be replaced. Because of the noise created by exiting air, mufflers are often required; you will want to check these for damage or clogs. Water and dirt fall out of the air in the reservoir tank of a pneumatic system, so they need periodic cleaning and draining to insure proper operation and prevent damage to the components in the system. The last thing you want is dirty wet air forced through the valves and components of your system, as this is a fast track to breakdowns.

For mechanical systems, the important PM task is greasing and lubrication. Unless the bearings are sealed, they need the grease replenished at some point with the correct type of grease. Many robotic systems use white lithium grease for the

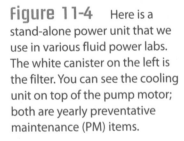

accumulators
Devices used to store hydraulic pressure and then release it back into the system as needed

Figure 11-4   Here is a stand-alone power unit that we use in various fluid power labs. The white canister on the left is the filter. You can see the cooling unit on top of the pump motor; both are yearly preventative maintenance (PM) items.

11-1   Hydraulic Accumulators

Some accumulators use a rubber bladder filled with nitrogen gas inside a metal housing to store pressure. As the systems forces oil into the accumulator, the gas bladder is compressed and exerts pressure on the oil in the accumulator. If the system pressure drops, the pressure from the bladder forces the oil back into the system, helping with drops in system pressure. With this type of system, any loss of pressure in the bladder equates to a loss of stored pressure in the system. Since the accumulators create flow when the system is overtaxed, lower pressure in the bladder could mean slow operation of hydraulic motors or cylinders, which can influence robot movement or tooling performance.

Another common accumulator uses a solid weight with seals to force fluid back into the system. With this type of accumulator, the PM is to check the seals and make sure the drain, for oil that gets past the seal, is working properly. No matter the type of accumulator, there will be some form of preventative check to perform for verification of proper operation.

Figure 11-5   This pneumatic regulator does not have a lubricator; however, it does have a filter bowl that needs to be drained and cleaned from time to time as well as a muffler to keep an eye on.

bearings and gears because of its high-speed properties and the tight tolerances of the systems. Speaking of gears, the common check is to make sure everything is meshing properly and that there is no excessive wear or broken gear teeth. For belt and chain drives, the PM includes making sure that any stretching is within a given tolerance, the systems are still transmitting power without slipping, and everything is in alignment. This is also a good time to make sure nothing is rubbing against the metal skin of the robot or contacting things such as electrical wires and fluid power hoses. Catching this kind of contact early allows the problem to be dealt with easily and minimizes any damage, whereas letting the problem continue is almost sure to result in a major robot repair.

These common PM tasks are in no way a complete list of everything you can expect. The specifics of your robot's PM depend on what it does, how it does it, and where it does it. Harsh environments and heave use tend to shorten the time between checks or replacement. Infrequent use or optimal conditions may extend the amount of time you can get out of various components, but make sure not to push your luck in these cases. If you find something is wearing out more quickly than the recommended time between checks, go ahead and change the time interval for replacing that part. If you have consistent problems with something that is not on the list, feel free to add it to your PM schedule. As the end user, the responsibility for properly maintaining the robot falls squarely on your shoulders. That said, make the changes you see fit to improve the overall performance of the system.

Unfortunately, some companies do not recognize the value of preventative maintenance or simply do not want to take the robot out of production long enough for the work needed. In these cases, you must do the best you can; however, there are a few arguments you can make in favor of preventative maintenance.

1. **We can schedule preventative maintenance.** Some industries build time into their production schedules for machine failure, but no one has a crystal ball that tells them when the equipment will fail... Oh wait, they do! That is the whole point of a preventative maintenance schedule. This means that they could schedule a time before the breakdown to take care of the problem. While PMs may not prevent all breakdowns, they will certainly help to reduce the number of repair events.

2. **Preventative maintenance saves money.** I am sure you have heard the saying "Time is money." In industry, managers have this concept pounded into their heads day in and day out and are acutely aware of this whenever machines stop working. A few hours of preventative maintenance now could save many hours of downtime later. On top of that, one failing part can cause damage to or the failure of other parts in a kind of domino effect. In those cases, anything damaged by the original failure is added cost in parts, time, and labor.

3. **Warranty voided.** Many robots come from the factory with a warranty that covers the replacement of any components that fail within a given time. Failure to keep up with preventative maintenance, which is the user's responsibility, could very well make the warranty null and void.

4. **Shipping and handling.** Another cost factor when it comes to repairs is having the right parts at the right time. It costs money to maintain a large stock of replacement parts, so many companies try to keep only the essentials on hand. This can lead to extended downtime and high shipping costs when ordering parts in response to machine failure. With a planned PM, there is time to order the parts at cheaper, slower, shipping rates. In addition, if the company knows that they will need a large number of certain parts, they can order them all at once and perhaps get a bargain on part cost, shipping, or both.

These are just a few of the arguments that you might make for the case of performing PMs. Some companies simply have not examined the pros and cons, while others will only see the reasoning of your arguments after they have lost a large amount of time and money on reactionary maintenance only. In these cases, just do the best you can and keep pointing out the benefits of a preventative maintenance program. If your company is a believer in PMs, make sure that you do them properly and hold up your end of the bargain. Share your ideas and observations, as we often become stuck in a habit and forget to take time to reevaluate the situation. If you are responsible for the robot's upkeep, the frustration you prevent may well be your own!

Precautions to Take before You Begin Repairs

As pointed out in the previous chapter, before you fix a system, you have to know what is wrong with it. The whole point of the troubleshooting phase is to determine the problem, but that is not all the prep work that you should do. An important consideration before beginning repairs is to make sure the area is safe for you to work. LOTO is a big part of this.

Lockout/tagout (LOTO) is a methodical process of removing all power from a device so that there is no active or latent power; this is also known as a zero energy state, making it safe for you to work on and around the equipment. Sometimes during the troubleshooting phase, you have to perform checks with the power on because there is simply no other way to troubleshoot the machine; however, as soon as you have a plan of action, it is time to power everything down so that you can work safely. To get to a zero energy state, turn off all electrical, hydraulic, and pneumatic power coming to the machine and bleed off any residual pressure, discharge any capacitors, and either lower portions of the equipment that could fall or block them in place. As part of the LOTO process, we use specialized devices called lockouts that hold the power source in the blocked or de-energized state. No matter how you turn the power source off, there is a lockout designed to keep it turned off, and each lockout has a place for a lock or hasp that can hold multiple locks. Each person who works on the equipment must put his or her own uniquely keyed lock on each lockout for the machine. This is the only way to ensure that someone else does not turn on the power while you working on the equipment. There are only two people who can legally unlock that lock: the person who put the lock on and his or her supervisor. The supervisor must talk with that person to verify that he or she not in the danger zone before removing the lock.

Once you believe that you have all the power locked out, bled off, and accounted for, try to turn the system on. This helps to ensure that the machine is in a zero energy state before you begin repairs and verifies that you have done everything right before entering the danger zone. Make sure to check any gauges on the system for fluid power pressure and use a multimeter to verify that there is no electricity present. Once you have performed all these checks and proved that there is no power left in the system, neither kinetic nor potential, you can safely begin repairs. Each person working on the robot must have their own lock on the lockout devices, for their safety. Make sure that you have any necessary tags in place as well. Common warnings include "DO NOT RUN" or "Maintenance working on machine." Make sure that your lock has your name on it so that everyone knows you are working on the machine. When a lock has no name, it is very difficult to verify that the owner is clear of the danger zone.

The following checklist gives you a step-by-step procedure for the LOTO process.

- Notify affected individuals that you are about to shut the machine down. This includes operators, nearby employees, and management.

- Stop the machine cycle, if necessary.

- Turn off or remove all external power supplies and lock them in the off position using lockout devices and a lock with your name on it.

- Place appropriate information tags on the equipment, such as "Do Not Run—Under Maintenance."

lockout/tagout (LOTO)
The process of removing and draining all power sources from a piece of equipment to reach a zero energy state and then ensuring that they cannot be reactivated by the use of lockout devices and personalized locks

zero energy state
When a machine has no active or latent power

lockouts
Devices used to hold the power source in the blocked or de-energized state

- Verify a zero energy state. Make sure to account for capacitors, compressed springs, items that could fall, stored fluid pressures, or other potential energy sources.

- Perform repairs.

- Once finished, remove all tools and any blocking devices or other items that you added to the machine for safety reasons.

- Once everything is clear, each person working on the robot removes his or her own lock; the last person can return power to the equipment.

Failure to follow this procedure can put not only your life, but the lives of your coworkers, in danger (see Figure 11-6). History is full of tales of injury or death from improper lockout. This is not something to take lightly. In fact, you can perform all the steps of LOTO and still end up in a life-or-death situation. (Read Food for Thought 11-2 for a tale of a near miss that happened to my best friend.) You should always be aware of your surroundings when working on equipment, paying particular attention to any signs that someone is powering up the system. LOTO protects you from harm, but it does not mean that you can check your brain and situational awareness at the door.

Once you complete LOTO, there are a few other preparatory things that you can do to help with the maintenance process. An important step is to gather all the necessary technical manuals, schematics, and other information necessary to complete repairs. Having these on hand saves time later when you need to determine the next step in a repair or confirm the wiring of a portion of the circuit. It is likely that you already gathered the materials you need during the troubleshooting phase, so just make sure you have them at hand. Be sure to place them somewhere out of harm's way, otherwise damage might occur and some of the materials may be difficult to replace. A common place for such materials is the top of a roll-around toolbox or a workstation that is near the system but far enough away so that no fluids can splash or parts can fall on them. Make sure to wipe grease and grime off your hands before handling these resources as well. I can assure you that a greasy thumbprint has impeded repairs more than once.

Figure 11-6 When working with large robots that move large payloads, it's crucial for your safety that you make sure that nothing can fall on you during the LOTO process!

food for thought

11-2 LOTO Awareness

This LOTO incident happened to my best friend, James E. Stone, who passed away in 2012 after losing his battle with cancer. The incident occurred during his years as a maintenance technician and, in particular, while he was working at an injection molding company. For those of you who are not familiar with the injection molding process, the machine presses two halves of a mold together, injects either molten plastic or metal into the mold, and then rapidly cools the part before the two halves separate, exposing the newly created part to the world. These machines range from bulky to massive; the one in this story was the massive type for metal parts. In fact, it was so large that my friend Jim could easily fit between the two halves when the machine was open.

Now that I have the stage set, I can begin recanting the details of that day. Jim had some maintenance to do on the molds so he locked out the piece of equipment, verified the zero energy state, and crawled into the mold area with tools in hand to get to work. He had been working on the machine for a while when an operator showed up and decided to start the machine. After hitting the start button and getting no response, the operator noticed that the main electrical disconnect was turned off and that someone had stuck this aluminum tag thing, Jim's lockout, through a hole and put a lock on it. The operator studied the device long enough to determine that he would have to pry it off, but apparently did not get the meaning of "Maintenance Working On Machine." Most lockout devices are made of thick plastic or thin aluminum, so they are not indestructible. It is more of a hassle to remove them than impossible.

So, after a bit of work, the operator had bent and warped the lockout device in such a way that he could remove it and throw the arm on the side of the box, returning power to the injection-molding machine. This is where the real problem began for Jim. During the several minutes that the operator was working on his nefarious mission, my friend was working in the molds and had no idea what was going on. Jim's first clue was when he heard the hydraulics for the machine cycling up. Knowing the piece of equipment as Jim did, he knew that part of the power-up process for this machine was to close the two halves of the mold and then open them back up, ensuring that everything was working correctly. The halves of the mold for this machine close with tons of force and Jim was right in the danger zone!

It was only a combination of quick reflexes, panic, and odd luck that saved Jim's life that day. As he scrambled to get clear of the machine, he happened to reach up and grab a large bar running from one side of the mold to the other. The bar he grabbed was the main power bus for the two halves of the mold and energized with 480V of three-phase power. The grab was the panicked reflexes, the fact that he grabbed the bus bar was the odd luck. As soon as Jim grabbed said bus bar, it electrocuted him and caused such a violent muscle contraction that it threw Jim clear of the machine just as the halves closed. In fact, the timing was so close that part of the material of his pant leg was caught between the two halves of the mold. He was literally saved by an event that most maintenance technicians work their entire careers to avoid!

So there was Jim, lying on the ground, feeling the effects of a severe electrical shock, bleeding from the ears and nose, and riding out the fight or flight adrenaline reaction. Once he realized that he was alive and functional he had a . . . discussion . . . with the operator of a rather physical nature and then informed the management of the facility that if it was going to hire that caliber of employee, he quit. His supervisor did try to talk him into staying, but Jim had had enough of that place.

This just goes to show that you can do everything right and still find yourself in a life-or-death situation. Learn from Jim's experience and keep your wits about you at all times when working on or repairing equipment! I can assure you that this advice has saved me from injury more than once over my years of repairing equipment and robots, making the difference between near misses and life-altering accidents.

Gather the tools that you know you will need before you begin. Many people try to make do with the tools they have on hand rather than going somewhere to get the right tool. This can lead to damaged parts, stripped bolts, and busted knuckles. If you have your tools in a roll-around toolbox, then wheel those over before you begin. If you have a tool pouch or a carry-around toolbox, then make sure that you stock it with the tools you need for the repairs at hand. Many times, you can find a manual with an exploded view of the parts you need to work on, and this will give you an idea of the tools you need. If you need specialty tools, such as gear pullers, torque wrenches, or forklifts, get those as well. If by chance you find that you need a tool that you do not have with you, it is best to take the time and go get it rather than run the risk of making do with something that *might* work.

If you know that you need to replace certain parts, such a bearings or motors, make sure that you have them on hand before you begin. I remember several times when I was halfway through a repair only to find out that we did not have all the parts needed. It may take days or weeks for a part to arrive, depending on what it is and where it comes from. This delay turns into a chance for other technicians to rob parts from the machine to repair other machines, nuts and bolts seem to wander off, and when the part does finally come in, someone else may finish the repair. All of this equates to added time and frustration in the repair. In my experience, it is much better to wait to start the repair until the parts are on hand than to stop halfway through.

You may wonder why we perform LOTO before we gather all the tools and parts. The reason is that this gives the machine more time to dissipate any stored energy. Many drive systems use capacitors and are set up in such a manner that the stored power dissipates slowly over time. In my experience, it takes 5 to 10 minutes, on average, before it is safe to work on the system, making this hands-off time ideal for getting tools and other necessary materials gathered up. Remember to double-check with a meter before you begin repairs. Leaving the machine to allow dissipation of energy means that it has not reached a confirmed zero energy state; you must confirm all the energy is gone before you begin repairs.

Repair Tips

Now that you have gathered all the parts, found the necessary information, and brought the equipment to a zero energy state, you are ready to begin repairs or preventative maintenance. There are several things that you can do during this process to make it run more smoothly. While you may not need or be able to use all the tips from this section for each job, the more you take advantage of, the greater the benefit that you will reap.

"One picture is worth a thousand words," according to Fred R. Barnard (Titelman 2000). I have found this a very true statement when it comes to maintenance. Before you remove a part or unwire a system, use your cell phone or another device with a camera to snap a few pictures. This will give you a visual roadmap back to the original system configuration and help with any questions that you may have. I have used this many times myself to help me remember how the system was wired, where everything went, how something looked before I took it apart, how hoses or cables were routed, and so on. This technique has saved me many hours of trouble, especially in those cases where days, weeks, or months have passed from the time I started a repair to when I finished it. These pictures also help others who must finish what you started or in case you need to explain what is going on with a system (see Figure 11-7a through Figure 11-7e).

Keep all the parts together. To fix most equipment, you need to remove covers, take off brackets, remove parts, and in general end up with a pile of stuff that needs to go back on the machine once you have finished the repairs. Keeping everything together makes the assembly process much easier. When deciding where to place the parts, be aware of things such as areas where people are still working, any

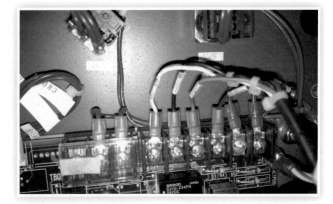

Figure 11-7   I took this series of photos before unhooking a FANUC R-J2 controller in order to move our lab robot from our old building to the new campus. It was nearly a year later before I had the chance to hook the system backup, but these pictures allowed me to complete the task as if it had only been a day or two!

openings that parts could fall into, areas where parts could roll under the machine, places where the parts might fall on you or others, and anything else that could cause problems. For small parts such as nuts, bolts, screws, and washers, you may want to have a small container in which to place them in so they do not roll away. Another trick is to lay the parts out in the order in which they came off the robot. By placing parts in a linear fashion, you create an easy-to-reverse timeline of what goes on next.

Get help before you need it, not after you have created an emergency. Many times I have seen maintenance technicians bite off more than they could handle and get themselves into dangerous situations because they would not ask for help. There is no trophy for taking foolish risks just to prove that you can do the job all by yourself. If you need help lifting something, ask (see Figure 11-8). If you are not sure how to proceed with a task, see if someone you work with has experience and can help. If you need someone to use a forklift or crane to remove or install a heavy portion of the equipment, then make sure the person knows how to run the equipment. The only prize for trying to be a one-person show when two would be better is injury, scars, and possibly death.

figure 11-8 While the robot in this picture is not small by any means, take a moment to look at the tooling it uses during operation. Now imagine trying to hold onto this while removing the bolts holding it in place. It is situation like these when you need help from your coworkers.

When you enlist help, make sure everyone is on the same page. One of my favorite parts from the *Lethal Weapon* series is when Danny Glover and Mel Gibson are arguing over whether it is one, two, three, go or one, two, and go on three. While this is one of the running jokes of the movies, it brings to light a problem that happens when people are working together. If two people are lifting a heavy motor, but they are not working in unison, there is a good chance for injury and accidents. If one person is about to lose his or her grip and does not tell the other, then disaster may follow. If both people are thinking about going to a different place with the heavy object or have different ideas of what various hand signals mean, then mayhem is almost certain to occur. The whole point of teamwork is to work as a team, not as a group of individuals all trying to do their own thing. Take into consideration the physical differences as well. If it is a two-person lift and one person is over six feet tall and can lift 300 lbs while the other is just under five foot tall and can only lift 100 lbs, then there is a good chance the team lift will fail. Carrying anything at an angle puts more weight on the person on the lower end. If the shorter person drops whatever it is due to muscle fatigue or it being more weight than he or she can handle, then the taller person may hurt himself or herself trying to catch the load or end up under it. These are all considerations when working as part of a team.

Take good notes. Often during repairs, we learn better ways to do things, track down information that did not come with the equipment, or make other discoveries that can help later. This information is very beneficial six months or a year later, when you run into a similar problem but forget exactly what you did to fix it. This is one of the main purposes behind the technician's journal, and is the perfect place to put technical sheets found on the Internet or in parts boxes, those pictures you took earlier, and anything you found helpful or learned from the troubleshooting process. The more you save today, the less you have to search for tomorrow.

I remember copying the binder of my mentor when I first started my industrial maintenance career. It took a couple of hours and included stuff that made no sense to me whatsoever, but I copied it all. Years later, after my mentor had moved on to another company, I still used that book and found, from time to time, a nugget of knowledge that had meant nothing to me on the day I copied it, but was invaluable when I finally understood what it was for. Of course, I added to this book over the years from my own experiences. When I left that company to pursue a different path, I passed down that binder of knowledge to one of the newer techs, in the hope that it would serve him as it had me.

Break the mold. We tend to get stuck in a rut and do things the same way over and over again, because that is the way we learned to do it. Just because that is the way it has always been done does not mean it is the best way! Sometimes, a fresh pair of eyes can see a different path that can save time, money, and frustration. You may need to create your own special tools to do what you want, but do not let that stop you. You will likely get resistance from those who are stuck in the rut; again, do not let that stop you. Sometimes your idea will fail. Sometimes your idea will make matters worse. But more importantly, sometimes a new way that makes it all worth it. When talking with your peers about trying something new, the argument that "This is the way we have always done it" is *not* a valid reason not to try something new. Valid arguments would be along these lines: There is something preventing your plan from working, it could cause unnecessary damage to parts or equipment, you would have to make major modifications to the equipment, etc.

These are a few of my favorite tips to help you when you begin repairs, and I highly encourage you to find other tips and tricks that help: Talk to those who have experience with the equipment and see what they recommend. See if you can get someone with years of experience to mentor you on the finer arts of maintenance. Search the Internet for information from reputable sources. In this digital age and with all the social media resources out there, information is easy to gather for those willing to put a little effort into it. When you discover a gem of knowledge, glean a new insight, or create a new process, make sure to record it in your technician's journal, whatever it is you are using for such, to preserve it for future use.

Parts Swapping versus Fixing the Problem

When it comes to repairs, swapping out broken parts for working parts is a big part of the process, but it is not the end of the process. In our modern world, where many things are disposable, it is common to replace whole assemblies of parts instead of trying to fix the part or component of the assembly that is causing problems. Years ago, maintenance technicians would pull a control card from the robot controller, find the specific damaged component or components, and replace them. Today, technicians pull the damaged card and put in a brand new card. Sometimes they send the old card back to the manufacturer for repair, but many times we simply toss or recycle the bad card. This lack of need for deeper troubleshooting has led to a trend where technicians swap out bad parts for good and then stop their troubleshooting, assuming that they have repaired the robot. Yes, the robot is ready to run once more, but have they truly solved the problem?

The truth of the matter is that everything faults out for a reason and simply changing parts may not be enough to remove the root cause of the problem. If the motor burns out because the robot is picking up too heavy a part, then changing the motor does nothing to fix the problem (see Figure 11-9). If a loose wire is sparking and causing a current problem that damages components on a control card, then replacing that card does nothing to address the loose wire. There are times that part swapping only addresses the symptoms and not the cause of the problems. To truly fix the robot or other equipment, we have to figure out why it failed in the first place. Sometimes this is due to the part in question's failing; in these cases, swap and go is the right path. For everything else, try to answer

Figure 11-9

A servomotor is something that you will likely change out at some point. Make sure you determine whether the motor just reached the end of its life or if something caused it to wear out prematurely.

the question, "Why did it fail?" Failure to do this can result in repeated damage to certain components and a possible cascade effect, where the second or third time that it happens, other components and systems are damaged as well.

If you cannot find a bigger reason why, make sure you perform all reasonable checks for functionality and record your results. If possible, compare these to another robot of the same type and see if the system is running normally. If you do not find any smoking guns or obvious reasons, make sure to pass along what you did find out to others that work on the machine, so that those involved have the background information should it happen again. Sometimes it takes two or three failures before the root cause is determined, similar to the way in which a doctor may need to run a series of test and try a couple of medicines before truly diagnosing the problem. This deeper level of analysis is what separates the novice technician from the experienced maintenance person.

Precautions before Running the Robot

Once repairs are complete, there are a few things you need to do to make sure that everything is ready to go before powering up, after the system has power, and during the first few cycles. These simple steps ensure that the system is ready to go and prevent the need to turn around and come right back to a robot that you had just declared functional. I can tell you from experience that it is easier to do it right the first time than to go back and explain why the robot you just fixed is back down.

Check the system for tools, spare parts, and foreign objects before putting on the covers. You would be surprised at the amount of damage a forgotten wrench or screwdriver can cause when you fire up the system. Once the covers are on, these foreign objects remain unnoticed until they become damage in the system and require repair.

Make sure that everyone is clear and all the covers are in place. This is part of the LOTO process; however, it is important to mention again, as you really do not want to be the person who started the robot with someone in harm's way. Having all the covers on before startup is for your safety. This protects you from electrical arcs, flying parts, rotating equipment, and dangerous events that may occur after a repair, especially if you were unable to find the problem's root cause. Even if all the locks are off the lockout, make sure everyone is clear of the work envelope before turning the power back on. The robot could start unexpectedly, and we have discussed the ills of what could happen. Make sure that any operators or other persons around the robot know you are powering up the system. This prevents them from being startled and possibly hurting themselves on something else.

After power up, check for any alarms or unusual action by the system. Alarms will let you know if the system's diagnostic functions detected any issues. Also, what for irregular action of the robot that can indicate deeper problems or improper repair of the system. Once you verify that the system is ready, place it in manual mode and move each axis to ensure proper operation. If you have a simple program, such as a homing program or alignment check program, try running one of those to see how the robot responds. If the robot stopped at a specific point in a program before repairs, load that program and step through it manually to see if it functions correctly. Also, verify that all the safety equipment and sensors of the robot are working. Sometimes sensors are damaged either by the breakdown of the robot or by accident during repairs.

The last and perhaps most important check is to load in the proper program and start the robot back into normal operation (see Figure 11-10). While it is running, keep your hand on the E-stop or stop button so that you can stop the system should any signs of problems arise. Murphy's law of maintenance says that those who fail to check proper operation in automatic are doomed to receive a callback to the robot. I learned this the hard way during my years of fixing machines; many times, everything seems fine during startup and all the checks, but the problem came back during normal operation. I always spend some time watching it run normally before I truly call anything repaired. After a couple of cycles of the program with no apparent issues, I usually turn the machine over to the operator and tell him or her to let me know if anything goes wrong.

Another great benefit of these steps is the fact that you have a front row, center seat for any problems that arise. There are times when no amount of data from the system or descriptions of what happened from the operator can replace your own experience with the system. As a technician, you view the system differently than someone who only runs or programs the systems, thus the reason why seeing the event can give deeper insight to the root cause than just hearing about it. Do not be afraid to spend some time with the equipment after startup just watching the system for those special clues that mean something to you. This is time well spent at the beginning that could save you hours later.

figure 11-10 When checking repairs or programs, do not forget to wear the proper safety equipment.

What to Do if the Robot Is Still Broken

At this point one of two things has happened. Best-case scenario, the robot is running normally once more and you can proceed with a few finishing details. Worst-case scenario, the robot is still broken and you have to decide what to do next. Obviously, if the system is not running properly, we are back in the troubleshooting phase of things, but we have some new data to factor in. You need to ask yourself these four questions:

- Is it better than it was?

- Is it worse than it was?

- Was there no discernible change?

- Is there a new problem?

Depending on when you say yes will indicate your next course of action and help with the troubleshooting process.

If the system is better than it was, you are on the right track, but need to do some more work. It may be that you only replaced some of the damaged components. In the case of offsets and adjustments, you will likely need to make a larger change in the same manner as the last. Make sure that you think about the systems that depend on whatever it is you worked on and check them for damage or misalignment as well. Keep following the track you are on and you should get the system back up and running.

If the system is worse than it was, there is one of two possible reasons. One, you are on the right track, but going in the wrong direction. Perhaps you made an offset that was positive instead of negative, or realigned the tooling at the wrong position. In these cases, you have the right system, just the wrong repair action. This is the case most of the time and is at least a positive outcome since you know what you worked on is related to the problem.

On rare occasion, we misdiagnose the problem so badly that we actually create new problems during the repairs! In these instances, do the best you can to undo what you just did and see how the system responds. If you return to your original problem, then you know whatever it was you did was unrelated. If your original problem does not return, then you have a worst-case scenario, where the misdiagnosis has created new problems that are likely masking the original problem. These situations can take a large amount of time, effort, and patience to work through. The best advice I can offer is to work through the problems one at a time, starting with the most severe issue or alarm first. I remember a few times in industry when I walked into the middle of this type of situation; the person working on the machine before me went down the wrong path and created several new problems that masked the original, muddying the troubleshooting waters. Most of the time, I had to figure out what that person had done and correct or undo his or her actions just to get back to the original problem. Regardless of the specifics, this is an advanced troubleshooting scenario; should you work your way through one, you need to realize that you have completed a rite of passage that all professional troubleshooters eventually experience.

If there is no discernible change, then whatever you worked on is not likely part of the problem. This often happens when there are several possible right answers to the problem and you picked the one that seemed the top contender at the time. If this is the case, simply reevaluate the other options and pick the next most-likely culprit. If you only had one corrective action in mind, go back to the troubleshooting phase and factor in the new information (that whatever you just tried had no effect). This will often help you narrow down where the problem is and get you back on track. If you are stumped, ask for help from those you work with or see if you can find new sources of information about your problem.

The other possibility is that the new or rebuilt part that you put in the robot is also bad. It does not happen often, but it does happen. Most manufacturers consider a 95 percent quality rate great; however, that means that 5 out of 100 parts have some kind of flaw. These flaws are often minor and not a real concern, but on occasion bad parts slip out of the plant and make it to the customer. When this happens, you can spend a large amount of time trying to decide what is really behind the problem when you already had the right solution. Again, this is a rare event and not a first answer to your problem of status quo. Just keep this in mind if you run out of other good options for what is wrong. I have used a part from a working system to test this theory before, but I ran the risk of damaging the working machines part if I was wrong and having two robots down instead of just one. Be careful trying this type of testing: having to tell the boss or owner that you now have two systems down instead of one or having to explain why you need three of a part, two in the original machine and one in a second machine, is rarely a pleasant conversation.

A new problem after repairs is one that I personally hate. There are several explanations for this, and none are particularly pleasant. The least of the evils is that the system had multiple failures and you have fixed some, uncovering other problems with the system that are as of yet unaddressed. If this seems logically possible, troubleshoot it like a new problem, but keep in mind there could be other causes behind the issue and you may need to backtrack. A good logical test here is to consider whether the problem could have been caused by the same fault. An example of this is when a power surge burns out the fuses in a drive and damages one of the control cards. When you fix the first card, you may discover a second damaged card. Another common multiple failure is the loss of home position for the axes of the robot with the robot powered down and something interrupts battery backup power.

Next step down the chain of evils is that you damaged something on the robot while repairing another portion. No one wants to believe that he or she has damaged the robot while trying to fix it, but it does happen. The logic test here is whether the new error involves a system that you could have damaged during repairs. For instance, if you changed the tooling on the robot and the only power source involved was pneumatic, then there is very little chance you shorted out a board in the main controller. On the other hand, it is quite easy to damage signal cables or other power cables while installing a motor, especially if the signal cables are fiber optic (literally made of glass) and run near the motor.

In my opinion, the worst-case scenario is when you have misdiagnosed the robot and create a completely new set of problems while fixing what you believed was the problem. Yes, we mentioned this already in the "made it worse" discussion, but it bears a deeper look. The problem here is mistaking improper troubleshooting conclusions for a multiple failure situation, and that can lead to a long road of improper logic. To avoid this, be vigilant for elements of the original problems that are still present. The mistaken repair may mask some but usually not all of the core problems. If the new error seems totally unrelated to the original problem, but you have some essence of the original problem as well, that could be the clue you are looking for. This is a good time to get a second pair of eyes on the problem or to talk with someone else who is familiar with repair and the system in question. If you have a nagging feeling that you are on the wrong track, retrace your actions and see if you would do the same thing again. If all else fails, return the robot to the state it was in when you first started and see what happens.

For instance, let us start with an original problem of the robot not reaching a specific spot in the work envelop due to a singularity alarm. You know other

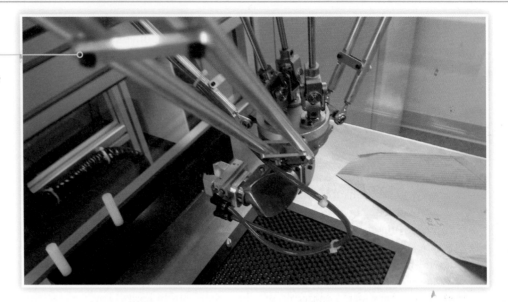

Figure 11-11   This image shows the three minor axes that manipulate the gripper. The three small diameter shafts are rotated by motors in the top of the robot and these in turn operate the gearing that moves the tooling.

robots of the same make and model have no problem reaching this spot in their work envelope. To correct the issue we reset the soft limits, allowing the robot to move further in the work envelope. When we head for the new spot this time, we still get the singularity alarm as well as hitting a hard limit of the system. We now have a new problem as well as the same old problem, indicating that resetting the soft limits was not the way to go. We would want to reset the soft limits back to the original settings and try something new. This exact troubleshooting scenario happened in my robotics class when my students tried to correct the problem by resetting the soft limits. In the end, we re-mastered the minor axes of the robot, the singularity issue went away, and the students got their first taste of trying something and creating new problems on top of the old (see Figure 11-11).

Experience is the best teacher for these various tough situations. If you feel as if you are in over your head, ask for help from those you work with or call the manufacturer's technical hotline. I had an issue with the FANUC training robots that we have at my college; it took about six weeks and three different technicians before we got to the root of the issue and corrected the problem. Some problems are just tough to solve because of all the possible things it could be, the various systems involved, and, in this case, we had to work around shipping schedules and the time I could devote to the repairs. When you come across these tough situations, make sure you document what you discover to help yourself and others in the future.

The Robot Is Running, Now What?

Once we have confirmed that the robot is running and all systems are go, we are ready to walk away and tackle the next challenge, right? If there is a pressing need or problem, then yes, we can address it at this time, but eventually we will need to take care of a few things. If you push the tasks below back to a later time, make sure you put something in place so that you do not forget to come back to them. The main tasks to keep in mind are the following:

- Put everything back in its place.

- Finish the paperwork.

- Deal with the parts used.

Some of the specifics for these tasks depend on the systems involved and where you perform maintenance. If you are repairing someone's telepresence system, you likely have a different set of requirements than the technician working in the industrial setting, so make sure you know the particulars of your situation for the follow up tasks.

Put everything back in its place. You have probably heard "put that back where you got it" many times over the years, as this is a common phrase used by bosses and parents alike. Often times, the special equipment needed for specific jobs is shared by all of those performing maintenance, so it is paramount that the tools are stored where those who need them can find them. If you use a specialty tool, but fail to return it, you can expect the next person who needs it to be rather upset with you. Putting tools and equipment where they belong also saves time the next time you need them to perform your job. If you have ever spent half an hour looking for a screwdriver to do 30 seconds of work, then you know what I mean. Everything in its own designated place is also a big part of the lean manufacturing model that many companies follow these days.

Finish the paperwork. Many times when you finish a repair, you need to fill out some related paperwork. This may be as simple as writing down how long you worked on the machine or as complex as a detailed write-up of what was wrong and how you fixed it. You may need to turn a completed checklist in to your boss or add data to a computer program that tracks the equipment. This is also a good time to update your technician's journal for future reference. It is much better to do this while everything is fresh in your mind instead of waiting a week or two, when new problems and stresses can erase the memories of what you did. If you are working for yourself, you may need to create a bill for the customer and other accounting-type activities to ensure compensation and to keep your business running smoothly. If any of the parts you replaced are under warranty, then you can expect some paperwork involved with that too. It really is true; the job is never truly finished until all the paperwork is done.

Deal with the parts used. Any parts used during the repairs will require your attention. Some you may have to turn in for repair by the manufacturer or return for warranty claim. If the part is beyond repair and not a required return for core costs, then make sure you dispose of it properly. Where applicable, recycle the parts to reduce waste and offset some of the cost of new parts. If you junked out the old part, then there is a good chance you need to order a replacement. Doing this after completing the repair is the best policy to ensure that you will have the right parts on hand for the next repair. Some companies have a parts room or other such entity that is responsible for this; in those instances, you are off the hook as long as you follow the proper in-house procedures.

You may have other things to take care of once the robot is running, depending again on where you work or the robot you are working on. Make sure that you understand what falls under your responsibility when maintenance is completed. I also recommend swinging by the robot later on to check operation and ensure that everything is running correctly. This is a great way to avoid damage to the system if you did not take care of the root cause or when the issue damaged other systems, but not fatally. Talk with the person running the robot and see if he or she noticed anything unusual about its operation or behavior. The person(s) who work with the system day in and day out will notice something odd more quickly than someone who only works with the system on occasion. These simple steps can pay great dividends down the road when you need to work on it again.

review

There are many other areas we could drill down into, but this chapter should give you a good overview of what maintenance entails and some things to watch for. Nothing compares to time in the field, but you must have a basic understanding before you start. I encourage those of you who are interested in repairing and building robots to delve deeper into the maintenance field and all the various subjects that entails. As we explored repairing the robot, we hit on the following topics:

- **Preventative maintenance.** This is all about the maintenance tasks we perform to prevent the robot from breaking down and increasing operational time.

- **Precautions to take before you begin repairs.** Here we covered LOTO and other important tasks to get ready to fix a robot.

- **Repair tips.** Here I shared some tips and tricks I have picked up over my years working in various maintenance roles.

- **Part swapping versus fixing the problem.** This section of the chapter discussed the difference between changing parts and truly fixing the machine.

- **Precautions before running the robot.** This covered what to do before and after powering the robot up.

- **What to do if the robot is still broken.** These tips help determine what to do if problems persist and ways to filter the information gained.

- **The robot is running, now what?** This was about all the finishing details that complete a repair.

key terms

Accumulators	Lockout/tagout (LOTO)	Reseat
Lockouts	Preventative maintenance (PM)	Zero energy state

review questions

1. What are the common time frames for preventative maintenance tasks?

2. How do we determine when components need preventative maintenance?

3. List some of the common preventative maintenance tasks for electrical systems.

4. List some of the common preventative maintenance tasks for hydraulic systems.

5. List some of the common PM tasks for pneumatic systems.

6. What are the four arguments that you can make in favor of doing preventative maintenance?

7. How do we get a piece of equipment to a zero energy state?

8. List the step-by-step process for lockout.

9. What are some of the dangers of stopping mid-repair to wait on parts?

10. What can you do to help repairs or preventative maintenance go smoothly once you are ready to start?

11. What is one of the major problems with parts swapping versus troubleshooting?

12. What should you do if you do not track down the root cause of the problem?

13. What is the checklist of tasks once you have finished repairs?

14. Once you have powered up the system after a repair, what are some simple tests that you can perform to check operation?

15. What are the four questions to ask if the system is not fixed after repairs?

16. What is the best fix for misdiagnoses of the problem that causes new problems and how do you know if you have a worst-case scenario?

17. What are a common and an uncommon cause of no discernible change in the system after repairs?

18. What are the possible reasons behind a new problem after repairs?

19. What are the three things to remember once we fix the system and confirm proper operation? Give an example of each.

Justifying the Use of a Robot

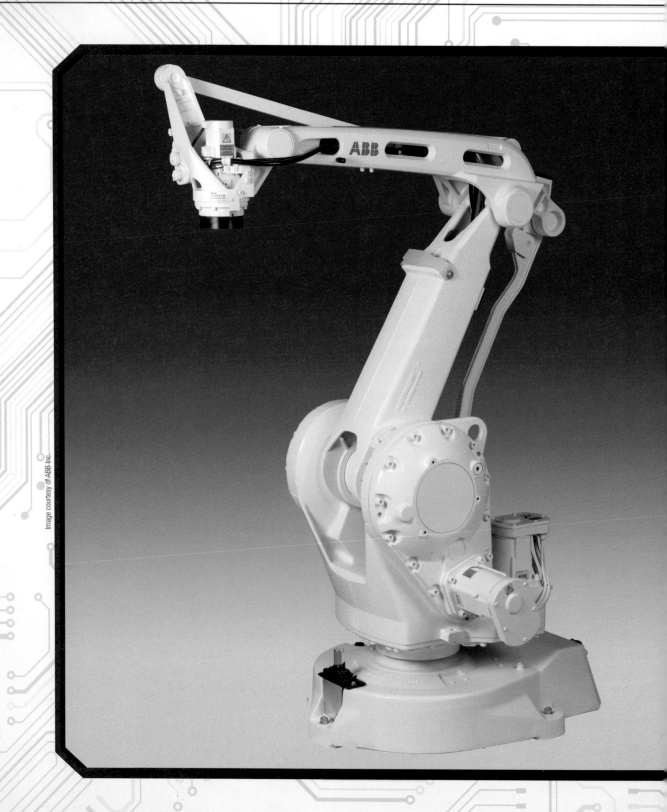

Image courtesy of ABB Inc.

overview

We have explored the historical events that gave us the modern robot, some of the reasons why we use robots, how to work with them safely, the parts and systems that make up the modern robot, the basics of programming and operation, and how to troubleshoot and repair the robot. To complete our exploration, we will examine the justifications, or reasonable reasons, for using robots. The justifications are as varied as the types of robots available and the areas in which we use them. With this in mind, we will look at some of the broad justifications used over the years to validate the use of a robot. During our exploration, we will cover the following topics:

- Robot versus human labor
- Return on investment (ROI)
- Precision and quality
- Use of consumables
- Hazardous environments
- Nonindustrial justifications

Robot versus Human Labor

In Chapter 1, we talked about the four Ds of robotics. These provide a great place to start when justifying the use of a robot. Robots are the perfect answer to many of the dull, dirty, difficult, or dangerous tasks people perform every day. If we damage the robot, we can get replacement parts, while a person who has been harmed requires medical care. A robot does not feel pain, it only knows what its sensors tell it, and even then, the information is purely data. When a person receives damage, there is pain and suffering involved, not to mention the possibility of scaring, surgery, rehabilitation, and permanent loss of mobility or function. Robots do not have emotions, so there is no fear of doing a dangerous job, and they do not get bored by dull tasks or frustrated by difficult tasks. Robots simply continue to perform the task until told to stop or something breaks, whereas a person may become feed up over time and quit.

> **justifications**
> Reasonable reasons for using robots

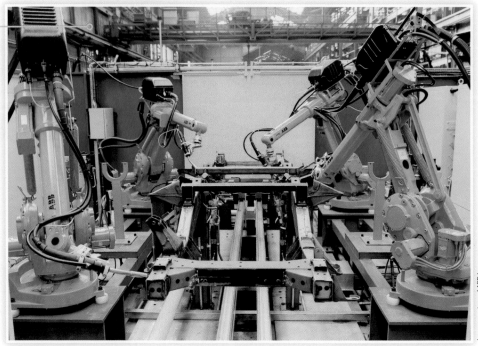

figure 12-1 Repetitive jobs, such as those found on assembly lines, are perfect tasks for robotic systems.

The same lack of emotion that gives robots a leg up in the four Ds also helps to justify their use in other situations as well. Robots are methodically constant where we mere mortals are not. A robot performs its tasks in the same manner until something mechanical changes or we modify the program; this consistency translates into efficiency in production as well as improved quality (see Figure 12-1). On the other hand, humans have emotions, and as much as we like to deny it, our emotions do color our actions. We have Monday mornings where we would rather be anywhere but work. We have Friday afternoons where all we can think about is the coming weekend. We get sick enough to be off our game, but not sick enough to stay home. If we are unhappy because the boss chewed us out or because something is going on in our personal lives, our heart is just not in our work. If we are excited about something, we tend to let our thoughts wander or find excuses to leave our post and tell others about our big news. In other words, our own emotions work against us, creating distractions that influence the quality and quantity of our work. While we are off our game, the robot is steadily doing its job and being productive.

Wages also come into play when we talk about using a robot rather than a person. I have seen statistics that show that a midsized robot costs about 72 cents an hour to run when electricity, maintenance, programming, etc., are taken into account (see Figure 12-2). At the writing of this book, the minimum wage in America is $7.25 an hour. In other words, it costs an employer the same to run robot for 10 hours or pay a worker for 1 hour. Factor in that most factories pay well over minimum wage and add in the cost of the various benefits employees receive, and you can see the cost savings. In many cases, this type of savings is what allows a manufacturer in the United States to compete with manufacturers in such countries as Mexico, Africa, China, and others who have a large amount of workers willing to work for low pay. Keep in mind that the 72 cents an hour mentioned above is the operational cost and does not include the initial cost of the system or the cost of damaged parts. We will talk more about initial cost in the return on investment (ROI) section of this chapter.

While talking about the cost of human labor versus running the robot, I want to take a moment to address the concern of lost jobs for human workers. Over the years, I have heard workers, students, employers, and others debate and talk about how the use of robots leads to fewer jobs in the industrial world. At times when jobs are tough to find, such as during the recession of 2007–2009 and the subsequent recovery, discussions about the use of automation happen frequently and can get rather heated. When we talk about this in my class, I make the following points.

- **First, if the business is not profitable, it will not stay in business**. Yes, the use of robots may eliminate some positions at a company and lead to fewer total people employed, but this is better than the business closing its doors and employing no one. Moreover, the use of automation and robotics may mean the difference between a company staying in the United States or moving operations to another country.

- **Second, every robot needs a support staff**. At the very least, every robot in industry needs someone to run it, someone to program it, and someone to fix it. Usually three different people complete these three jobs. If a company has a large number of robots, it is a safe bet that they will need multiples of all three (see Figure 12-3).

- **Third, many tasks are better suited to humans than robots**. Robots are great at black and white tasks, but still struggle with the gray areas of decision making. It is these gray areas that people excel in and why we still need people in the automated

Image courtesy Yaskawa America, Inc. Motoman Robotics Division

Figure 12-2   A robot's consistency often outperforms even the fastest human workers due to the robot's ability to work 24 hours a day, seven days a week at the same pace.

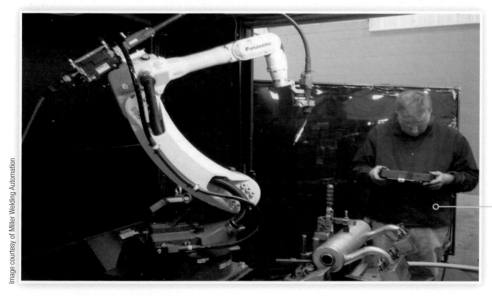

Image courtesy of Miller Welding Automation

Figure 12-3   People are still a vital part of the robot equation, doing the tasks the robot cannot.

factory. The robot can load the machine while the operator tweaks programs, checks quality, double-checks order quantities, or any of the other thinking tasks necessary in modern industry. Someday we may figure out a way to write programs that deal with the gray areas in a better way, but until then, we need the decision-making capacity of humans.

As we continue to advance the robot, we will find new and exciting ways for it to perform tasks that people would rather not. As we free the operator from the mundane tasks of industry, we allow him or her the time needed to perform other functions such as monitoring quality, adjusting programs, and figuring out better ways to accomplish various goals. Moreover, the more we use robots in the industrial world, the greater the number of good paying technical jobs for people interested in robots, like you!

Return on Investment (ROI)

return on investment (ROI)
A measure of the amount of time it takes a piece of equipment to pay for itself

Return on investment (ROI) is a measure of the amount of time that it takes a piece of equipment to pay for itself. We calculate this by taking the operating cost of the equipment and subtracting it from the amount of money it saves. Once we know the amount of profit or money made using the machine, we divide the total cost of the equipment by this number to determine how long it takes to pay for itself. If we figure it on a part-by-part basis, we would determine how many parts it needs to make to pay for itself and then turn that into a time scale using some correlation between parts produced and the time needed to produce them. If we are working on a cost savings versus human labor, then we simply figure the per-hour savings and use that to get the total hours needed to pay for the robot. Once we have that figure, we simply divide by the number of hours it runs per year or week to figure out the time for payback. The following examples illustrate how to calculate ROI.

Example 12-1

For this example, we will figure out the ROI for a robot that costs $107,500 new, which includes the cost of shipping and installation. For a cost savings, we will use $0.50 per part after subtracting the operating cost of the robot, at a rate of 50 parts per hour. Lastly, we will use 2,040 hours per year, as this is the standard 40-hour week times 51 weeks. Remember, the robot does not get sick days, paid vacations, or holidays, but we do factor in a week of downtime due to the facility closed for holidays.

FIRST:
Take the per-part cost savings of $0.50 and divide this into the total cost.

$$\frac{\$107{,}500}{\$0.50} = 215{,}000 \text{ parts}$$

NEXT:
Take the total number of parts needed to pay for the robot and divide this by the rate of 50 per hour to determine the total number of hours needed.

$$\frac{215{,}000}{50} = 4{,}300 \text{ hours}$$

LASTLY:

Take the 4,300 hours and divide this by the 2,040 hours for a year that we figured earlier.

$$\frac{4,300 \text{ hr}}{2,040 \text{ hr}} = 2.11 \text{ years, or just over 2 years}$$

Example 12-2

Let us look at another example. This time, we will use the same data, but instead of a per-part cost savings, we will figure it purely against a person's wages. The robot still costs $107,500 new, which includes the cost of shipping and installation. For a cost savings, we will use the $0.72 cents per-hour operating cost above versus the hourly wage of our fictional worker at $19.75 per hour when benefits and wages are added together. Lastly, we will use 2,000 hours per year, as this is the standard 40-hour week times 50 weeks (since our worker gets a week off each year for vacation and all the holidays mentioned previously). I want to make it a straight comparison for this example and not give the robot an extra week of work against its human counterpart, though this would be a week of extra profit as the robot is producing parts for that week while the company pays the worker who is *not* producing parts.

FIRST:

We take the $19.75 hourly wage of the human worker and subtract from that the $0.72 cost of hourly production for the robot.

$19.75 − $0.72 = $19.03

NEXT:

Take the total cost of $107,500 and divide this by $19.03 to see how many hours the robot must run to pay for itself.

$$\frac{\$107,500}{\$19.03} = 5,648.975$$

We will round this up to 5,649 hours so we are working with whole hours.

LASTLY:

We take the 5,649 hours and divide this by our 2,000-hour working year for an employee at this company.

$$\frac{5,649 \text{ hrs}}{2,000 \text{ hrs}} = 2.82 \text{ years, or almost three years}$$

Keep in mind that these are rough figures and do not factor in real work events, such as machine downtime for preventative maintenance, unexpected plant closures, programming, or breakdowns. Any parts used to maintain or repair the robot would lengthen the amount of time for payback, as this would be an added cost. If anything changed over this time, such as the cost of electricity or having to buy new tooling, that would also factor in. In reality, it would likely take longer than our math shows in both cases when you add these factors in, but for a general idea of the amount of time until the robot is profitable, these figures are sufficient. (See Figures 12-4 through 12-7 for various examples of robots with their initial costs listed.)

figure 12-4   NAO robot: Initial cost about $8,000, which is much cheaper than the $15,000 price tag of the earlier models.

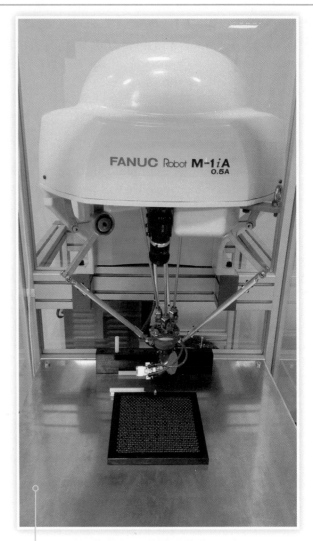

figure 12-5   FANUC M-1iA training system: initial cost about $33,000.

Another factor in ROI is how long the robot will run after it pays for itself. Some systems come with a designated useful life from the manufacturer that tells you how long you can expect the robot to run before it should be replaced. This is just a recommendation. There are numerous accounts (like the ABB robot that we looked at in Chapter 1 that has been running for almost 40 years now) of robots running long past the time stipulated by the manufacturer. The longer a system runs after it has paid for itself, the more money it makes for the company. Because of this, most companies want their equipment to have a ROI time of two years or less. In this way, they can get the most out of the equipment before they must replace it or make large investments in repairs and preventative maintenance.

In some cases, the cost of the system far outweighs the return on investment. We have a few options in those cases. One, we forget the robot and continue to do it the way it is currently done. Two, we can try to find a cheaper robot that will perform the task. Or three, we figure out a new way to use the robot. Remember, robots work in the same manner over and over again with little human input. This means that with some thought and preparation, we can set up the system in such a manner that it can run all night while the human workers are away. Perhaps

Figure 12-6 Panasonic-Miller welding robot PA55: initial cost about $55,000.

Figure 12-7 Panasonic-Miller PA 102S, initial cost: over $100,000.

there is a way that the robot could do several jobs instead of just one or two, thus increasing the amount of savings the system creates. In the next two sections, we will cover some of the other areas that figure into the cost savings of a robot and thus more ways to help with the ROI.

Here's the simple truth: If we do not find a way to justify the cost of a robot, there is a very small chance of industry using it. Companies are in business to make money, and things that cost money with no tangible return are rarely funded. You can argue until you are blue in the face about the value of something, but until you can show the math, you are often just wasting time. If the ROI is more about saving injury to people, make sure to figure out a way to show this mathematically, in cold hard monetary facts, in addition to arguing that it is the right thing to do. This approach gives you the best chance for justifying the purchase of a robot.

Precision and Quality

As discussed in Chapter 1, robots have the ability to reach an exact point repeatedly with only the smallest of error, often within 0.0003 in. to 0.005 in. This is a level of precision that is difficult for a human to match once, much less each time it is required. For jobs requiring this level of precision, such as placing micro-sized components on an integrated circuit board or precision laser fabrication for aerospace, we need either a dedicated machine or a robot that is up to the task. When you add the robot's speed to this level of precision, it is clear why some tasks are better suited to a robot than a person.

The fact that a robot has this level of repeatability also translates to the quality of the work. Since it can follow the same path repeatedly with only the smallest amount of error, all you need is to edit the program until you get the quality of part you want. Then you can let the robot run (see Figure 12-8). People can and do produce quality parts all the time, but there is a greater risk that they will do something different from time to time that could compromise quality. When we catch bad parts inside the facility, it equates to lost time and materials. When bad parts make it to the customer, it can hurt future sales, as the customer shares that experience. Imagine what you would have to say about the manufacturer if you bought a new TV or cell phone and turned it on for the first time only to discover it would not work. It does not take too many of these types of stories spread by customers to decrease the sales of a product. This is why the quality work of the robot is so attractive to industry.

To go along with this train of thought, robots give operators with low skill levels or experience the ability to produce parts that require a great amount of precision and technical skill. For instance, if someone with 20 years in the welding field creates a program to weld parts the same way they would do it by hand, then anyone who knows how to run the robot can create parts with proper welds. This

Figure 12-8 Welding thick metals, or heavy plate as it is called in the industry, requires a great deal of skill and experience when done by hand. When we program a robot to perform welding, we give operators who are new to heavy plate the chance to make parts with quality welds while they learn all the specifics of this type of welding.

creates situations where companies need only a few workers with the right years of experience and training to program the robots while operators with a lower level of training run them. In a time when those with years of experience are retiring, this has become the operating philosophy at more than one company. They can train the operators in the basics of running the robot and quality inspection in a week or two and have quality parts flowing down the line. This leaves their experienced workers free to write programs, make changes to programs as needed, and perform other tasks to improve the overall quality of the products.

Precision and quality savings figure into the overall ROI of a robot. Each time a company scraps out a part, it costs money. Each time a bad product leaves the factory, there is a good chance there will be a dip in customer satisfaction and potential sales. Lack of precision requires larger spaces for various components and raises the overall cost of the product. Therefore, by increasing the precision of processes and part placements while ensuring that these actions occur in the same manner each time, the company reaps a cost savings that figures into the overall cost per-part savings of the robot. With the improvements in robots and their precision levels, some companies are replacing worn-out dedicated machines with precision robotic system and saving money from day one, as the robot is cheaper than the cost of a new dedicated machine. As long as the length of service is comparable, there really is no need to worry about the ROI in this instance.

Use of Consumables

In addition to quality and precision, a robot often saves on the amount of raw materials used in the production process (see Figure 12-9). The same factors that allow the robot to increase quality and precision in production also tend to optimize the use of materials, thus minimizing waste. In many applications, subtle changes in the speed of the process translate into savings in dollars or cost of materials used. For example, when a person paints a car, he or she usually uses some form of spray device and sweeping motions to make sure that the entire vehicle is covered. If the person moves the sprayer too fast, the paint may be thin in certain areas and not stand the test of time. If he or she moves too slowly, the end result is a thicker coat, which wastes paint and could lead to cracking or other problems down the road. If we use a robot in the application, we can program a constant motion rate that equates to the optimal paint thickness, saving on paint used while increasing the overall quality.

Besides tweaking the program and thereby optimizing any materials used in the process, robots can work with materials that are hazardous to humans. Many chemicals and materials help to save costs in the manufacturing process but cause harm or death to humans. A human worker who

Figure 12-9 A FANUC robot hard at work painting parts as they pass by on the conveyor. Notice how the parts get a nice even coat while the conveyor system beneath is free from overspray or waste.

personal protective equipment (PPE)
The various items worn on the body to negate the hazards of a work task or area

will interact with these materials often requires personal protective equipment (PPE), which amounts to various items worn on the body to mitigate the hazards of a work task or area. (Such items include safety goggles, chemical-resistant gloves, special clothing, and other safety equipment.) The PPE has to protect the worker from the hazards of the task, can be less than comfortable for the wearer, and represents an added cost in the manufacturing process. In addition, if the PPE fails, the worker will be exposed to a hazard of some kind. This adds up to added risk and expenses above and beyond the cost of the materials used. Since a robot is a machine and not a person, these materials often present no added danger to the robot. In the cases where a chemical or material could damage the robot, often a simple change in the construction of the robot is all it takes to remove the hazard. This is another dangerous tasks that robots do for us, but it can also be a cost savings for the company.

The precision use of consumables was the driving force behind the first industrial robot, built by the DeVilbiss Company in 1941 (see Chapter 1). To this day, industries continue to look for ways to use an ounce less here or a coat less there, increasing the overall cost savings by reducing the amount of materials used. Every cost savings is another plus to the ROI.

Hazardous Environments

One of the great benefits that robots have over people is that they are machines and not organic systems. This gives the robot the ability to not only survive, but continue to work without issue in environments that are physically taxing, dangerous, or deadly to their human counterparts. Sometimes this is all the justification we need for using a robot, especially in situations where the ROI math does not add up. Even if we overlook the moral obligations, there are expenses associated with injury or death of employees. On-the-job injuries can cost a company thousands if not millions of dollars, so avoiding even one of these events can mean the difference between a company making a profit or going bankrupt. Even in cases where the medical expenses are not that high, the employer still has to find someone to do the work while the injured employee recovers. In addition, the injury could make the companies workers' compensation insurance premiums go up. This all equates to added expense. While this is not a common factor of the ROI math, some cases warrant the addition of these figures in the calculations (see Figure 12-10).

In instances where worker death is highly probable, there should be no debate. Does the company hire someone knowing that there is a fair chance that the person could die performing the work, or do they purchase a robot that can be replaced? The obvious answer is to get a robot! Even if the management at the company had no concern for human life, the cold hard facts about cost should make the decision clear. A death on company property, due to working conditions, immediately means an inspection by the Occupational Safety and Health Administration (OSHA). OSHA's job is to ensure that everyone has a safe and healthy working environment. If someone dies on the job due to work-related factors, OSHA immediately inspects the facility and investigates the death to figure out why it happened and to prevent anyone else from being hurt or killed. If OSHA were to shut a plant down until corrective actions had been completed, that alone could cost the company thousands of dollars in fines, not to mention lost production time. Then there is the possibility of a wrongful death lawsuit by the victim's family, which is another large expense. Insurance rates for the company are likely to increase, which is more money out of the company's pocket. Moreover,

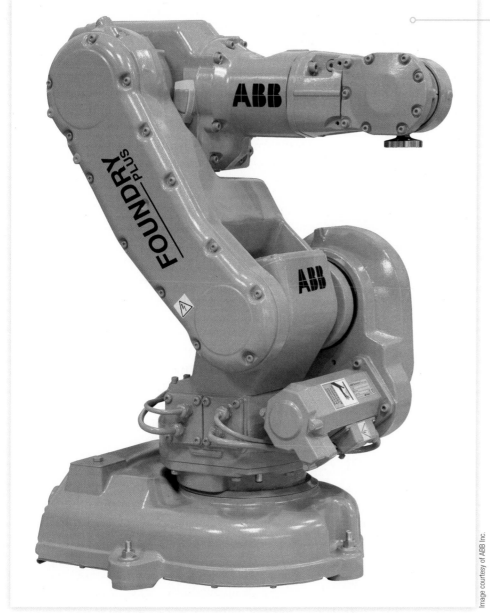

Figure 12-10 Foundry operations often have multiple areas where human habitation is hazardous, if not downright lethal. The use of robots in these areas makes good sense!

we cannot forget the impact that such an incident would have on the morale of the workers at the facility, which has costs all its own. Low moral generally decreases the production rates of workers, reduces the quality of parts produced, and leads to increased turnover (employees leaving the company), which necessitates the hiring and training of new workers. All of these factors should convince even the hardest-hearted company that no job is worth the worker's life when a robot can take on the risks instead.

Robots can go into some hazardous environments that people simply cannot. With our current technology and equipment, it is impossible to send a person below a set depth in the ocean, as the pressure would crush him or her. To gather information or do research at these depths, we need a machine. Without protective equipment and oxygen, humans cannot live in space, so deep space missions require a robot. Current technology can only shield people from certain levels of radiation, and even then, for only for a finite amount of time. We have miles of buried pipeline that are too small for adults to fit through that require regular

turnover

A measurement of employees leaving the company

inspection. In all these instances and many more, the only viable option is to use a robot to perform the tasks that we cannot.

No matter why it is hazardous to humans, it generally makes good business sense to use a robot. With the maturing of robot technology and the advancement in sensors over the last decade, the old arguments of "A robot can't do the job as well as a person" has gone out the window in most applications. While there is a cost associated with robot repair and replacement, it pales in comparison to the pain and suffering of humans performing the same task. It is this reasoning that has put robots into space, under the sea, cleaning Chernobyl, destroying buildings, inspecting pipes, and performing a multitude of tasks in industry.

Nonindustrial Justifications

When we look outside of the factory walls and industrial-grade applications of robotics, we find many of the same justifications hold true. The specifics depend on the application of the robot and the person or organization funding the robot. Sometimes there is no better reason than we simply want a robot. This has been the driving force behind many of the toys out there as well as many of the robots built in the hobby realm. Sometimes we simply want to see what we can do with a robot and thus use it for experimentation. This is the justification for many of the systems developed at various universities as well as the robotics created by enthusiasts the world over. Sometimes robots fit a specific need outside of industry; thus, this provides their justification in the world. Let us look at some of the broad categories of robots used outside of industry and their justifications.

Entertainment

When used for entertainment, the only real justification that a robot needs is to bring the user joy. Consider the Robosapiens, LEGO NXTs, and other systems designed to spark the imagination while providing hours of fun (see Figure 12-11).

Figure 12-11 Here is my personal collection from WowWee. Some will become part of future experiments of my own, but most will remain as they are for simple operational enjoyment.

The only justification needed for these systems is that the buyer wants one and has enough cash to make that happen. The ROI comes from the hours of entertainment, fiddling, building, programming, and whatever else comes to the mind of the user. The best part is that should the owner tire of the robot, he or she can always try to sell it to someone else interested in robots. If you do not believe me, do a quick search for WowWee robots on eBay.

If we look at Hollywood and the various theme parks around the United States, we find robots wowing the masses and bringing surreal situations to life. Hollywood has used robots for years now to play roles in movies as well as in the actual process of filming (see Figure 12-12). The expensive camera shot once performed by helicopter is now the job of the quadcopter, with a gyroscopic rig to keep the camera steady. Robotics bring dinosaurs to life as well as scare audiences with creatures and monsters, creating a realistic look and "feel" that is hard to duplicate with computer graphics. Robots also draw crowds at conventions and expos so that companies can get their message out to the masses in an effort to boost sales. The ROI here is the money the companies rake in when people pay for the entertainment provided by these systems or buy a product after listening to the robotic sales pitch. As long as people are interested, robots will continue to have a place in entertainment.

Service

When it comes to service robots, there are three general categories: general task, personal care, and risk mitigation. Service robots perform tasks that people cannot or would rather not do for themselves. In these instances, the robot takes its human master's place, with all the responsibilities that entails. A decade ago, this was a nearly nonexistent field outside of industry, but today it is growing and thriving in a manner that rivals the early years of industrial robotic applications.

General task service robots cover everything from iRobot's Roomba, which sweeps floors, to the Lawnbot, which keeps the grass in check. These robots perform a specific task or chore, and the time they save their human owners is often the justification for their use. Many people would happily pay the cost of a decent vacuum cleaner to purchase a robot that performs the task automatically. As the

service robots
Robots that perform tasks that people cannot do for themselves or would rather not perform

Figure 12-12   A great example of an industrial robot used in conjunction with Hollywood props to create a real-world feel.

Image courtesy of ABB Inc.

price goes down on these robots, they are finding more and more homes in which to ply their trade. Someday, homeowners may have the option of purchasing a robot that takes care of most of the mundane tasks around the home. However, for now, they must buy robots to handle one task at a time.

Personal care robotics is an exploding field right now, as roboticists find new and exciting ways to use robots and robotic exoskeletons to improve the quality of life for those with debilitating conditions. The robotic prosthetics we have discussed earlier fall in this category as well. Some of these robots retrieve items for their owners; others allow them to have a virtual presence in locations that would be otherwise off-limits. Here is a popular example of a personal care robot: A young child with a severe allergy to nuts uses a telepresence robot to interact with his classmates at school. Instead of being homeschooled in isolation, this child can experience the sights and sounds of his classroom while talking with kids his age and asking his teacher questions.

Exoskeletons are robotic systems worn over the body that enhance the wearer's stamina or, in many cases, give mobility back to those who have lost the use of their legs. ReWalk has made great strides in this field by allowing people who are wheelchair-bound to walk upright once more. There are also robotic chairs out there that put the motorized wheelchair to shame when it comes to stairs or other height obstacles. These robots improve the quality of life for the user; that is really the only justification one needs. Of course, it does not hurt if an insurance company or similar organization covers part of the cost.

Risk mitigation is just what the name implies: we send in the robot to take the risk so that people do not have to. These robots help detonate bombs, perform surveillance for law enforcement and the military, sweep the sea and land for mines, and carry out many other tasks that are hazardous to humans. The justification is to prevent injury or death to humans in industry. One thing to note is that many of these robots are mobile and require the helping hand of a human controller. To overcome the problem of line of sight, many of these systems have onboard cameras and other sensors so that the operator knows what is going on around the robot even when it is out of sight. While this extra equipment could easily seem like added cost at first, this information often proves invaluable to the people using the robot, adding another justification for this type of robot's use.

Research

Without research, we would not have the robots of today nor the advancements for the robots of tomorrow. The kind of robot or robots needed for research will depend on where the researcher(s) hope to take robotic knowledge (see Figure 12-13). To study BEAM robotics, one has to build many simple robots and see how they behave. To study

Exoskeletons
Robotic systems worn over the body that enhance the wearer's stamina or, in many cases, give mobility back to those who have lost the use of their legs

Figure 12-13 The NAO robot was specifically designed to allow educational facilities to experiment with complex humanoid robots and see where they can take the technology.

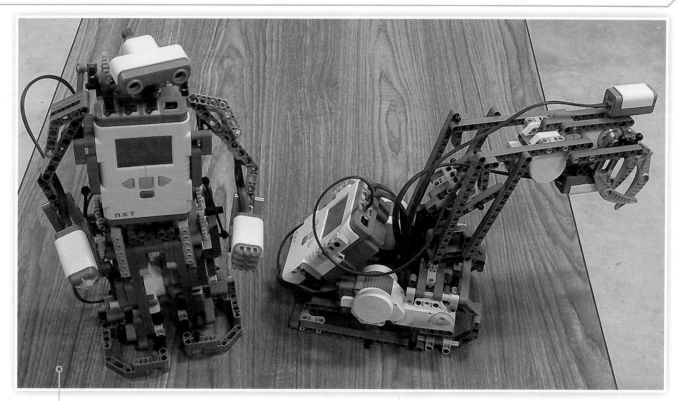

Figure 12-14   Believe it or not, you can do a large amount of engineering and mechanical experimentation with a good stock of parts for a LEGO NXT Mindstorms system. A lot of gearing and other mechanical principles go into creating these two robots. Luckily, for those using the NXT systems, there is a wealth of build instructions on the Internet to kick-start the learning process.

swarm robotics, one must first have enough robots to meet the experimental designation for a swarm. To see how humanoid-type robots could help in industry or the home, researchers must have a humanoid robot. This is how schools, colleges, and universities justify the funds they spend on everything from LEGO NXT kits to NAO or Baxter robots (see Figure 12-14). An added benefit is that a one-time purchase of robots and equipment can allow hundreds of interested students the chance to learn, explore, and expand the realm of robotics.

Sometimes the point of research is to create a system that the world has never seen before. This application of robotics can be very expensive. Someone has to buy the parts with which to build the robot—and the higher the torque or precision of the robot, the higher the part cost. If the group needs specialty parts, this could mean costs in prototyping and designing the parts as well as getting them made. Since these researchers are creating a robot, they have to either find a controller that works for their application or create that as well, which is another expense to factor in. The cost of replacement parts for anything damaged during the building or testing process also figures into the total bill. This can lead to a hefty price tag for an advanced robot built from scratch. This means that there must be a clear goal deemed worthy by those

NASA

Figure 12-15 This record-setting event of the first human and robot handshake in space may well be the first glimpse of a wonderful future of human–robot cooperation.

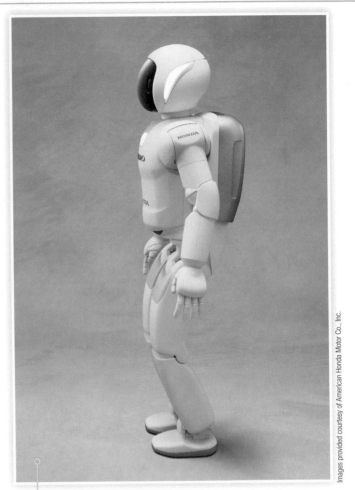

Images provided courtesy of American Honda Motor Co., Inc.

Figure 12-16 Honda has invested heavily in the development of the ASIMO robot, playing a key role in the advancement of robot motion and the robot's interaction with the world around them.

controlling the funds in order for the project to get approval. The justification for research systems is often the advancement of the robotics field, as students build this cutting-edge system during their courses. The students get real-world hands on experience building the system and the institution gets the prestige of advancing robotics.

Sometimes industry invests in the research field in hopes of reaping the rewards down the road. Two great examples of this are Honda's ASIMO (see Figure 12-16) and NASA and GM's partnership on the Robonaut (see Figure 12-15). Both of these systems are humanoid-type robots that are pushing the boundaries of robot design and operation to create systems that are more like you and me. As they improve this technology, we may well see a day where the humanoid-type robot is common in industry and our everyday lives. It is hard to predict where this technology will end up, but I personally feel that the money spent on these systems and the research they represent will pay great dividends in the robots of tomorrow.

review

The justification of a robot, no matter if it is for industry or some other application, depends on the where, what, and why for using a robot in the first place. In a perfect world, we would use robots because they are a good fit and not have to look at budgets or cost, but we live in an imperfect world. There are many places where we could use a robot but simply do not because of cost. If you find yourself presented with one of these situations, it is my sincere hope that the information from this chapter will help you to plead your case and acquire the funds needed to purchase a robot. This chapter by no means covers all the possible justifications for using robots, but you should now understand the components of justifying a robot as well as picking up a few tools to use in your justification efforts. Our exploration of justifying the use of a robot included the following topics:

- **Robot versus human labor**. We covered the differences between using a robot and using a person to perform tasks, from the justification standpoint.

- **Return on investment (ROI)**. Return on investment is all about the robot paying for itself and the math that goes along with it.

- **Precision and quality**. This section added a few more points to add to the justification toolbox.

- **Use of consumables**. Here, we talked about how robots can save money by using less of the materials needed to produce things.

- **Hazardous environments**. This section covered how the dangers associated with a job can help to justify the use of a robot.

- **Nonindustrial justifications**. Here we looked at some of the reasons for using robots outside of industry and how we justify the expense of these systems.

key terms

Exoskeletons

Justifications

Personal Protective Equipment (PPE)

Return on investment (ROI)

Service robots

Turnover

review questions

1. What are the four Ds of robotics?
2. How does the damage to a robot compare to the damage of human workers?
3. How do people and robots compare when it comes to job performance and emotions?
4. How does the cost of operating a robot compare to the cost of human labor?
5. What are three main points to remember when talking about the loss of human jobs to robots?
6. How do we figure ROI?
7. What are the options if the cost of the robot outweighs the return?
8. How does using a robot help to save on consumables such as paint or welding wire?

9. What are the downsides of PPE for human workers?
10. What are the costs associated with a worker's injury on the job?
11. What is the potential cost and impacts of a worker fatality for the company involved?
12. List some of the places robots can go that humans cannot.
13. What is the ROI for commercial entertainment systems?
14. What are the three general categories of service robots and what does each type do?
15. What are some of the costs associated with building a robot that has never been seen before?

glossary

A

Abacus   A counting device that uses beads to help the user keep track of numbers and make complex mathematical calculations easier

Absolute optical encoder   An encoder with enough emitters and receivers to give each position of the encoder its own unique binary address

Accumulators   Devices used to store hydraulic pressure and then release it back into the system as needed

Algorithm   A systematic procedure for solving a problem or accomplishing a task

Alphanumerical format   Information in the form of letters and numbers

Alternating current (AC)   Voltage where the electrons flow back and forth in the circuit

Amperage/amperes/amps   A measurement of how many electrons, or how much electricity, is flowing through a system

Amp-hours (Ah)   A measure of the number of amps a power supply can deliver over a specific time

Analog   Data with a range of values based on the sensor or device

Analog signal   A signal with a range of voltage or amperage that correlates to a set scale used to express changing values in the sensors and output devices

And   A logic function that requires two or more separate events or data states to occur before the output of the function occurs

Angular grippers   Grippers with fingers that hinge or pivot on a point to move the tips of the fingers

Anthropomorphic   The term used to describe things that move in the same manner as humans would, imparting to the motion the feeling of a living presence

Arc flash   An electrical catastrophe where two main legs of a system short out and draw enough amperage to create an electrical explosion complete with pressure waves, shrapnel, vaporized metal, and enough heat to instantly cause second- and third-degree burns

Arcs   A portion of a circle

Artificial Intelligence (AI)   Software/hardware that is capable of processing data similar to human thought, giving it the ability to deal with questions where there is no clear right answer, or where multiple answers would work as well as ultimately intuitive jumps in problem solving

Automata   Devices that work under their own power and are often designed to mimic people

Automatic mode   A condition where the robot runs the program without continued operator input based on the operating parameters of the program

Automatons   Self-operated machines, often designed to emulate something living

Axis   Each part of the robot that has controlled movement

B

Backlash   The distance from the back of the drive gear tooth to the front of the driven gear tooth

Ball screw   A large shaft with a continuous tooth carved along the outer edge with a nut or block that moves up and down the length of the shaft

Base   The portion of the robot that we attach to either the solid mounting surface or mobile unit

Battle bot   Robots used to compete in various forms of robotic combat

BCE   Before Common Era

Bevel gear   A gear that has its teeth cut along a tapered edge that would make a pointed cone if it were not flattened on the end

Binary   A base 2 number system with 0 or 1 for each position

Binary address   A unique set of 1s and 0s that the controller can understand and use

Biology Electronics Aesthetics Mechanics (BEAM)   The study of robotics involving simple systems without complex controllers

Bionic   A robotic system designed to mimic the human part it replaces that is controlled by nerve impulses, usually through the use of transmitters implanted into the user's body

Biosensor   A device that picks up the electrical activity associated with movement thru wires incorporated into the surface of the patient's skin as well as needle-sized electrodes implanted directly into the muscles

Bit   The smallest unit data can be broken into

Blunt-force trauma   An impact that does not penetrate the skin

Bottom-up approach   The robot evolution approach in which the inventor starts with a very simplistic input and control system and then sees what happens

Brainstorming   A group of people getting together to come up with a variety of ideas about a topic or problem

C

Calibration   A specified process that ensures that a precision system performs properly and provides for any adjustments needed

Call program/subroutine   A command that calls up subroutines or other programs to help reduce the lines of code in a program

Capacitive proximity switch   A switch that generates an electrostatic field; working on the same principle as capacitors, it uses the item sensed to complete the capacitive circuit as well as sense materials at various distances

Cautionary zone   The area where one is close to a robot, but still outside of the work envelope

CE   Common Era

Center of gravity   Where we consider the mass to be centered and all forces are in equilibrium

Character recognition   The ability to read written or printed letters, numbers, and symbols

Circuit   A path that electrons flow through

Circular   A motion described by no fewer than three points to create arcs or circles

Closed-loop system   A system that sends out the control pulse to initiate movement and then receives a return signal that confirms movement, often including the direction, speed, and distance moved

Common connection   Connections used to provide power or signal to contacts controlled by switches or relays

Compound gear   Two or more gears on the same shaft, often made from one solid piece of material

Computer Numerically Controlled (CNC)   The common descriptor used for machines that use computers and software to control system operation

Consistency   The ability to produce the same results or quality each time

Continuity   Unbroken electrical connection between two points

Continuous mode   A condition where the program runs as it would normally, provided that the proper inputs—such as the dead man switch and manual button or similar—are maintained

Controller   The brains of the operation and the part of the robot responsible for executing actions in a specific order and timing or under specified conditions

Conventional current flow theory   Electron flow from the positive terminal to the negative terminal

Coordinate systems   Another term for frames

Crash   The robot's unexpected contact with something in the work envelope

Critical alarms   Error conditions that prevent the machine from running until corrected

Cycle   One complete sine wave from zero to positive to zero to negative and back to zero of AC power

Cycle time   The time it takes to complete the program one time from start to finish

D

Danger zone   The area that a robot can reach and where all the robot's tasks take place

DC brushes   Items made of carbon to transfer electricity from the power wires going into the motor to the rotating portions of the motor

Dead man's switch   A switch, often found on the back of teach pendants, that must be held for the system to run in manual and stops all movement when released or pressed too hard

Decimal numbers   The base 10 numbering system that most of us are familiar with

Degree of Freedom (DOF)   Each axis of the robot, with more axes equating to more DOF and thus more ways in which the robot can move

Diametral pitch   The ratio of the number of teeth per pitch diameter that describes the size of the gear teeth

Digital   Data with a value of 0 or 1

Digital signal   A signal that has exactly two states, 1 or 0

Direct current (DC)   Voltage where the electrons flow in only one direction

Direct-drive systems   Systems that have the motor shaft connected directly to the robot for motion

Drive pulley   The pulley that is attached to the motor or force in a system

Drive system   A combination of gears, sprockets, chains, belts, shafts, and other power transmission equipment that transmits energy from a generation source, such as electric motors, to where useful work is performed

Driven pulley   The pulley attached to the output or load of a system

E

Eddy current   A flow of electrons created by the magnetic field moving across a ferrous metal item that generate their own magnetic field

Electricity   The flow of electrons from a place of excess to a place of deficit that we route through components to do work

Electromotive force (EMF)   Another name for voltage that is a measure of the potential difference between two points that causes electron flow in circuits

Electron flow theory   Electron flow from the negative terminal to the positive terminal

Emergency   A set of circumstances or a situation that requires immediate action and often involves events (or potential events) that have caused injury to people and/or severe damage to property

Encoders   Devices tied to the shaft of the motor that provide information about the motor's movement, such as direction, speed, and location in the rotation, by breaking the complete rotation of the motor into a specific number of measurable units

End   This command stops the scanning of the program and triggers the system's normal end-of-program responses

End-of-Arm-Tooling (EOAT)   Tools, devices, or equipment at the end of the robotic manipulator used to manipulate parts or perform the programmed tasks of the robot

E-stop   An emergency stop used to shut down most of the powered operations of the machine and stop all motion as quick as possible

Exclusive Or   A logic filter that sends true logic when only one of the elements is true, but not more than one on one shaft

Exoskeletons   Robotic systems that strap onto and around the user's body to enhance strength, endurance, and in some cases mobility

Expanded metal guarding   Metal that is perforated and then stretched to create diamond-shaped holes with eighth-inch pieces of metal around it used to enclose equipment while providing a clear line of site

Exploded view   Assemblies of complex systems drawn in such a fashion as to show all the parts that make up the assembly and how they fit together

External axis   Axis or axes of motion that are not a part of the main robot often used to move parts, position tooling for quick change, or in some other way help with the tasks of the robot

F

Ferrous metals   Metals that contain iron

Fingers   The metal projections of the gripper that move to hold parts; often machined for the specific application that they are used for

Fixture   A device that holds parts in place for various industrial processes by clamping or securing them in some manner

Flat belt   A belt made up of a flat band of rubber reinforced with steel or fibers that transfers power from the drive pulley to the driven pulley

Flowchart   A group of boxes with either a question or a task to perform with directions on where to go next depending on the answer or result

Frames   A way of referencing movements and points in the work envelop that controls how the robot moves

Friction   The force resisting the relative motion of two materials sliding against each other

Fringe benefits   Health and dental insurance, retirement, life insurance, Social Security, and other items that the company pays part or all of the cost on behalf of an employee

Fuse Bank   Multiple fuses concentrated in one area for ease of wiring, checking, and replacement

Fuzzy logic   Situations where there is more than one right answer or no clear-cut difference between several plans of action, where a computing system must choose one

 G

Gantry   A simple two- or three-axis machine/robot designed to pick up parts from one area and place them in another

Gantry base   A linear base with a finite reach

Gear ratio   The ratio used to determine what happens with the driven gear in reference to the drive gear in terms of torque and speed

Gear train   Two or more gears that are connected together and used to transmit power

Global function   Variables, subroutines, and other code or data accessible by any program created on the robot

Global Positioning System (GPS)   A system that determines geographical position based on the time it takes to receive signal from three or four separate satellites in orbit around Earth

Green power   Power derived from a source that is easily renewable that often has little or no negative environmental impact

Grippers   Tooling that applies some force in order to secure parts or objects for maneuvering

Ground   The wire that provides a low resistance path for electrons to flow when the electrons escape the system due to insulation or component failure

Grounded point   A point somehow connected to the earth

Guards   Devices or enclosures designed to protect us from the dangers of a system

H

Hall effect sensor   A sensor used to track rotation via interaction with a magnetic field that causes voltage flow in a semiconductor

Harmonic drives   Specialized gear systems that use an elliptical wave generator to mesh a flex spline with a circular spline that has gear teeth fixed along the interior

Helical gear   A gear similar to a spur gear that has teeth set at an angle along the edge instead of parallel

Helix   A smooth space curve

Hertz (Hz)   Sine wave cycles per second

Hexadecimal   A base 16 numbering system that uses digits 0–9 and letters A–F for each position

Hydraulic power   The use of a noncompressible liquid given velocity and then piped somewhere to do work

Hypoid bevel gear   Similar to the spiral bevel gear; however, if you draw a line from the shaft set at an angle, it will not meet the shaft of the other gear

I

Idler gear   A gear on a dedicated shaft used to change the direction of the output gear and/or to link two gears together when there is a large distance between the gears

If then   An advanced logic filter that allows you to set a complex set of conditions to occur before a desired output function happens

Impact   A robot's contact with an object in the intended movement path

Incremental optical encoder   A disk that has either holes for light to pass through or special reflectors to return light, an emitter, a receiver, and some solid-state devices for signal interpretation and transmission

Inductive proximity switch   A switch that uses an oscillating magnetic field to detect ferrous metal items and send out signals accordingly

Intermittent   A condition where a system will work fine for a while and then fault out without any clear-cut answer as to why

International Standards Organization (ISO)   An organization that develops, updates, and maintains sets of standards for use by industries of the world

J

Jaws   Another name for tooling fingers

Joint   Point-to-point motion where all the axes involved move either as fast as they can or at the speed of the slowest axis, which may result in nonlinear robot motion

Joint frame   A reference system that moves one axis at a time, with the positive and negative directions determined by the setup of each axis's zero point

Jump to   This command advances the reading of the program to the designated line, avoiding the execution of the lines skipped

Justifications   Reasonable reasons for using robots

L

Laser photo eye   A sensor that uses a concentrated beam of light known as a laser to sense the presence or absence of objects, often over large distances, by the reflection of the laser light to a receiver

Laser welders   Systems that use intense beams of light to create the high temperatures needed to join metal

Lean manufacturing   An initiative that is all about cutting production costs by minimizing wasted time and materials

Legend   Pictures of the symbols used in a drawing with details of what each symbol represents

Light curtain   A sensor that houses the emitter and receiver separately to create an infrared sensing barrier, often used to sense people in or entering a dangerous area

Limit switch   A device activated by contact with an object that changes the state of its contacts when the object exerts a certain amount of force

Linear   Motion where the controller moves all the axes involved at a set speed to ensure straight-line motion

Live   Electrically charged equipment

Local function   Data accessible by only one program

Lockout/tagout (LOTO)   The process of removing and draining all power sources from a piece of equipment to reach a zero energy state and then ensuring that they cannot be reactivated by the use of lockout devices and personalized locks

Lockouts   Devices used to hold the power source in the blocked or de-energized state

M

Macro   A single instruction or specified button used to generate a sequence of instructions or other outputs

Major axes   The axes of the robot that get the tooling and minor axes of the robot into the general area that work is performed, usually the first through third axes

Manipulator   Available in all shapes and sizes, what the robot uses to interact with and affect the world around it by activating and positioning the End-of-Arm tooling

Manual   The books that come with a piece of equipment or system that give you deeper details on how it works

Manual mode   A condition in which the operator maintains control of the robot and can stop its action as quickly as he or she can react

Mastering   The process of setting or defining the home/zero position for each moveable axis

Math functions   Commands that let you add, subtract, multiply, and divide in varying levels of complexity for data manipulation

Mentoring   Knowledge shared by those more experience with their trainees through teaching, coaching, and helping the trainees with job experiences as they happen

Mesh   The mating of gear teeth to transmit power through physical contact

Metal mesh   The common name for guarding that uses either welded or expanded metal strips with small openings, allowing visual checks, to protect people from something potentially hazardous

MIG welders   A machine that uses wire feed through the system and high-current electricity to join two pieces of metal

Minor axes   The axes of the robot that position and orient the tooling of the robot, usually the fourth through sixth axes

Miter gear   Bevel gears with equal numbers of teeth and the shafts at a 90-degree angle

Mobile base   A mounting system that gives the robot mobility options, often wheeled or tracked in nature

Motor encoder   A device that directly monitors the rotation of a motor shaft and turns that information into a meaningful signal

Muffler   A device used with pneumatic power to reduce the velocity and noise generated by venting used air

Multiple failures   When more than one component is at the core cause of a machine breakdown

N

NAnd   The opposite of the "And" command, in that all the inputs must be false before the output is true

Networking   Two or more systems sharing information over some form of connection

Neutral wire   The wire that provides a return path for the electrons and allows for a complete circuit

Noise   Mathematical difficulties in properly calculating the torque required at the start of motor motion for robot movement

Nor   A logic sorting filter in which one or all input conditions must be false before the output occurs

Normal force   The force that an object pushes back with when acted on by a force and that is necessary to prevent damage to the object

Normally closed (NC)   Contacts that allow electricity through when in their normal or de-energized state

Normally open (NO)   Contacts that do not allow electricity through when in their normal or de-energized state

Not   The opposite of the "And" command, in that all the input conditions must be false before the output is triggered

Numerically Controlled (NC)   Numerically controlled machines use punch cards or magnetic tapes with information encoded on them to control the operation of equipment

O

Occupational Safety and Health Administration (OSHA)   The federal agency charged with the enforcement of safety and health legislation in the United States

Octal   A base 8 numbering system that uses 0–7 to represent each position

Odd-shaped parts   Parts with unique shapes and proportions that are not symmetrical in nature

Open-loop system   A system that assumes that everything is working correctly because there is only limited information on a few positions or no feedback at all to confirm everything is working as directed

Or   A logic filter where at least one, or more than one, of two or more input events must be true for the output to occur

P

Parallel grippers   Tooling with fingers that move in straight lines toward the center or outside of the part

Parameters   Data that designates the type of equipment, signals, and information that the controller needs to have for proper communication and control of various devices

Paraplegic   A person who has lost all mobility in his or her lower limbs

Payload   The specification of a robotic system that informs the user how much weight the robot can safely move

Peripheral system   System or equipment that performs tasks related to or involving the robot but is not part of the robot

Personal protective equipment (PPE)   The various items worn on the body to negate the hazards of a work task or area

Photoelectric proximity switch (photo eye)   A switch that sends out a specific wavelength of light and uses a receiver to detect that specific wavelength of light when returned by a reflector or the detected item

Pick-and-place operation   The common description used for robotic operations where parts are picked up from one location and then placed or loaded into another

Pinch point   Any place where a robot could trap a person against something solid

Pitch diameter   The diameter of an imaginary circle used to design a gear

Pneumatic power   Fluid power that uses air to generate force

Polarity   The condition found in DC circuits where each component has a defined positive and negative terminal that must be maintained in the circuit for proper operation

Precision   The exact or accurate performance of a task within given quality guidelines

Presence sensors   Sensors that detect when a person is inside the danger zone that are tied into the robotic safety system to prevent automatic operation

Pressure angle   How the forces interact between two gears, at what angles, and how a gear tooth is rounded or shaped

Pressure sensor A device that detects the presence or absence of a predetermined amount of force

Preventative maintenance (PM) The practice of changing out parts and doing repairs in a scheduled fashion (based on previous equipment and component performance) before the equipment breaks down or quits working

Program A system of sorting and direction function created by the user to define the operation of equipment based on the various monitored input conditions

Programmable Logic Controller (PLC) A specialized computer with a logical program to direct equipment action based on inputs and other information filtered by the program

Programming languages The rules governing how we enter the program so that the robot controller can understand the commands

Prosthetic An artificial replacement for a missing or damaged organic part that often closely resembles the part it is replacing in form and function

Proximity switch A solid-state device that uses light, magnetic fields, or electrostatic fields to detect various items without the need for physical contact

Punch card control system A control system that uses a card made of a sturdy substance with a series of carefully placed holes that can control the sequence and operation of machines

Q

Quadriplegic A person who is paralyzed in the torso and limbs (in other words, from the neck down)

R

Rack and pinion Systems that consist of a spur gear and a rod or bar that has teeth cut along the length

Reduction drive Systems that take motor output and via mechanical means reduce or alter it

Relay Device that uses a small control voltage to generate a magnetic field and make or break connections between field devices using contacts

Relay logic Control system that uses a device known as a relay to create various logic-sorting situations, which would in turn control the operation of the system

Remote Center Compliance (RCC) A simple way for tooling to respond to parts that are not always in the same position by mechanical means of flexing to adapt

Repeatability The ability to perform the same motions within a set tolerance

Reseat The process of removing electronic cards from the controller and putting them back in to insure proper connection

Resistance The opposition to the flow of electrons in the circuit and the reason why electrical systems generate heat during normal operation

Return on investment (ROI) A measure of the amount of time it takes a piece of equipment to pay for itself

Revolutions per minute (RPM) Commonly used to measure the speed of rotating systems

Right-hand rule When facing in the same direction as the robot, your thumb will be pointing in the positive direction for the Z-axis, your forefinger in the positive direction for the X-axis, and your middle finger in the positive direction for the Y-axis, when your thumb is pointing straight up, your forefinger is pointing straight ahead and your middle finger is pointing to the left side of your body

Robot A machine equipped with various data gathering devices, processing equipment, and tools for operational flexibility and interaction with the systems environment, which is capable of carrying out complex auctions under either programmed control or direct manual control

Robot program The list of commands that run within the software of the robot controller and dictate the actions of the system based on the logic sorting routine created therein

Robota Drudgery, or slave-like labor

Root Mean Square (RMS) A mathematical average of the sine wave

RS232 A communication standard where certain wires are designated to transmit specific signals

Rungs The lines going horizontally across the page of a schematic

S

Safe zone The area where a person can pass near a robot without having to worry about making contact with the system

Safety factor The margin of error we build into a process or system to ensure that accidents or dangerous situations do not occur

Safety interlock A system where all the safety switches have to be closed or made for the equipment to run in automatic mode

Schematics Drawing of the electrical flow or logic of a system with no regard to the component placement or design in an effort to show system operation clearly

Selective Compliance Articulated Robot Arm (SCARA) Robots that blend linear Cartesian motion with articulated rotation to create a new motion type

Service robots Robots that perform tasks that people cannot do for themselves or would rather not perform

Servomotor A continuous rotation type of motor with built-in feedback devices called encoders, often designed for use with variable frequency drives

Shock   When a person becomes a part of the electrical circuit and has electrons flowing through his or her body

Short circuit   A direct path between a charged wire/component and a ground, neutral, or another powered wire/component

Show-Stoppers   Devices built for the purpose of getting people to stop and take notice, typically used to draw attention to products or booths at trade shows

Single-phase AC   AC power that has a single sine wave provided to the system via one hot wire and returned on one neutral wire

Singularity   A condition in robotics where there is no clear-cut way for the robot to move between two points that often results in unpredicted motions and operational speeds

Skew gear   Another name for helical gear

Slag   Molten metal splatter that has hardened in the welding gun tip

Slippage   Loss of rotation between the drive and driven elements, typically used to reference the loss in belt driven systems

Solid-mount base   A nonmobile base to which a robot is attached with bolts and other fasteners and from which it works

Solid-state device   A device made up of a solid piece of material that manipulates the flow of electrons without any moving parts

Spiral bevel gear   A type of bevel gear with the tooth curved to reduce noise and smooth out its operation

Sprocket   A circular metal device similar to a gear, with teeth spaced to fit inside the links of a chain and used to transmit force

Spur gear   A gear made from a round or cylindrical object with teeth cut into the edge

Staging points   Positions that get the robot close to the desired point but are a safe distance away, allowing for clearance and rapid movement

Step mode   A manual mode in which the robot executes one line of program code each time a specific button or button combination is pressed

Stepper motor   A motor that moves a set portion of the rotation with each application of power

Sub chart   A chart that takes a portion of a larger chart and breaks it into greater detail

Subroutines   A sequence of instructions grouped together to perform an action that the main program accesses for repeated use

Swarm robotics   The research field that focuses on ways to use a large number of simple robots to perform complex tasks

Synchronous belt   A belt that has teeth at set intervals that is used to transfer power without slipping

T

Tactile   The ability to sense pressure and impact

Teach pendant   A device that allows the operator to view alarms, make manual movements, stop the robot, select which program to run, change or create a program, and carry out any of the other day-to-day tasks required of those who run robots

Telepresence   The interaction with others from a remote location using technology

Territories   Areas of an environment that are favored

Third party   Systems, software, or other items designed by someone other than the manufacturer of the equipment or customer

Three-phase AC   AC power that has three sine waves that are 120 degrees apart electrically

Tooling base   The part of the tooling that attaches to the robot and holds the mechanisms for movement

Top-down approach   The robot evolution approach in which one starts with something very complex, such as people or independent thought, and then tries to create something as close to that as possible with the technology available or innovative new designs

Torque   A measure of the work potential of rotational force

Tourniquet   A tightened band that restricts arterial blood flow to wounds on the arms or legs to stop severe bleeding

Tracking the power   The process of verifying that the components that should have power do indeed have power

Transmission   A group of gears, often including compound gears and multiple gear ratios, that are used to transmit power

Troubleshooter   A skilled person who determines the cause of a fault and makes the corrections necessary to get the system or process working properly once more

Troubleshooting   The logical process of determining and correcting faults in a system or process

Tunnel vision   When a person focuses on a specific way of thinking or seeing the problem and ignores everything else

Turnover rate   How often workers quit a job or working for an employer

U

Ultrasonic sensor   A sensor system that emits sounds above the normal hearing range of humans and then determines the distance of items by the amount of time it takes for the sound to return to the receiver

User frame   A specially defined Cartesian-based system where the user defines the zero point and how the positive directions of the axes lie

V

Vacuum   Any pressure that is less than the surrounding atmospheric pressure at that location

Vacuum gripper   A device that works by creating a pressure that is less than the atmospheric pressure of 14.7 psi and is often used to pick up delicate or large flat items

Variable frequency drive (VFD)   A control system that takes AC power and converts it to pulsing DC power to control the torque and speed of motors

V-belt   Rubber belts with steel or fiber reinforcement that are shaped like a V with the point cut off that are used to transmit force

Velocity   A measure of how fast something is moving

Ventricular fibrillation   A condition where the heart quivers instead of actually pumping blood

Vision system   A system that uses cameras and software to process images and provide that information to the robot

Voltage   A measurement of the potential difference or imbalance of electrons between two points and the force that will cause electrons to flow

W

Wait   This command creates timed pauses in the program or has the robot wait for a specified set of conditions before continuing with programmed actions

Weave   Straight line or circular motion that moves from side to side in an angular fashion while the whole unit moves from one point to another

Welding guns   Tube-like tools used to direct the charged wire of welding operations as well as the shield gas used to prevent weld oxidation

Wiring diagrams   Drawings that show how the power moves through the system with components drawn to some degree as they actually appear

Work cell   A logical grouping of machines that perform various operations on parts in a logical order during the production process

Work envelope   The area that a robot can reach during operation

World frame   A Cartesian system based on a point in the work envelop where the robot base attaches

Worm gear   A cylinder that has at least one continuous tooth cut around it

X

XOr   A logic filter where the output is only activated when any one of the inputs is true, but not more than one

Z

Zero energy state   When a machine has no active or latent power

Zerol bevel gear   A gear with the same curved tooth of the spiral bevel gear, but without angled sides

bibliography

Adept Technology. *About the Company*. 1996. http://www.adept
.com/company/about (accessed May 27, 2012).

Ament, Phil. "Fascinating Facts About the Invention of the Abacus
by the Chinese in 3000 BC." *The Great Idea Finder*. March 6,
2006. http://www.ideafinder.com/history/inventions/abacus
.htm (accessed May 12, 2012).

Angle, Colin. "Gengis, a Six Legged Autonomous Walking Robot."
DSpace@MIT. March 1989. http://dspace.mit.edu/bitstream
/handle/1721.1/14531/20978065.pdf?sequence=1 (accessed
May 26, 2012).

"Animation Notes #1 What Is Animation?" *Center for Animation &
Interactive Media*. February 16, 2010. http://sec1aep.wiki.hci
.edu.sg/file/view/01+Introduction+to+Animation.pdf (accessed
May 12, 2012).

Babbit, Seward S., and Henry Aiken Crane. United States Patent
484870. June 13, 1892.

Baik, Seung Hyuk. "Robotic Colorectal Surgery." *Yonsei Medical
Journal*. December 31, 2008. http://www.ncbi.nlm.nih.gov
/pmc/articles/PMC2628019/ (accessed July 24, 2014).

Bares, John. "Dante II." *Carnegie Mellon, the Robotics Institute*. n.d.
http://www.ri.cmu.edu/research_project_detail.html?project
_id=163&menu_id=261 (accessed May 26, 2012).

BBC News. "Fish-Brained Robot at Science Museum." *BBC News*.
November 27, 2000. http://news.bbc.co.uk/2/hi/science
/nature/1043001.stm (accessed May 26, 2012).

—. "World's First 'Bionic Arm' for Scot." *BBC News*. August 25, 1998.
http://news.bbc.co.uk/2/hi/health/154545.stm (accessed May 26,
2012).

Berkeley, Edmund C. "Small Robots—Report." *Blinkenlights Archae-
ological Institute*. April 1956. http://www.blinkenlights.com
/classiccmp/berkeley/report.html (accessed May 26, 2012).

"Blaise Pascal (1623–1662)." *Educalc.net*. 2002–2012. http://www
.educalc.net/196488.page (accessed May 12, 2012).

Bonev, Ilian. "The True Origins of Parallel Robots." *ParalleMIC*.
January 24, 2003. http://www.parallemic.org/Reviews
/Review007.html (accessed May 12, 2012).

Boyle, Alan. "The Human Behind This Year's Hot Robot." *Science on
msnbc.com*. November 23, 2004. http://www.msnbc.msn.com
/id/6567169/ns/technology_and_science-science/t/human
-behind-years-hot-robot/#.T8O8xMWuWHQ (accessed
May 28, 2012).

Buckley, David. "1868 - Zadoc P. Dederick - Steam Man." *davidbuckley
.net*. August 21, 2007. http://davidbuckley.net/DB/
HistoryMakers/1868DederickSteamMan.htm (accessed
May 20, 2012).

—. "History Making Robots." *davidbuckley.net*. August 22, 2007.
http://davidbuckley.net/DB/HistoryMakers/1890EdisonTalking
Doll.htm (accessed May 20, 2012).

Canright, Shelley. "What Is Robonaut?" *NASA*. February 1, 2012.
http://www.nasa.gov/audience/forstudents/5-8/features
/what-is-robonaut-58.html (accessed May 28, 2012).

Capek, Karel. *R. U. R.* New York: Doubleday, 1890–1938.

CBSNews.com staff. "For the Person Who Has Everything… A
Personal Robot." December 10, 1992. http://www.cbsnews.com
/news/for-the-person-who-has-everything-a-personal-robot/
(accessed July 24, 2014).

"Charles Babbage." *Charlesbabbage.net*. n.d. http://www
.charlesbabbage.net/ (accessed May 12, 2012).

Chemical Heritage Foundation. "Joseph John Thomson." Chemical
Heritage Foundation. 2010. http://www.chemheritage.org
/discover/online-resources/chemistry-in-history/themes/atomic
-and-nuclear-structure/thomson.aspx (accessed May 26, 2012).

Ciampichini, Lauren. *Robotics Insights Newsletter*. May 2012.

COGNEX. *Company History*. 2012. http://www.cognex.com
/company-history.aspx (accessed May 27, 2012).

Computer History Museum. "Timeline of Computer History." *Computer History Museum*. 2006. http://www.computerhistory.org
/timeline/?category=rai (accessed May 26, 2012).

Control Engineering. "Cincinnati Milacron T3 Robot Arm."
WordPress. July 26, 2009. http://lakshmimenon.wordpress
.com/2009/07/26/cincinnati-milacron-t3-robot-arm/ (accessed
May 26, 2012).

Copeland, Jack. *Colossus: The First Electronic Computer*. Oxford
University Press, 2006.

Cruz, Frank da. "Herman Hollerith." *Columbia University
Computing History*. March 28, 2011. http://www.columbia
.edu/cu/computinghistory/hollerith.html (accessed May 12,
2012).

Dalakov, Georgi. "Friedrich von Knauss." *History-Computer.com*.
May 4, 2012. http://history-computer.com/Dreamers/Knauss
.html (accessed May 12, 2012).

—. "Juanelo Torrian." *History-Computer.com*. May 4, 2012. http:
//history-computer.com/Dreamers/Torriano.html (accessed
May 12, 2012).

—. "Jacques de Vaucanson." *History-Computer.com*. May 4, 2012.
http://history-computer.com/Dreamers/Vaucanson.html
(accessed May 12, 2012).

—. "Joseph-Marie Jacquard." *History-Computer.com*. May 4, 2012.
http://history-computer.com/Dreamers/Jacquard.html
(accessed May 12, 2012).

—. "Pierre Jaquet-Droz." *History-Computer.com*. May 12, 2012.
http://history-computer.com/Dreamers/Jaquet-Droz.html
(accessed May 12, 2012).

—. "The Robots of Westinghouse." *History-Computer.com*. May 15,
2012. http://history-computer.com/Dreamers/Elektro.html
(accessed May 20, 2012).

Doyle, Arthur Conan. *BrainyQuote.com, Xplore Inc*. 2012.
http://www.brainyquote.com/quotes/quotes/a/arthurcona134512
.html (accessed February 18, 2012).

Edsinger-Gonzales, Aaron, and Jeff Weber. "Domo: A Force Sensing
Humanoid Robot for Manipulation Research." *International
Journal of Humanoid Robotics*, 2004.

Einstein, Albert. *BrainyQuote.com, Xplore Inc.* 2012. http://www.brainyquote.com/quotes/quotes/a/alberteins133991.html (accessed February 18, 2012).

FamousPeople. "Isaac Asimov." *TheFamousPeople.com.* n.d. http://www.thefamouspeople.com/profiles/isaac-asimov-158.php (accessed May 20, 2012).

Fanuc. *FANUC's History.* 2011. http://www.fanuc.co.jp/en/profile/history/index.html (accessed May 27, 2012).

Fell-Smith, Charlotte. *John Dee (1527–1608).* London: Constable & Company LTD, 1909.

Fundazioa, Elhuyar. "Humanoid Robot Works Side by Side with People." *ScienceDaily.* May 22, 2012. http://www.sciencedaily.com/releases/2012/05/120522084322.htm (accessed May 28, 2012).

"George Boole." *Encyclopedia Britannica.* 2012. http://www.britannica.com/EBchecked/topic/73612/George-Boole (accessed May 12, 2012).

Goodreads. *Steven Wright > Quotes > Quotable Quote.* 2013. http://www.goodreads.com/quotes/141045-if-at-first-you-don-t-succeed-then-skydiving-definitely-isn-t (accessed November 9, 2013).

Guizzo, Erico, and Evan Ackerman. *How Rethink Robotics Built Its New Baxter Robot Worker.* October 2012. http://spectrum.ieee.org/robotics/industrial-robots/rethink-robotics-baxter-robot-factory-worker (accessed January 13, 2013).

Hapgood, Fred. "Living Off the Land." *Smithsonian.com.* July 2001. http://www.smithsonianmag.com/science-nature/phenom_jul01.html?c=y&page=1 (accessed May 27, 2012).

Harper, Chris. *Current Activities in International Robotics Standardisation.* 2012. http://europeanrobotics12.eu/media/15090/2_Harper_Current_Activities_in_International_Robotics_Standardisation.pdf (accessed July 14, 2012).

Hollis, Ralph. "BALLBOTS." *Scientific American,* 2006: 72–77.

Honda Motor Co. Ltd. "Asimo Technical Information." *Asimo, the World's Most Advanced Humanoid Robot.* September 2007. http://asimo.honda.com/downloads/pdf/asimo-technical-information.pdf (accessed May 27, 2012).

Hornyak, Tim. "The Face That Launched a Thousand Robots." *The Japan Times.* August 20, 2008. http://www.japantimes.co.jp/text/nc20080820a1.html (accessed May 12, 2012).

Hrynkiw, Dave, and Mark W. Tilden. *Junkbots, Bugbots & Bots on Wheels.* Osborne: McGraw-Hill, 2002.

Humanoid Robotics Institute. *WABOT - WAseda roBOT-.* n.d. www.humanoid.waseda.ac.jp/booklet/kato_2.html (accessed May 26, 2012).

Independence Hall Association. "The Electric Ben Franklin." *UShistory.org.* July 4, 1995. http://www.ushistory.org/franklin/info/kite.htm (accessed May 26, 2012).

Intuitive Surgical, Inc. *History.* http://www.intuitivesurgical.com/company/history/ (accessed May 27, 2012).

Kanade, Takeo, and Haruhiko Asada. Robotic Manipulator. United States of America Patent 4,425,818. January 17, 1984.

Kauderer, Amiko. "Historic Handshake for Robonaut 2." *NASA.* February 15, 2012. http://www.nasa.gov/mission_pages/station/main/r2_handshake.html (accessed May 28, 2012).

Kawasaki Robotics (USA). *Company History.* 2012. http://www.kawasakirobotics.com/Kawasaki-Robotics-History (accessed May 27, 2012).

Klobucar, Jack. "ReconRobotics Introduces Throwable, Mobile Reconnaissance Robot with Night Vision Capabilities." *ReconRobotics.* September 30, 2008. http://www.reconrobotics.com/contact/press_news_9-30-08.cfm (accessed June 12, 2012).

Kotler, Steve. "Man's Best Friend." *Discover Magazine.* December 1, 2005. http://discovermagazine.com/2005/dec/robot-robot (accessed May 28, 2012).

KUKA. "KUKA History." *KUKA.* December 14, 1998. http://www.kuka-robotics.com/en/pressevents/news/NN_981214_KUKA-History.htm (accessed May 28, 2012).

Lahanas, Michael. "Ctesibius of Alexandria." *Wordpress.com.* October 3, 2010. http://hydraman7.files.wordpress.com/2010/03/ctesibiusorktesibiosofalexandria.pdf (accessed May 12, 2012).

—. "Heron of Alexandria." *Mlahanas.de.* n.d. http://www.mlahanas.de/Greeks/HeronAlexandria.htm (accessed May 12, 2012).

LeBouthillier, Arthur Ed. "The Robotu Builder." *Robotics Society of Southern California,* Volume 11, Number 5. May 1999. http://www.rssc.org/sites/default/files/newsletter/may99.pdf (accessed May 20, 2012).

Ledford, Heidi. "Injured Robots Learn to Limp." *Nature.* November 16, 2006. http://www.nature.com/news/2006/061113/full/news061113-16.html#B1 (accessed May 28, 2012).

Lerner, Evan. "Penn Researchers Build First Physical 'Metatronic' Circuit." *Penn News.* February 22, 2012. http://www.upenn.edu/pennnews/news/penn-researchers-build-first-physical-metatronic-circuit (accessed May 28, 2012).

Levy, Steven B. "No Battle Plan Survives Contact with the Enemy." *Lexician.* November 1, 2010. http://lexician.com/lexblog/2010/11/no-battle-plan-survives-contact-with-the-enemy/ (accessed August 11, 2013).

Lowensohn, Josh. "Timeline: A Look Back at Kinect's History." *CNET.* February 23, 2011. http://news.cnet.com/8301-10805_3-20035039-75.html (accessed June 21, 2012).

Malone, Bob. "George Devol: A Life Devoted to Invention, and Robots." *ieee Spectrum.* September 26, 2011. http://spectrum.ieee.org/automaton/robotics/industrial-robots/george-devol-a-life-devoted-to-invention-and-robots (accessed May 20, 2012).

Marsalli, Michael. "McCullogh-Pitts Neurons." *The MIND Project.* n.d. http://www.mind.ilstu.edu/curriculum/modOverview.php?modGUI=212 (accessed May 20, 2012).

Merriam-Webster, Incorporated. "robot." *Merriam-Webster.* 2012. http://www.merriam-webster.com/dictionary/robot (accessed July 27, 2012).

Metropolis. Directed by Fritz Lang. Performed by Brigitte Helm, Gustav Frohlich, Alfred Abel, and Rudolf Klein-Rogge. 1927.

MIT news. "MIT Team Building Social Robot." *MIT News.* February 14, 2001. http://web.mit.edu/newsoffice/2001/kismet.html (accessed May 26, 2012).

Mo. "16th-Century Mechanical Artificial Hand." *Scienceblogs.com.* July 14, 2007. http://scienceblogs.com/neurophilosophy/2007/07/16th_century_mechanical_artifi.php (accessed May 12, 2012).

Mortensen, Tine F. "Timeline 1990–1999." *LEGO.* January 9, 2012. http://aboutus.lego.com/en/lego-group/the_lego_history/1990/ (accessed May 27, 2012).

Munson, George E. "The Rise and Fall of Unimation, Inc." *Robot.* December 2, 2010. http://www.botmag.com/the-rise-and-fall-of-unimation-inc-story-of-robotics-innovation-triumph-that-changed-the-world/ (accessed May 26, 2012).

Nadarajan, Gunalan. "Islamic Automation: A Reading of al-Jazari's *The Book of Knowledge of Ingenious Mechanical Devices.*" *MediaArtHistories.org.* 2007. http://www.mediaarthistory.org/wp-content/uploads/2011/04/Gunalan_Nadarajan.pdf (accessed May 12, 2012).

NASA. "The Continuing Adventures of Deep Space 1." *NASA Science News.* September 19, 2001. http://science.nasa.gov/science-news/science-at-nasa/2001/ast19sep_1/ (accessed May 26, 2012).

Needham, Joseph, and Colin A. Ronan. *The Shorter Science and Civilisation in China: An Abridgement of Josep Needham's Original Text, Volume 4.* Cambridge University Press, 1994.

Nilsson, Nils J. "Shakey the Robot." *SRI International.* April 1984. http://www.ai.sri.com/pubs/files/629.pdf (accessed May 20, 2012).

Nocks, Lisa. *The Robot: The Life Story of a Technology.* Westport: Greenwood Publishing Group, 2007.

O'Connor, J. J., and E. F. Roberston. "William Oughtred." *School of Mathematics and Statistics University of St. Andrews, Scotland.* December 1996. http://www-history.mcs.st-andrews.ac.uk /Biographies/Oughtred.html (accessed July 1, 2012).

O'Connor, J. J., and E. F. Roberston. "Gottfried Wilhelm von Leibniz." *School of Mathematics and Statistics University of St. Andrews, Scotland.* October 1998. http://www-history.mcs.st-and .ac.uk/Biographies/Leibniz.html (accessed May 12, 2012).

O'Connor, J. J., and E. F. Robertson. "Alan Mathison Turing." *School of Mathematics and Statistics University of St. Andrews, Scotland.* October 2003. http://www-groups.dcs.st-and.ac.uk /history/Biographies/Turing.html (accessed May 12, 2012).

—. "Augusta Ada King, Countess of Lovelace." *School of Mathematics and Statistics University of St. Andrews, Scotland.* August 2002. http://www.gap-system.org/~history/Biographies/Lovelace .html (accessed May 12, 2012).

—. "Wilhelm Schickard." *School of Mathematics and Statistics University of St. Andrews, Scotland.* April 2009. http://www-history .mcs.st-andrews.ac.uk/Biographies/Schickard.html (accessed May 12, 2012).

Oxford University Press. "robot." *Oxford Dictionaries.* 2012. http:// oxforddictionaries.com/definition/english/robot (accessed July 27, 2012).

ReWalk. *UCPN Offering Gait Training for Ground-Breaking ReWalk Device.* 2012. http://bionicsresearch.com/ucpn-offering-gait -training-for-ground-breaking-rewalk-device/ (accessed September 8, 2012).

Roberge, Pierre R. "Sir William Grove (1811–1896)." *Corrosion Doctors.* August 1999. http://www.corrosion-doctors.org /Biographies/GroveBio.htm (accessed May 12, 2012).

Robotic Industries Association. "Robot Terms and Definitions." *Robotics Online.* 2012. http://www.robotics.org/product -catalog-detail.cfm?productid=2953 (accessed July 27, 2012).

Robotics Trends. "Robot to Robot: Dragon and Space Station Meet Up." *Robotics Trends.* May 25, 2012. http://www.roboticstrends .com/security_defense_robotics/article/robot_to_robot_ dragon_and_space_station_meet_up/ (accessed June 21, 2012).

Rosheim, Mark Elling. *Leonardo's Lost* Heidelberg: Springer Berlin, 2006.

Saenz, Aaron. "Dairy Farms Go Robotic, Cows Have Never Been Happier (video)." *Singularity Hub.* November 16, 2010. http:// singularityhub.com/2010/11/16/dairy-farms-go-robotic- cows-have-never-been-happier-video/ (accessed June 21, 2012).

Salton, Jeff. "Nao — A Robot That Sees, Speaks, Reacts to Touch and Surfs the Web." *gizmag.* December 6, 2009. http://www.gizmag .com/nao-all-rounder-robot/13445/ (accessed May 28, 2012).

Scassellati, Brian, and Ganghua Sun. "A Fast and Efficient Model for Learning to Reach." *International Journal of Humanoid Robotics,* 2005: 391–413.

Seiko Epson Corp. "Monsieur." *Epson.* 2012. http://global.epson.com /company/corporate_history/milestone_products/23_monsieur .html (accessed May 27, 2012).

Sharlin, Harlod I. "William Thomson, Baron Kelvin." *Encyclopedia Britannica.* n.d. http://www.britannica.com/EBchecked/topic /314541/William-Thomson-Baron-Kelvin/13896/Later-life (accessed May 12, 2012).

Simkin, John. "Richard Arkwright." *Spartacus Educational.* September 1997. http://www.spartacus.schoolnet.co.uk /IRarkwright.htm (accessed May 12, 2012).

Steele, Bill. "Researchers Build a Robot That Can Reproduce." *Cornell Chronicle.* May 11, 2005. http://www.news.cornell.edu /stories/may05/selfrep.ws.html (accessed May 27, 2012).

Takanishi Laboratory. *History of Waseda Talker Series (from WT-1 to WT-7R).* February 2, 2012. http://www.takanishi.mech.waseda .ac.jp/top/research/voice/wt_series.htm (accessed May 28, 2012).

Tate, A. "Edinburgh Freddy Robot (Mid 1960s to 1981)." *The University of Edinburgh.* May 16, 2011. http://www.aiai.ed.ac.uk /project/freddy/ (accessed May 26, 2012).

Texas Instruments. "Jack Kilby." *Texas Instruments.* 1995. http:// www.ti.com/corp/docs/kilbyctr/jackstclair.shtml (accessed May 26, 2012).

"Timeline of Flight." *The Library of Congress.* July 29, 2010. http:// www.loc.gov/exhibits/treasures/wb-timeline.html (accessed May 12, 2012).

"Timeline of Robotics 1 of 2." *The History of Computing Project.* November 19, 2007. http://www.thocp.net/reference/robotics /robotics.html (accessed May 12, 2012).

Titelman, Gregory. *America's Popular Proverbs and Sayings.* New York: Random House Reference, 2000.

—. *"America's Popular Proverbs and Sayings."* New York: Random House Reference and Seaside Press, 2001.

Toshiba-Cho, Komukai. "Hisashige Tanaka." *Toshiba.* 1995. http://toshiba-mirai-kagakukan.jp/en/learn/history/toshiba _history/spirit/hisashige_tanaka/index.htm (accessed May 12, 2012).

Virk, Gurvinder S. "Robot Standardization." February 4, 2013. http://www.clawar.org/downloads/Medical_Workshop /1.%20GSVirk-Robot%20standardization.pdf (accessed July 14, 2012).

Vojovic, Ljubo. "Nikola Tesla: Father of Robotics." *Tesla Memorial Society of New York.* July 10, 1998. http://www.teslasociety.com /robotics.htm (accessed May 12, 2012).

Vujovic, Ljubo. "Tesla Biography." *Tesla Memorial Society of New York.* July 10, 1998. http://www.teslasociety.com/biography.htm (accessed May 12, 2012).

Wiederhold, Gio. "Robots and Their Arms." *Stanford University Info-lab.* 2000. http://infolab.stanford.edu/pub/voy/museum /pictures/display/1-Robot.htm (accessed May 26, 2012).

Wilson, Jim. *Robonaut2, the Next Generation Dexterous Robot.* April 28, 2010.

Yamafuji, Kazuo. "Celebrating JRM Volume 20 and Three Epoch-Making Robots from Japan." *Journals of Robotics and Mechatronics.* 2008: 3.

Yaskawa Motoman. *YASKAWA MOTOMAN, Our History.* 2010. http://www.motoman.eu/company/about-yaskawa/our-history/ (accessed May 28, 2012).